Polymer–Clay Nanocomposites

Wiley Series in Polymer Science

Series Editor:
John Scheirs
ExcelPlas Australia
PO Box 163
Casula, NSW 2170
AUSTRALIA

Modern Fluoropolymers
High Performance Polymers for Diverse Applications

Polymer Recycling
Science, Technology and Applications

Metallocene–Based Polyolifins
Preparation, Properties and Technology

Forthcoming titles:

Dendritic Polymers

Polymer–Clay Nanocomposites

Edited by

T. J. PINNAVAIA
Department of Chemistry, Michigan State University, East Lansing, MI, USA

and

G. W. BEALL
Missouri Baptist College, St. Louis, MO, USA

WILEY SERIES IN POLYMER SCIENCE

John Wiley & Sons, Ltd
Chichester · New York · Weinheim · Brisbane · Singapore · Toronto

Other Wiley Editorial Offices

John Wiley & Sons, Inc., 605 Third Avenue,
New York, NY 10158-0012, USA

WILEY-VCH Verlag GmbH, Pappelallee 3,
D-69469 Weinheim, Germany

Jacaranda Wiley Ltd, 33 Park Road Milton,
Queensland 4064, Australia

John Wiley & Sons (Asia) Pte Ltd, Clementi Loop #02-01,
Jin Xing Distripark, Singapore 129809

John Wiley & Sons (Canada) Ltd, 22 Worcester Road,
Rexdale, Ontario M9W 1L1, Canada

Library of Congress Cataloging-in-Publication Data

Polymer–clay nanocomposites / edited by T.J. Pinnavaia and G.W. Beall.
p. cm. — (Wiley series in polymer science)
Includes bibliographical references and index.
ISBN 0-471-63700-9 (alk. paper)
1. Nanostructure materials. 2. Polymeric composites. 3. Clay. I. Pinnavaia, Thomas J.
II. Beall, G.W. III. Series.

TA418.9. N35 P65 2001
620.1'92—dc21   00-043272

British Library Cataloguing in Publication Data

A catalogue record for this book is available from the British Library

ISBN 0-471-63700-9

Typeset in Times by Techset Composition Ltd, Salisbury, Wiltshire
Printed and bound in Great Britain by Biddles Ltd, Guildford, Surrey
This book is printed on acid-free paper responsibly manufactured from sustainable forestry,
in which at least two trees are planted for each one used for paper production.

Contents

II NANOCOMPOSITE SYNTHESIS AND PROPERTIES

Contributors

P. Aranda,
Instituto de Ciencia de Materiales de
Madrid,
CSIC,
Cantoblanco E-28049, Madrid,
Spain

A. Balazs,
Department of Chemical and
Petroleum Engineering,
1249 Benedum Engineering Hall,
University of Pitsburgh,
Pittsburgh, PA 15261,
USA

G. W. Beall,
Missouri Baptist College,
One College Park Dr.,
St. Louis, MO 63141,
USA

C. A. A. Bloomquist,
CHM 200 Argonne National
Laboratory,
9700 S. Cass Ave.,
Argonne, IL 60439,
USA

K. A. Carrado,
CHM 200 Argonne National

Laboratory,
9700 S. Cass Ave.,
Argonne, Il 60439,
USA

R. Csencsits,
CHM 200 Argonne National
Laboratory,
9700 S. Cass Ave.,
Argonne, Il 60439,
USA

M. P. Eastman,
Department of Chemistry,
Northern Arizona University,
Flagstaff, AZ 86011,
USA

K. Fujimoto,
Unitika Ltd,
23, Kozakura, Uji-Shi,
Kyoto, 611, Japan

J. W. Gilman,
National Institute of Standards and
Technology,
Polymer Building 224, Rm B258,
Mail Stop 8652, Office A265,
100 Bureau Drive,

Gaithersburg, MD 20899-8652,
USA

V. V. Ginzburg,
Department of Chemical and
Petroleum Engineering,
1249 Benedum Engineering Hall,
University of Pittsburgh,
Pittsburgh, PA 15261,
USA

T. Kashiwagi,
National Institute of Standards and
Technology,
Polymer Building 224, Rm B258,
Mail Stop 8652, Office A265,
100 Bureau Drive,
Gaithersburg, MD 20899-8652,
USA

S. Katahira,
Unitika Ltd,
23, Kozakura, Uji-Shi,
Kyoto, 611, Japan

M. Kato,
Toyota Central R & D Labs,
Nagakute, Aichi, 480-11,
Japan

Y. Komori,
Department of Applied Chemistry,
Waseda University,
Okubo-3, Shinjuku-ku,
Tokyo 169-8555,
Japan

R. Krishnamoorti,
Department of Chemical Engineering,
University of Houston,
4800 Calhoun,
Houston, TX 77204-4792,
USA

K. Kuroda,
Department of Applied Chemistry,
Waseda University,
Okubo-3, Shinjuku-ku,
Tokyo 169-8555,
Japan

Y. Lyatskaya,
Department of Chemical and
Petroleum Engineering,
1249 Benedum Engineering Hall,
University of Pittsburgh,
Pittsburgh, PA 15261,
USA

J. Massam,
Department of Chemistry,
Michigan State University,
East Lansing, MI 48824-1322,
USA

J. C. Matayabas Jr,
Polymers Technology Group,
Eastman Chemical Company,
Kingsport, TN 376621,
USA

A. Oya,
Faculty of Engineering,
Gunma-University,
Kiryu, Gunma,
376-8515
Japan

T. J. Pinnavaia,
Department of Chemistry,
Michigan State University,
East Lansing, MI 48824-1322,
USA

T. L. Porter,
Department of Astronomy and Physics,
Northern Arizona University,

Flagstaff, AZ 86011,
USA

E. Ruiz-Hitzky,
Instituto de Ciencia de Materiales de
Madrid,
CSIC,
Cantoblanco E-28049, Madrid,
Spain

S. Seifert,
CHM 200 Argonne National
Laboratory,
9700 S. Cass Ave.,
Argonne, IL 60439,
USA

A. S. Silva,
Department of Chemical Engineering,
University of Houston,
4800 Calhoun,
Houston, TX 77204-4792,
USA

C. Singh,
Department of Chemical and
Petroleum Engineering,
1249 Benedum Engineering Hall,
University of Pittsburgh,
Pittsburgh, PA 15261,
USA

T.-Y. Tsai,
Industrial Technology Research
Institute,
321 Kuang Fu Road, Section 2,
Hsinchu, Taiwan 300,
ROC

S. R. Turner,
Polymers Technology Group,
Eastman Chemical Company,
Kingsport, TN 376621,
USA

A. Usuki,
Toyota Central R & D Labs,
Nagakute, Aichi, 480-11,
Japan

R. A. Vaia,
Air Force Research Laboratory,
Materials Science Group Leader,
Polymer Branch,
AFRL/MLBP, Bldg 654,
2941 P Street, Rm 336,
WPAFB, OH 45433-7750,
USA

Z. Wang,
Department of Chemistry,
Michigan State University,
East Lansing, MI 48824-1322,
USA

L. Xu,
CHM 200 Argonne National
Laboratory,
9700 S. Cass Ave.,
Argonne, IL 60-439,
USA

K. Yasue,
Unitika Ltd,
23, Kozakura, Uji-Shi,
Kyoto, 611,
Japan

M. Yoshikawa,
Unitika Ltd,
23, Kozakura, Uji-Shi,
Kyoto,611,
Japan

E. Zhulina,
Department of Chemical and
Petroleum Engineering,
1249 Benedum Engineering Hall,
University of Pittsburgh,
Pittsburgh, PA 15261,
USA

Series Preface

The Wiley Series in Polymer Science aims to cover topics in polymer science where significant advances have been made over the past decade. Key features of the series will be developing areas and new frontiers in polymer science and technology. Emerging fields with strong growth potential for the twenty-first century such as nanotechnology, photopolymers, electro-optic polymers etc. will be covered. Additionally, those polymer classes in which important new members have appeared in recent years will be revisited to provide a comprehensive update.

Written by foremost experts in the field from industry and academia, these books place particular emphasis on structure–property relationships of polymers and manufacturing technologies as well as their practical and novel applications. The aim of each book in the series is to provide readers with an in-depth treatment of the state-of-the-art in that field of polymer technology. Collectively, the series will provide a definitive library of the latest advances in the major polymer families as well as significant new fields of development in polymer science.

This approach will lead to a better understanding and improve the cross fertilization of ideas between scientists and engineers of many disciplines. The series will be of interest to all polymer scientists and engineers, providing excellent up-to-date coverage of diverse topics in polymer science, and thus will serve as an invaluable ongoing reference collection for any technical library.

John Scheirs
June 1997

Preface

Polymer–clay nanocomposites are formed through the union of two very different materials with organic and mineral pedigrees. Linus Pauling first elucidated the structures of clay minerals using early X-ray crystallographic methods and thereby provided the basis for subsequent extensive studies of these materials in the disciplines of colloid science, rheology, mineralogy, catalysis, tribology, civil engineering, ceramic engineering and, especially, soil science. The smectite clays, in particular, are unique in the fact that they exist in nature in turbostratic units that are hydrophilic and can be broken down into one nanometer thick platelets. They find many and diverse industrial applications, including use as water-based thickeners, cation exchangers, oil-well drilling fluids, cat litter, cosmetics, paper coatings, paint additives, metal casting greensands, catalysts, pharmaceuticals, animal feed additives, landfill liners, foundation leak barriers and adsorbents for wine clarification, vegetable oil refining, water purification and the efficient application of agricultural chemicals. Smectite clays can also be modified with various organic chemicals to render them hydrophobic. These so-called organoclays have been traditionally used as rheological modifiers for oil-based paint, grease and ink, as well as in cosmetics, pharmaceuticals, waste water treatment and poison ivy rash prevention. Indeed, organoclays are also used extensively in the developing field of polymer–clay nanocomposites, the subject of the present volume.

In contrast to the clay minerals industry, the polymer industry has been built with synthetic chemicals as its basis, although this has begun to change recently with the introduction of polylactic acid and other polymers derived from biomass. The origin of modern polymers can be traced to the pioneering work by Wallace H. Carothers at Du Pont, where his crowning achievement was the synthesis of nylon as a superior replacement for silk. As new polymers were rapidly introduced in the 1950s and 1960s, plastics earned an undeserved reputation as being a cheap substitute for wood or metal. This point of view has largely faded owing to the enormous and varied use of plastics in all kinds of products that we encounter in our daily lives. This can best be seen in the food packaging industry where traditional glass, metal and paper packaging has been and continues to be displaced by plastic. The automobile industry also has embraced plastics to lower the weight of cars and improve manufacturing methods. However, plastics have certain limitations in both the food packaging and the automobile industries. In the packaging industry, for instance, certain foods (e.g. tomato products and beer) are sensitive to oxygen and cannot be stored in plastic containers owing to the oxygen permeability of the plastic. In the auto industry, low stiffness and tensile strength and the tendency to warp or creep under heat load have limited the use of plastics in automotive applications. If need is indeed the mother of invention, then nanocomposites are a great example of the fulfilling of a need.

Polymer–clay nanocomposites have their origin in the pioneering research conducted at Toyota Central Research Laboratories where these two divergent organic and mineral materials were successfully integrated. Fittingly, the first practical application of a nanocomposite was in the use of a nylon–montmorillonite clay nanocomposite as a timing belt cover on a Toyota Camry automobile. This nanocomposite exhibited large increases in tensile strength, modulus and heat distortion temperature without a loss in impact resistance. The composite also had lower water sensitivity, permeability to gases and thermal coefficient of expansion. All of these property improvements could be realized without a loss of clarity in the polymer. It has been further found that nanocomposites impart a level of flame retardancy and UV resistance not present in the pure polymer. Since the initial work by Toyota on thermoplastics, nanocomposite technology has also been successfully applied to thermoset polymers. These initial research successes have prompted a large amount of research in industrial and university settings over the past decade. Many of these efforts are now coming to fruition with the introduction of new products.

This volume presents a broad view of the state of polymer–clay nanocomposites from a theoretical as well as a practical viewpoint. We have arranged the fifteen contributions according to the following four sections: I. Polymer–Clay Intercalates; II. Nanocomposite Synthesis and Properties; III. Special Properties and Applications; IV. Structure and Rheology. Section I emphasizes materials that are akin to intercalation compounds insofar as they generally have fixed compositions or compositions that have a relatively high clay fraction. The chapters in Section II describe the preparation and general properties of representative thermoplastic and thermoset nanocomposites. Two properties of polymer–clay nanocomposites that are of special practical and commercial significance, namely their fire retardant and barrier properties, are treated in Section III. Finally, Section IV provides much needed theoretical and experimental approaches to the elucidation of the structural and rheological factors that influence the performance and processing properties of polymer–clay nanocomposites. It is hoped that this volume will spur further research and advances in this fascinating field and ultimately lead to the widespread use of nanocomposites commercially.

Thomas J. Pinnavaia
Michigan State University, East Lansing, Michigan, USA

Gary Beall
Missouri Baptist College, St Louis, Missouri, USA

PART I
Polymer–Clay Intercalates

1

Layered Silicate–Polymer Intercalation Compounds

Y. KOMORI AND K. KURODA
Department of Applied Chemistry and Kagami Memorial Laboratory for
Materials Science and Technology, Waseda University, Tokyo, Japan

1 INTRODUCTION

Interactions between inorganic layered materials and organic substances have attracted increasing interest from both scientific and industrial perspectives [1–4]. In addition to the rapidly expanding field of exfoliated clay–polymer nanocomposites described throughout this book, intercalation of polymers into inorganic layered materials with retention of the layered nature is also an excellent way of constructing novel inorganic–polymer nanoassemblies [5]. There are two main ways of preparing layered inorganic–polymer systems. Figure 1 shows a schematic view of intercalation reactions of polymers into the interlayer spaces. Some polymers are directly intercalated, and others are intercalated by *in situ* polymerization of preintercalated monomers between the layers of host materials. Because the kind of polymers that are intercalated directly is limited, the route utilizing preintercalated monomers is important, though the control of molecular weight of the polymers formed *in situ* is normally difficult. A coprecipitation method is also effective in some cases and will be described below (Section 2.3).

Intercalation compounds of inorganic layered materials with polymers have the following unique characteristics.

(a) Various kinds of compounds are synthesized by the combination of a wide variety of both host materials and polymers.

Polymer–clay nanocomposites Edited by T. J. Pinnavaia and G. W. Beall
© 2000 John Wiley & Sons Ltd

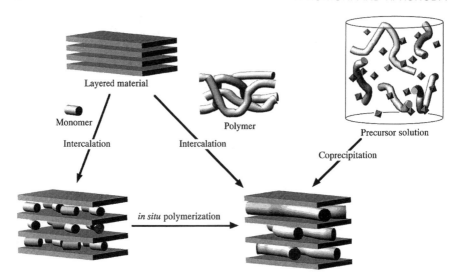

Figure 1 Schematic view of preparation methods for polymer intercalation compounds

(b) Polymers can be accommodated in the interlayer region with the retention of the structural features of layered hosts, which yield anisotropic arrangements of polymers in two dimensional nanoenvironments.

(c) Interlayer spaces are adaptable to the size of guest polymers.

(d) Guest species in the interlayer spaces are influenced by characteristic environments surrounded by adjacent host layers, and the properties are controlled by the interactions between hosts and guests in addition to the properties of hosts and guests themselves.

(e) Structural, chemical and thermal stabilities due to rigid inorganic frameworks are expected to work as a barrier or capsule for various vulnerable guest polymers.

Among the large number of inorganic layered materials that exhibit intercalation capabilities, layered silicates are one of the most typical because of the versatility of the reactions. In particular, the smectite group of clay minerals such as montmorillonite, saponite and hectorite have mainly been used because they have excellent intercalation abilities. Several reviews have recently appeared on polymer nanocomposites based on smectites [6–8]. Studies on the other types of layered silicate–polymer intercalation compound have been conducted to a much lesser degree because of the relative difficulties in the preparation of polymer-intercalated compounds. We believe, however, that extension of the kind of host materials leads to various layered silicate-based nanocomposites with compositional and structural variations that can be directed towards new applications.

In this review, we present some examples of intercalation compounds composed of layered polysilicates, kaolinite and layered double hydroxide as the hosts. Although layered double hydroxides are not silicates, they are often regarded as 'anionic clay' and can be treated in a manner similar to layered silicates. The structures of these layered materials are depicted in Figure 2.

The framework of layered polysilicates consists of SiO_4 tetrahedra with some silanol groups in the interlayer region, and exchangeable cations are present in the interlayer region. Kaolinite consists of SiO_4 tetrahedral sheets and $AlO_2(OH)_4$ octahedral sheets. Neither cations nor anions are present between the layers, and the

(a) layered polysilicate

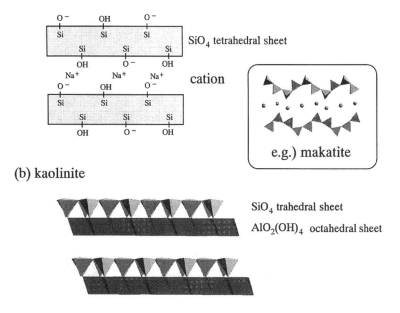

(b) kaolinite

(c) layered double hydroxide

Figure 2 Schematic structures of layered materials. Some ions, water molecules or hydrogen atoms are omitted for clarity

layers are linked to each other by hydrogen bondings between hydroxyl groups on the octahedral sheets and the oxide arrangement of the tetrahedral sheets. The structure of layered double hydroxide (LDH) consists of mainly brucite-like sheets. Anions are present between the layers of LDH. All these materials can be synthesized artificially, meaning that the compositions are free from impurities and, in some cases, variable. Their intercalation properties and the formation of intercalation compounds with polymers are quite different from those of smectites. In this review, methods for synthesizing intercalation compounds with polymers, some of which have been conducted in our laboratory, are described here. Some interesting ideas on the role of polymers in the interlayer spaces are also presented, which provides a wider view of clay–polymer nanosystems in addition to nano-composite applications.

2 VARIATION IN HOST MATERIALS FOR POLYMER INTERCALATION

2.1 LAYERED POLYSILICATE

2.1.1 Various Reaction Routes in Layered Polysilicates

Layered polysilicates include kanemite, makatite, octosilicate, magadiite and kenya-ite whose structures consist mainly of SiO_4 tetrahedra with different layer thick-nesses. These are naturally occurring minerals, except for octosilicate, but they can also be synthesized in relatively simple ways. The structures of kanemite [9], makatite [10] and octosilicate [9, 11] have been reported. As an example, the structure of makatite is schematically shown in Figure 2a (inset). Although the structures of magadiite and kenyaite have not been determined yet, the basic structures are composed of layered silicate networks and interlayer hydrated alkali metal cations. Their acidic analogues, layered polysilicic acids, are obtained by mild acid treatment of those polysilicates. Layered polysilicates and polysilicic acids have silanol groups in the interlayer regions, which leads to various organic modification by grafting organic functional groups in the interlayers. Among the layered polysilicates, the intercalation ability of magadiite has been relatively well investi-gated in view of its chemical and thermal stability [12, 13]. Magadiite is hydro-thermally synthesized from simple reactants such as SiO_2, NaOH and water [14]. A typical scanning electron microscopy (SEM) image of magadiite is shown in Figure 3a.

Compared with those of smectites, magadiite possesses some unique properties such as higher cation exchange capacity and, very importantly, the presence of silanol sites in the interlayer space. Thus, organic intercalation reactions of magadiite are carried out by (i) ion exchange with organic cations [12, 15], (ii) adsorption of polar organic molecules to ions by dipolar interactions and/or to silanol groups by hydrogen bonds [13] and (iii) silylation and esterification of interlayer silanol groups [16–22].

(a) Magadiite

(b) Kaolinite

Figure 3　SEM images of (a) magadiite, and (b) kaolinite

2.1.2　Layered Polysilicate–Polymer Systems

In an early stage of investigation on magadiite, Lagaly *et al.* mentioned that caprolactam and acrylamide were intercalated into H-magadiite and polymerized in the interlayer space, although the characterization of the products was not described [13]. Intercalation of poly(acrylonitrile) into magadiite has been reported by Sugahara *et al.* [23]. The procedure is as follows. First, a $C_{12}H_{25}N(CH_3)_3^+$-magadiite is prepared by ion exchange. Then, it is soaked in an excess of acrylonitrile monomer containing 0.7 wt % benzoyl peroxide as an initiator for

24 h. Finally, the product is heated at $50\,^{\circ}C$ in air for 24 h to polymerize the monomers in the interlayer space of the $C_{12}H_{25}N(CH_3)_3{}^+$-magadiite. Further heat treatment of the magadiite–PAN compound above $1300\,^{\circ}C$ is useful for the synthesis of non-oxide ceramics (ex. β-SiC) by carbothermal reduction.

Detailed investigation of polymer intercalation into layered polysilicic acid has been reported by Yanagisawa et al. [24]. Acrylamide molecules are intercalated by the reactions of H-magadiite and H-kenyaite with acrylamide saturated aqueous solutions and the monomers are polymerized between the layers by heat treatment at $200\,^{\circ}C$ for 30 min. A part of the intercalated acrylamide molecules is oligomerized during the reaction because of the high surface acidity of H-magadiite.

Recently, Wang et al. have investigated the exfoliation of magadiite in an epoxy matrix [25, 26]. Depending on the kind of onium ions, intercalated or exfoliated magadiite nanocomposites are obtained. The exfoliated nanocomposites are typically disordered, but a new type of exfoliated structure is also observed in which the nanolayers are regularly spaced over long distances (e.g. $\sim 8\,nm$). The tensile properties of the polymer matrixes are greatly improved by the reinforcement effect of the silicate nanolayers.

2.2 KAOLINITE

2.2.1 Intercalation of Kaolinite—Effective Use of Guest Displacement

The layered structure of kaolinite, $Al_2Si_2O_5(OH)_4$, is depicted in Figure 2b, and a typical SEM image of kaolinite is shown in Figure 3b. Kaolinite is a 1 : 1 type layered clay mineral (basal spacing 0.72 nm) which consists of SiO_4 tetrahedral sheets and AlO_6 octahedral hydroxyl sheets. Neither cations nor anions are present in the interlayer space. Because the interlayer regions are surrounded by hydroxyl groups of the octahedral sheets on one side and the oxide arrangement of the silicate sheets on the other side, guest species are generally arranged with specific orientations [27]. However, the intercalation reactivity of kaolinite is low owing to its inherent hydrogen bondings between the layers, and intercalation of kaolinite has been investigated to a much lesser degree than that of the smectite group of clay minerals. Nevertheless, since the discovery of intercalation of kaolinite with potassium acetate in 1961 [28], intercalation of various kinds of organic species has been investigated [29–31].

In order to overcome the obstacle that the direct intercalation to kaolinite is limited, several techniques have been developed. The most effective technique is a guest displacement reaction, in which preintercalated organic species in kaolinite can be displaced with various types of organic molecules. (Figure 4). When kaolinite–DMSO or kaolinite–NMF is used as the intermediate, several molecules, such as lactam [32], aminoalcohols [33], methanol [34], and ammonium acetate [35], can be intercalated. Multistep displacement is also possible by using these compounds as

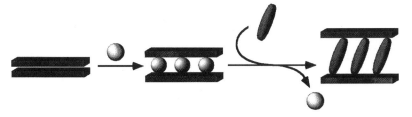

Figure 4   Guest displacement reaction

second intermediates. By utilizing a kaolinite–ammonium acetate compound as the second intermediate, further intercalation reactions of alkylamines [30], ammonium propionate [36, 37], acrylonitrile [35], and vinyl-2-pyrrolidone [38] have been realized. Recently, a kaolinite–methanol compound has been shown to be an excellent intermediate for the intercalation of alkylamines, *p*-nitroaniline, ε-capro-lactam and poly(vinylpyrrolidone) (PVP) [34, 39–42]. These intermediates have greatly expanded the intercalation chemistry of kaolinite towards organic guest species that cannot directly be intercalated.

2.2.2   Kaolinite–Polymer Systems

The first report on the intercalation of a polymer, namely poly(acrylonitrile) (PAN), into kaolinite was achieved by polymerization of acrylonitrile monomers between the layers [35]. A kaolinite–ammonium acetate compound is used as an intermediate and dispersed in acrylonitrile. By displacing intercalated ammonium acetate, the monomers are incorporated between the layers, the basal spacing of the product being 1.3–1.4 nm. A polymerization reaction is achieved by heating at 220 °C for 1 h in air. The hydrogen bonds between the hydroxyls of kaolinite and probably the CN groups of PAN are not affected after heating. Even after heating at 400 °C, the layers are expanded. Because the starting kaolinite–ammonium acetate compound decomposes at much lower temperature, the presence of PAN between the layers is strongly implied by this thermally stable behavior.

Evidence of *in situ* polymerization is clearly indicated by the ^{13}C NMR spectra of the products formed with poly(acrylamide) (PAAm) [43]. An acrylamide (AAm) monomer is first intercalated by the displacement reaction between a kaolinite–NMF compound and a 10 % acrylamide aqueous solution. The basal spacing of the resulting compound is 1.13 nm. On the basis of the IR and ^{13}C CP/MAS NMR data, the replacement of NMF by acrylamide is confirmed. The formation of hydrogen bonds with the hydroxyls of kaolinite is also corroborated. When the kaolinite–AAm compound is heated to 300 °C for 1 h, the basal spacing increases to 1.14 nm, and the C=C bonds disappear, indicating the polymerization of acrylamide. The heat-treated kaolinite–AAm compound is resistant to 30 min washing with water, whereas the untreated one collapses after the same treatment, which is consistent with acrylamide polymerization between the layers of kaolinite. PAAm is hydrogen bonded to kaolinite, but in a manner different from the hydrogen bonding of acrylamide.

Furthermore, the addition polymerization of vinyl compounds has been applied to the intercalation of poly(vinylpyrrolidone) [38]. The intercalation of an N-vinyl-2-pyrrolidone (VP) monomer is achieved by treating a kaolinite–ammonium acetate compound with a VP aqueous solution. The basal spacing decreases from 1.42 to 1.26 nm, and the IR and ^{13}C CP/MAS NMR spectra show the displacement of ammonium acetate by VP. Although the basal spacing did not change after heat treatment at 200 °C for 1 h, the polymerization of VP between the layers of kaolinite is confirmed by the IR and ^{13}C CP/MAS NMR spectra, which show the disappearance of vinyl groups and become similar to those of poly(vinylpyrrolidone).

An *in situ* polymerization is not always effective for organic monomers. For example, acrylic acid is not as polymerized as expected in the interlayer space of kaolinite by the heat treatment of the product at 250 °C for 1 h [44]. In this case, kaolinite–acrylic acid is prepared by a displacement reaction using kaolinite–DMSO as an intermediate. Thus, residual DMSO molecules which are not displaced with acrylic acid may block the polymerization of acrylic acid.

A novel *in situ* polymerization reaction with the formation of an amide bonding has been investigated in our laboratory. Intercalation of β-alanine has been achieved by using a kaolinite–ammonium acetate compound as an intermediate. After heat treatment, β-alanine is polymerized between the layers of kaolinite [45].

Direct intercalation of organic polymers has been reported by Tunney and Detellier [46]. Poly(ethylene glycol) (MW = 3400 and 1000) is incorporated into kaolinite by displacing DMSO from kaolinite–DMSO. Direct intercalation of poly(ethylene glycol) is performed from its polymer melt in the temperature range 150–200 °C. On the basis of the interlayer expansion of 0.4 nm, the intercalated oxyethylene units are in flattened monolayer arrangements.

A refined guest displacement method is quite effective for kaolinite to intercalate polymers at room temperature. Direct intercalation of PVP has recently been reported by the guest displacement method using a kaolinite–methanol compound [39]. By the reaction of the methanol–kaolinite with PVP dissolved in methanol, a kaolinite–PVP intercalation compound is formed, as indicated by an increase in the basal spacing to 1.24 nm. The ^{13}C CP/MAS NMR and IR spectroscopic results show the characteristic hydrogen bonding between the carbonyl groups in PVP and hydroxyl groups in kaolinite. These reports suggest that the ability of kaolinite to intercalate polymers is greater than recognized previously.

2.3 LAYERED DOUBLE HYDROXIDE

2.3.1 Intercalation Property of Layered Double Hydroxide

Layered double hydroxides (LDHs) are minerals and synthetic materials with positively charged brucite-type layers of mixed metal hydroxides [2, 47]. The schematic structure of LDH is shown in Figure 2c. Exchangeable anions located in the interlayer spaces compensate for the positive charge of brucite-type layers.

The most characteristic intercalation property of LDH is anionic exchange reactions. The methods for preparing LDH intercalation compounds depend on the kinds of guest species. Conventional anion exchange using an aqueous solution of guest species has been used widely. Compared with smectites showing cation exchange, ion exchange reaction in LDHs is not easily achievable because of their high selectivity for carbonate anions and the large anion exchange capacity. Therefore, CO_2 should be excluded during the sample preparation. Intercalation compounds have also been prepared via direct synthesis in the presence of guest species. The treatment of a mixed metal oxide solid solution, obtained by the heat treatment of LDH–carbonate, with an aqueous solution of guest species results in its structural reconstitution into an LDH–guest intercalation compound.

2.3.2 LDH–Polymer Systems

A hydrotalcite (Mg–Al–LDH)–PAN intercalation compound has been prepared by the *in situ* polymerization method [48]. Before intercalation of the acrylonitrile monomer (AN), chloride ions in the interlayer space of hydrotalcite (HT) are exchanged for dodecylsulfate (DS^-) ions, because the larger basal spacing is advantageous for the intercalation of AN. HT(DS^-) is soaked in AN containing benzoyl peroxide as an initiator. After the removal of excess AN, the product is heated at 50 °C for 24 h in air to polymerize AN between the layers of HT. Further investigations of the cyclization of PAN and the conversion to aluminum nitride have also been reported.

Intercalation of acrylate anions into a hydrotalcite-like compound by anion exchange has been reported by Tanaka *et al.* [49]. The reactivity of HT with a couple of interlayer anions (CO_3^{2-}, Cl^-, and NO_3^-) has been investigated. Utilizing HT–CO_3^{2-} is unsuccessful for exchanging interlayer anions. When HT–NO_3^- is employed, the reaction proceeds to form an HT–acrylate compound. The formation of the compound from HT–Cl^- is observed only under the reaction conditions using a concentrated acrylate solution. When the compound is heated at 80 °C with the addition of an initiator, the interlayer acrylate anions are polymerized to form an HT–polyacrylate intercalation compound.

LDH–poly(acrylic acid), poly(vinylsulfonate) and poly(styrenesulfonate) intercalation compounds have been directly synthesized by coprecipitation from solutions containing the desired polymers [50, 51]. Deaerated solutions of mixed metal nitrates and NaOH are simultaneously added to basic solutions of the polymers. The resulting nanocomposites contain the LDH sheet structure separated by 0.76–1.60 nm, which is large enough to accommodate polymer bilayers between the sheets.

The direct synthesis approach has been used to intercalate poly(α, β-aspartate) into LDH [52]. Synthesis of bioinorganic nanocomposites is achieved by coprecipitation involving simultaneous formation of the inorganic layers and intercalation of anionic species. In this report, *in situ* thermal polymerization of preintercalated aspartate monomers is also described as a second route for synthesizing LDH–poly(α, β-aspartate) hybrid materials. The direct synthesis approach by coprecipitation has been applied to the preparation of nanocomposites containing molecularly dispersed organic polymers in Ca–Al–LDH [53, 54]. Crystal growth of Ca–Al–LDH has been investigated in the presence of poly(vinyl alcohol), poly(dimethyldiallyl ammonium chloride) and poly(dibutyl ammonium iodide). The particles of poly-(vinyl alcohol)–LDH nanocomposite are spheroidal aggregates of thin plate crystals, whereas a polycationic polymer leads to rod-like particles. Although this finding is not directly related to polymer intercalation, it is an intriguing interaction between LDH and polymers.

An intercalative polycondensation of aniline molecules in swollen hydrotalcite-like compounds containing intrasheet oxidant Cu^{2+} centers has been reported by Challier and Slade [55]. The oxidizing host matrixes with terephthalate or hexacya-noferrate(II) ions acting as pillars are effective for interlamellar oxidative polymerization of aniline.

3 SOME EXAMPLES OF THE ROLE OF POLYMERS IN INTERLAYERS

3.1 MODIFICATION OF INTERLAYER ENVIRONMENT

Layered materials have been investigated for the organization of photofunctional molecules [2]. The presence of intercalated polymers provides a unique interlayer environment for photofunctional molecules or probes and induces a unique photo-luminescent behavior. The role of cointercalated poly(vinylpyrrolidone) (PVP) in the sterically limited interlayer space of swelling fluorotetrasilicic mica (TSM), for example, has been investigated using tris(2,2′-bipyridine)ruthenium(II) ($Ru(bpy)_3^{2+}$) as a probe [56, 57]. The luminescence maxima of intercalated $Ru(bpy)_3^{2+}$ gradually blue-shift with a decreased loading of $Ru(bpy)_3^{2+}$, reflecting the change in the polarity and/or rigidity of the microenvironment of $Ru(bpy)_3^{2+}$, presumably caused by cointercalated PVP. Moreover, $Ru(bpy)_3^{2+}$ self-quenching is suppressed even at its high concentration loading. It is supposed that cointercalated PVP prevents self-aggregation by surrounding $Ru(bpy)_3^{2+}$ in a close contact in the sterically limited interlayer spaces, as schematically shown in Figure 5.

3.2 CARBOTHERMAL REDUCTION

The reactions between layered materials and intercalated organic polymers have an advantage for specific carbothermal reduction processes because layered materials and organic substances can contact at a molecular level. A montmorillonite (Mont)–

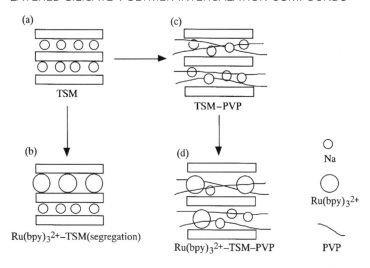

Figure 5 Schematic structure of (a) Na-TSM, (b) Ru(bpy)$_3^{2+}$-TSM intercalation compound, (c) TSM–PVP intercalation compound, and (d) Ru(bpy)$_3^{2+}$–TSM–PVP intercalation compound

poly(acrylonitrile) (PAN) intercalation compound was applied as a precursor for the carbothermal reduction for the first time in 1984. By the heat treatment of the precursor above 1150 °C in flowing N$_2$, β-sialon, SiC and AlN are mainly formed [58, 59]. When the precursor is heated in an Ar flow, β-SiC is formed at 1200 °C [60], and α-SiC, β-SiC and Al$_4$Si$_2$C$_5$ at 1700 °C [61]. Compared with the reactions of a Mont–carbon mixture, it has been clarified that the presence of PAN in the interlayer space leads to the formation of nitrides and SiC and suppresses the crystallization of oxide.

Furthermore, the carbothermal reduction of an alkylammonium-exchanged Mont–PAN intercalation compound in an N$_2$ flow has been investigated (Figure 6) [62]. The presence of alkylammonium ions induces both an increase in the amount of PAN in the interlayer and a lowering of the reaction temperature. Only β-sialon has been obtained at 1100 °C without the formation of SiC and AlN. The effect of the chain length of the alkylammonium ions has also been investigated [63]. The reactivity of precursors possessing longer alkylammonium ions is lower in spite of more organic content, indicating that the reactivities of the precursors are subjected to the structural conditions in the interlayer space rather than the carbon contents.

This method has been applied to other layered materials such as magadiite and LDH. When a magadiite–PAN compound is fired above 1300 °C in an Ar flow, only β-SiC is formed without the formation of any crystalline SiO$_2$ phases [23], indicating that PAN suppresses the crystallization of SiO$_2$ during the reaction. On the other hand, heat treatment in an N$_2$ flow induces the formation of Si$_3$N$_4$ above

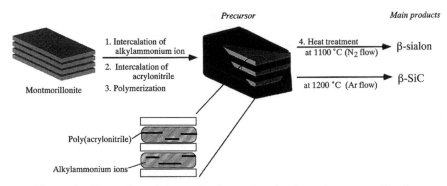

Figure 6   Examples of the carbothermal reduction of montmorillonite

1400 °C [64]. As a precursor for AlN, an LDH–PAN compound has been investigated [48]. The formation of LDH–PAN is described in Section 2.3. By heat treatment at 1600 °C in N_2, single phase submicron AlN grains have been obtained.

3.3   ORDERED GRAPHITE

When PAN itself is heated under an inert atmosphere, it produces carbon with three-dimensional networks with intertwined ribbons of stacked graphite sheets. However, by utilizing the two-dimensional structure of clay minerals, as shown in Figure 7, it is possible to synthesize ordered graphite film [65–67]. A film composed of clay–acrylonitrile intercalation compound is prepared and subjected to γ-ray irradiation to polymerize the acrylonitrile monomers between the layers of clay. The clay–PAN complex film is heated at 700 °C under an N_2 flow. The PAN is carbonized to form a clay–carbon complex film. The carbon is then released from clay by acid treatment and heated above 2500 °C. As a result, highly oriented carbon is formed owing to the orientation of the two-dimensional carbon precursor produced between the layers of clay. Other organic polymers such as poly(furfuryl alcohol) and poly(vinyl acetate) give similar results [68]. Oriented clay films such as Na-taeniolite produce higher crystallized and oriented graphite film [69].

Figure 7   Schematic representation of the preparation of a highly oriented graphite film by utilizing the interlayer space of smectite

3.4 CONTROL OF TRANSFORMATION OF SILICATE LAYERS

Studies of the thermal behavior of polymer-intercalated layered materials show that the intercalated polymers affect the thermal properties of the layered host. For metakaolinite, the dehydroxylated form of kaolinite obtained by heat treatment, the structure has been controversial because of its very low crystalline nature. Thus, the thermal transformation of a kaolinite–poly(acrylamide) (PAAm) intercalation compound has been investigated by heating in the range from 460 to 620 °C under a nitrogen atmosphere [70]. Carbonaceous materials remain between the layers in this temperature range. On the basis of the ^{29}Si NMR and ^{27}Al NMR spectra of the kaolinite–PAAm compounds heated above 500 °C, it has been concluded that the carbonaceous materials between the layers contribute to the retention of the two-

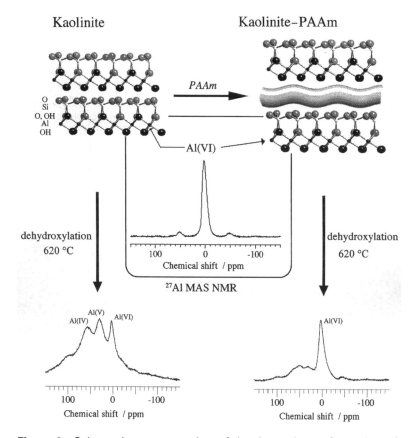

Figure 8 Schematic representation of the thermal transformation of kaolinite, and kaolinite–PAAm intercalation compound. Al(IV), and Al(V) appear after heat treatment of kaolinite, while the generation of Al(V), and Al(IV) is suppressed for the kaolinite–PAAm intercalation compound

dimensional network of the silicate sheets and suppress both the appearance of five-coordinated Al and the dehydroxylation reaction (Figure 8). Consequently, the formation of five-coordinated Al in the structure of metakaolinite is associated with the silicate sheets not only in the same layers but also in the adjacent layers. This new finding suggests that polymers can resist the skeletal transformation of inorganic layered materials by heat treatment.

The thermal properties of LDH intercalated with polymers has also been reported [71]. The presence of intercalated poly(vinyl alcohol) retards the decomposition of the inorganic nanolayers of Ca–Al–LDH and prevents the formation of crystalline hydroxide phases at high temperatures. The prevention of hydroxide crystallization is possibly caused by the intimate bonding contact between inorganic crystal layers and the intercalated polymer. Furthermore, the intercalation compound heated to 1000 °C transforms into an inorganic solid which has a different phase composition to the LDH treated to the same temperature.

4 CONCLUSIONS

Several examples of polymer intercalation compounds of layered polysilicate, kaolinite and layered double hydroxides have been reviewed. Since layered poly-silicates have versatile reactivities, such as ion exchange, adsorption and silylation, further applications with polymers are expected. Several reports for kaolinite–polymer intercalation compounds have demonstrated that kaolinite has higher intercalation ability toward a wider variety of guest species than accepted previously. Applications for kaolinite–polymer intercalation compounds are expected because of the high crystallinity, high aspect ratio and confined interlayer environment provided by the host. LDH hosts have an anion exchange ability and should accommodate polymers with negative charges. These intercalates will be surveyed in more detail in the future. Unique preparation methods are particularly useful for the preparation of LDH–polymer systems. Some examples of possible applications indicate that the interactions between inorganic layers and intercalated polymers are essential for constructing novel inorganic–organic nanoassemblies. The stereoregularity of inter-calated polymers should be examined in more detail. By utilizing the intercalation reactions of these hosts, it should be possible to design materials for a wide variety of applications.

5 REFERENCES

1. Bruce, D. W. and O'Hare, D. *Inorganic Materials*, 2nd edition, John Wiley & Sons, 1996.
2. Ogawa, M. and Kuroda, K. *Chem. Rev.*, **95**, 399 (1995).
3. Ogawa, M. and Kuroda, K. *Bull. Chem. Soc. Jpn*, **70**, 2593 (1997).
4. Ruiz-Hitzky, E. and Aranda, P. *An. Quím. Int. Ed.*, **93**, 197 (1997).

5. Theng, B. K. G. *Formation and Properties of Clay–Polymer Complexes*, Elsevier, New York, 1979.
6. Giannelis, E. P. *Adv. Mater.*, **8**, 29 (1996).
7. Lagaly, G. *Appl. Clay Sci.*, **15**, 1 (1999).
8. LeBaron, P. C., Wang, Z. and Pinnavaia, T. J. *Appl. Clay Sci.*, **15**, 11 (1999).
9. Gies, H., Marler, B., Vortmann, S., Oberhagemann, U., Bayat, P., Krink, K., Rius, J., Wolf, I. and Fyfe, C. *Microporous Mesoporous Mater.*, **21**, 183 (1998).
10. Annehed, H., Fälth, L. and Lincoln, F. J. *Z. Kristallogr.*, **159**, 203 (1982).
11. Vortmann, S., Rius, J., Siegmann, S. and Gies, H. *J. Phys. Chem. B.*, **101**, 1292 (1997).
12. Lagaly, G., Beneke, K. and Weiss, A. *Am. Miner.*, **60**, 642 (1975).
13. Lagaly, G., Beneke, K. and Weiss, A. *Am. Miner.*, **60**, 650 (1975).
14. Kosuge, K., Yamazaki, A., Tsunashima, A. and Otsuka, R. *J. Ceram. Soc. Jpn*, **100**, 326 (1992).
15. Kosuge, K. and Tsunashima, A. *Langmuir*, **12**, 1124 (1996).
16. Ruiz-Hitzky, E. and Rojo, J. M. *Nature*, **287**, 28 (1980).
17. Ruiz-Hitzky, E., Rojo, J. M. and Lagaly, G. *Colloid Polym. Sci.*, **263**, 1025 (1985).
18. Yanagisawa, T., Kuroda, K. and Kato, C. *React. Solids*, **5**, 167 (1998).
19. Yanagisawa, T., Kuroda, K. and Kato, C. *Bull. Chem. Soc. Jpn*, **61**, 3743 (1988).
20. Ogawa, M., Miyoshi, M. and Kuroda, K. *Chem. Mater.*, **10**, 3787 (1998).
21. Ogawa, M., Okutomo, S. and Kuroda, K. *J. Am. Chem. Soc.*, **120**, 7361 (1998).
22. Okutomo, S., Kuroda, K. and Ogawa, M. *Appl. Clay Sci.*, **15**, 253 (1999).
23. Sugahara, Y., Sugimoto, K., Yanagisawa, T., Nomizu, Y., Kuroda, K. and Kato, C. *Yogyo-Kyokai-Shi*, **95**, 117 (1987).
24. Yanagisawa, T., Yokoyama, C., Kuroda, K. and Kato, C., *Bull. Chem. Soc. Jpn*, **63**, 47 (1990).
25. Wang, Z., Lan, T. and Pinnavaia, T. J. *Chem. Mater.*, **8**, 2200 (1996).
26. Wang, Z. and Pinnavaia, T. J. *Chem. Mater.*, **10**, 1820 (1998).
27. Thompson, J. G. and Cuff, C. *Clays Clay Miner.*, **33**, 490 (1985).
28. Wada, K. *Am. Miner.*, **46**, 78 (1961).
29. Weiss, A., Thielepape, W., Göring, G., Ritter, W. and Schäfer, H. *Int. Clay Conf.*, **1**, 287 (1963).
30. Weiss, A., Thielepape, W. and Orth, H. in Proc. Int. Clay Conf., Jerusalem, 1966, Vol. 1, p. 277.
31. Theng, B. K. G. *The Chemistry of Clay–Organic Reactions*, Adam Hilger, London, 1974.
32. Sugahara, Y., Kitano, S., Satokawa, S., Kuroda, K. and Kato, C. *Bull. Chem. Soc. Jpn*, **59**, 2607 (1986).
33. Tunney, J. J. and Detellier, C. *Can. J. Chem.*, **75**, 1766 (1997).
34. Komori, Y., Sugahara, Y. and Kuroda, K. *J. Mater. Res.*, **13**, 930 (1998).
35. Sugahara, Y., Satokawa, S., Kuroda, K. and Kato, C. *Clays Clay Miner.*, **36**, 343 (1988).
36. Seto, H., Cruz, M. I. and Fripiat, J. *Am. Miner.*, **63**, 572 (1978).
37. Seto, H., Cruz-Cumplido, M. I. and Fripiat, J. J. *Clay Miner.*, **13**, 309 (1978).
38. Sugahara, Y., Sugiyama, T., Nagayama, T., Kuroda, K. and Kato, C. *J. Ceram. Soc. Jpn*, **100**, 413 (1992).
39. Komori, Y., Sugahara, Y. and Kuroda, K. *Chem. Mater.*, **11**, 3 (1999).
40. Komori, Y., Sugahara, Y. and Kuroda, K. *Appl. Clay Sci.*, **15**, 241 (1999).
41. Komori, Y., Matsumura, A., Itagaki, T., Sugahara, Y. and Kuroda, K. *Clay Sci.*, **11**, 47 (2000).
42. Kuroda, K., Hiraguri, K., Komori, Y., Sugahara, Y., Mouri, H. and Uesu, Y. *Chem. Commun.*, **22**, 2253 (1999).
43. Sugahara, Y., Satokawa, S., Kuroda, K. and Kato, C. *Clays Clay Miner.*, **28**, 137 (1990).
44. Sugahara, Y., Nagayama, Y., Kuroda, K., Doi, A. and Kato, C., *Clay Sci.*, **8**, 69 (1991).

45. Itagaki, T., Komori, Y., Sugahara, Y. and Kuroda, K. in preparation.
46. Tunney, J. J. and Detellier, C. *Chem. Mater.*, **8**, 927 (1996).
47. Cavani, F., Trifiró, F. and Vaccari, A. *Catal. Today*, **11**, 173 (1991).
48. Sugahara, Y., Yokoyama, N., Kuroda, K. and Kato, C. *Ceram. Int.*, **14**, 163 (1988).
49. Tanaka, M., Park, I. Y., Kuroda, K. and Kato, C. *Bull. Chem. Soc. Jpn*, **62**, 3442 (1989).
50. Oriakhi, C. O., Farr, I. V. and Lerner, M. M. *J. Mater. Chem.*, **6**, 103 (1996).
51. Wilson Jr., O. C., Olorunyolemi, T., Jaworski, A., Borum, L., Young, D., Siriwat, A., Dickens, E., Oriakhi, C. and Lerner, M. *Appl. Clay Sci.*, **15**, 265 (1999).
52. Whilton, N. T., Vickers, P. J. and Mann, S. *J. Mater. Chem.*, **7**, 1623 (1997).
53. Messersmith, P. B. and Stupp, S. I. *J. Mater. Res.*, **7**, 2599 (1992).
54. Messersmith, P. B. and Giannelis, E. P. *Chem. Mater.*, **6**, 1719 (1994).
55. Challier, T. and Slade, R. C. T. *J. Mater. Chem.*, **4**, 367 (1994).
56. Ogawa, M., Inagaki, M., Kodama, N., Kuroda, K. and Kato, C. *Mol. Cryst. Liq. Cryst.*, **216**, 141 (1992).
57. Ogawa, M., Inagaki, M., Kodama, N., Kuroda, K. and Kato, C. *J. Phys. Chem.*, **97**, 3819 (1993).
58. Sugahara, Y., Kuroda, K. and Kato, C. *J. Am. Ceram. Soc.*, **67**, C-247 (1984).
59. Sugahara, Y., Kuroda, K. and Kato, C. *J. Mater. Sci.*, **23**, 3572 (1988).
60. Sugahara, Y., Sugimoto, K., Kuroda, K. and Kato, C. *Yogyo-Kyokai-Shi*, **94**, 38 (1986).
61. Sugahara, Y., Sugimoto, K., Kuroda, K. and Kato, C. *J. Am. Ceram. Soc.*, **71**, C-325 (1988).
62. Sugahara, Y., Kuroda, K. and Kato, C. *Ceram. Int.*, 14, 1 (1988).
63. Sugahara, Y., Kuroda, K. and Kato, C. *Clay Sci.*, **7**, 17 (1987).
64. Sugahara, Y., Nomizu, Y., Kuroda, K. and Kato, C. *J. Ceram. Soc. Jpn. Int. Ed.*, **95**, 115 (1987).
65. Kyotani, T., Sonobe, N. and Tomita, A. *Nature*, **331**, 331 (1988).
66. Sonobe, N., Kyotani, T., Hishiyama, Y., Shiraishi, M. and Tomita, A. *J. Phys. Chem.*, **92**, 7029 (1988).
67. Sonobe, N., Kyotani, T. and Tomita, A. *Carbon*, **26**, 573 (1988).
68. Sonobe, N., Kyotani, T. and Tomita, A. *Carbon*, **29**, 61 (1991).
69. Kyotani, T., Mori, T. and Tomita, A. *Chem. Mater.*, **6**, 2138 (1994).
70. Komori, Y., Sugahara, Y. and Kuroda, K. *J. Mater. Chem.*, **9**, 3081 (1999).
71. Messersmith, P. B. and Stupp, S. I. *Chem. Mater.*, **7**, 454 (1995).

2

Electroactive Polymers Intercalated in Clays and Related Solids

E. RUIZ-HITZKY AND P. ARANDA

Instituto de Ciencia de Materiales de Madrid, CSIC, Cantoblanco, Madrid, Spain

1 INTRODUCTION

The combination of clays and functional polymers interacting at atomic level constitutes the basis for preparing an important class of inorganic–organic nano-structured materials [1]. Among these nanocomposites, the intercalation of electro-active species into the interlayer space of 2 : 1 phyllosilicates is a way to construct novel hybrid supramolecular assemblies with particular electrical behavior [2–7]. In fact, the ability of clays to intercalate organic compounds has been largely applied to obtain a wide family of intercalation materials (organoclays) exhibiting relevant properties [8, 9]. In this context, the design and synthesis of nanocomposite electroactive materials by modification of inorganic solids with organic entities constitute a new strategy for preparing materials with synergic or complementary performances afforded by the electroactive species and, in this case, by the clay host solid interactions.

Clays exhibit unique electrical properties mainly related to their ionic conductiv-ity. Therefore, although the silicate sheets can be regarded as ceramic lamellae, and indeed as insulating materials, the presence of hydrated cations in the interlayer space ensures a significant conductivity of the electrical signal associated with the ions from the clay interlayer region. According to Fripiat and coworkers [10], the water molecules associated with the interlayer cations of M^{n+}-montmorillonite show a high degree of dissociation and, consequently, contribute to the ionic conductivity of the system [11]. Consider montmorillonite, a smectite clay mineral, whose

Polymer–clay nanocomposites Edited by T. J. Pinnavaia and G. W. Beall
© 2000 John Wiley & Sons Ltd

structure (Figure 1) ideally consists of negatively charged layers built up by two tetrahedral silica sheets and a central octahedral sheet of magnesia or alumina [12]. Isomorphous substitutions originate a negative charge in the silicate layers which is compensated by cations (*exchangeable cations*) as extraframework ions located in the interlayer space. Homoionic clay samples are easily prepared by equilibrating the smectite with aqueous salt solutions of different metal cations. The mobility of such cations ensures electrical charge conductivity, the conductivity strongly depending on the hydration state of the clay system. The intercalation of neutral species could affect the hydration sphere of interlayer cations and even their mobility. This is the case, for instance, for nanocomposites resulting from the intercalation of crown ether complexing agents in clays [13]. The coordination between the interlayer cation and the intercalated molecule significantly modifies the ion mobility and, indeed, the electrical conductivity and other electrical parameters such as the capacitance of the material [5, 14]. Other molecular or ionic electroactive species could be intercalated in smectite clays, giving microparticulate materials with electrical behavior governed by the nature of both the clay and the intercalated species, as well as by the specific host–guest interactions [15].

The intercalation of electroactive polymers opens a way to novel organic–inorganic hybrid systems showing peculiar electrical properties [16]. This type of material can be prepared either by direct intercalation of the polymer or by preintercalation of the organic monomer precursor followed by its polymerization in the interlayer region. The choice of the guest polymer, as well as of the clay of

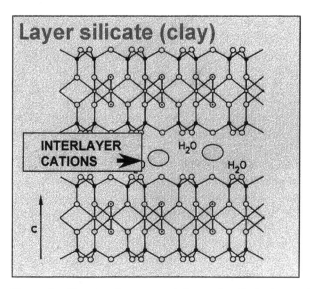

Figure 1 Structural representation of a 2:1 charged phyllosilicate (smectite group, such as montmorillonite and hectorite)

selected characteristics, gives rise to a large variety of possible clay-conducting polymer systems showing novel electrical properties of potential interest for and electrochemical devices.

2 CONDUCTING POLYMERS

Electroactive polymers are materials of increasing scientific and technical interest because of their important high-tech applications on rechargeable batteries, electrochromic displays, membranes and sensor devices, etc. Among these types of materials, two groups of electroactive polymers can be distinguished:

(a) *Ion-conducting polymers* (polymer electrolytes), where the electric current is ensured by ions associated with polymers such as poly(ethylene oxide) (PEO) [17–20].

(b) *Electronically conducting polymers* having conjugated bonds in the macromolecular network, as in polyacetylene (PA), polyaniline (PANI), polypyrrole (PPy), polythiophene (PTh), etc. [21, 22].

2.1 ION-CONDUCTING POLYMERS

Many compounds having oxyethylene ($-CH_2CH_2-O$) units show the ability to coordinate a large variety of cations including the alkaline metal ions. Cyclic polyoxyethylene compounds, so-called crown ethers, are complexing agents of such cations in solution [23, 24] and also in the solid state when intercalated into smectite clay minerals [13, 25–28] and other layered inorganic solids [29–33]. Linear polyoxyethylenes (PEO) having $-CH_2CH_2-O-$ flexible chains act as solvents of metal salts (Figure 2), giving solid solutions of so-called polymer electrolytes, with typical ionic conductivity values ranging from 10^{-4} to 10^{-5} S/cm at moderate temperatures ($<100\,°C$). This complexing ability, along with the electrical behavior of these systems, was first reported by Armand [34]. Later, numerous researchers devoted their activity to these topics [18]. One of the most salient features of these systems is the fact that, although crystalline PEO–salt complexes could be formed by coordination of the oxyethylene units with the corresponding cations, the conductivity behavior seems to be better related to the amorphous component than to the crystalline phases of the resulting polymer complex. In view of the fact that the state of aggregation is crucial to the ionic conductivity of polymer electrolytes, some inorganic additives, such as alumina or NASICON, were incorporated into the polymer electrolytes [35–37]. An alternative approach, consisting in the preparation of polymer electrolyte–clay nanocomposites, was first reported by Ruiz-Hitzky and Aranda [15], opening a new field of research. This novel class of solid electrolytes is considered in the present chapter.

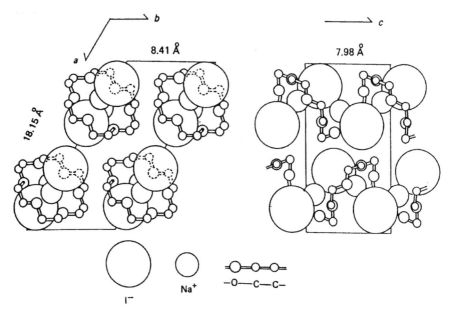

Figure 2 Crystal structure of a PEO–salt complex, showing the atomic disposition in a PEO–NaI complex. Reprinted from Chatani and Okamura [115]. Copyright 1987, with permission from Elsevier Science

2.2 ELECTRONICALLY CONDUCTING POLYMERS

Polymers having conjugated double bonds can act as electronically conducting materials when they are conveniently doped by coupling with charge-transfer agents. Polyacetylene (PA) was reported by MacDiarmid's group [38] as the first conducting polymer with nominal conductivity values close to that of metals [39–41]. Nevertheless, these *organic metals* exhibit an opposite temperature dependence of electrical conductivity to that of true metals as a consequence of a different conduction mechanism [42, 43]. In addition to polyenes, other important organic conducting polymers are polyaniline (PANI) [44, 45], polypyrrole (PPy) [46] and polythiophene (PTh) [47]. Among these polymers, PANI exhibits a large variety of properties related to its complex structure. In fact, PANI polymers may be described as a combination of four idealized repeating units (Figure 3). Each polymer presents a different structure, depending on the ratio of oxidized and reduced aromatic rings in neutral (base) or protonated (salt) states in the polymer chain [45]. The conducting form of PANI is the so-called 'emeraldine salt' which is the protonated (doped) form of the hemi-oxidized polymer (half-benzene/half-quinone rings in the chain) that allows the maximum conjugation.

Among other conducting polymers, PANI and PPy have been extensively used to prepare electroactive nanocomposites based on the ability of certain two-dimen-

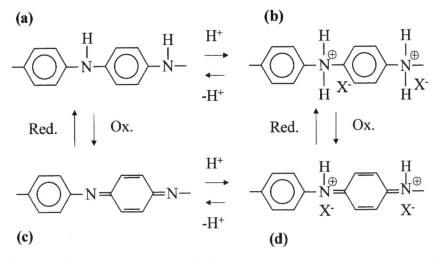

Figure 3   Chemical structure of PANI in different forms (oxidized, protonated, reduced and amine base)

sional inorganic solids to induce polymerization and to act simultaneously as doping agents in their interlayer space. This is the case for interlayer polymerization of pyrrole in FeOCl, reported by Kanatzidis [48]. From this discovery, an enormous amount of work has been invested by many research groups in order to develop new materials based on the intercalation of conducting polymers into different lamellar host solids [3, 4, 7]. Clay minerals containing transition metal ions, such as Cu^{2+}, deliberately introduced as exchangeable cations, induce interlayer polymerization of pyrrole and aniline, giving PPy–clay and PANI–clay nanocomposites respectively. This behavior is described below.

3   POLYMER ELECTROLYTE–CLAY NANOCOMPOSITES

The ionic conductivity of conventional PEO–salt complexes is strongly affected by various factors such as the crystallinity of the material, the ion-pair formation (anion–complexed cation interactions) and the high mobility of certain counter-ions. These factors reduce the cationic conductivity and, therefore, the potential applications of such polymer–salt compounds as solid state electrolytes. The combination of PEO with microparticulated solids such as alumina or some ionic superconductors [35–37] constitutes an approach aimed at improving the electrical behavior of polymer electrolytes. The use of clays to obtain nanocomposites with controlled ionic conductivity is a fascinating alternative, because it could directly affect the mobility of the cations while avoiding the mobility of the anions (charged silicate layers).

Cyclic polyoxyethylene compounds, such as crown ethers and cryptands, are able to intercalate 2:1 clay minerals, forming stable interlayer cation–oxyethylene complexes [13, 25–28]. This is a topotactic process involving the replacement of the solvation shell of the interlayer cations by the macrocyclic compounds. Structural features (Figure 4), stability, ion-exchange properties and other behavior of these nanocomposite materials have been extensively investigated [13, 14, 25–28, 49–52], showing that the enthalpy of the interlayer complex formation is directly correlated with the structure of the complex [28, 51].

In 2:1 clay silicates, the ionic conductivity associated with the interlayer cations exhibits an anisotropic character owing to the planar behavior of the clay host microparticles. The conductivity becomes very low upon elimination of water as the layers collapse. Appreciable ionic conductivity of dehydrated smectites is only observed at temperatures above 600 K [53, 54]. Intercalation of crown ethers into these phyllosilicates avoids the collapse of the layers, providing simultaneously an

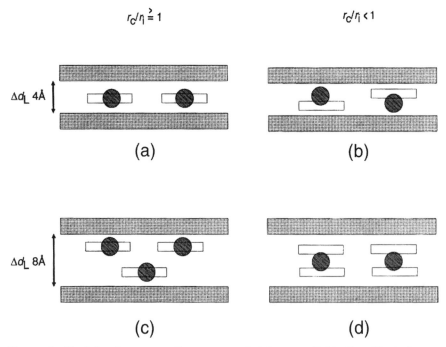

Figure 4 Models of macrocyclic compounds lying parallel to the silicate layers proposed on the basis of the ratio of intramolecular cavity r_c to cationic radius r_i, after Casal *et al.* [27]. Reproduced with kind permission of the Mineralogical Society of Great Britain and Ireland

adequate environment for the cations to keep their mobility. The ionic conductivity of crown ether–phyllosilicates is several orders of magnitude higher than that of the parent silicate. The conductivity increases with temperature until a maximum value which depends on the nature of the intercalated crown ether [14] being directly correlated with the enthalpy of interlayer complex formation (Figure 5). From these results it could be inferred that there is a clear tendency towards high values of conductivity at lower intercalation enthalpies [51], providing a basis for the design of new efficient composite polymer electrolyte materials derived from clays.

A new approach to obtaining oxyethylene–clay nanocomposites with enhanced ion conductivity was based on the intercalation of linear polyoxyethylenes into homoionic M^{n+}-smectites [15]. It is known that certain non-ionic polymers such as polyethylene glycol (PEG) and PEO, of different molecular weights, have been largely studied to control the flocculation properties of clay–water systems [55, 56]. Clay–polymer association involves polymer-exchangeable cation interactions that replace the cation hydration shell with oxyethylene units [57].

As reported by Ruiz-Hitzky and Aranda [15], higher molecular weight PEO $(6 \times 10^2–6 \times 10^5$ daltons) can be also intercalated (Table 1) into montmorillonite

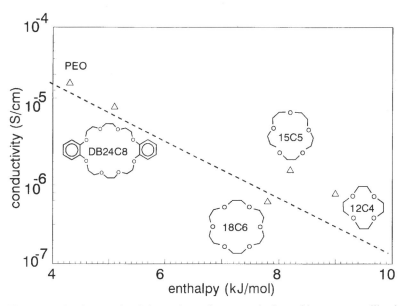

Figure 5 Ionic conductivity values for oxyethylene/Na-montmorillonite inter-calated compounds as a function of the formation enthalpy of the interlayer complexes, after Ruiz-Hitzky et al. [5]. Reproduced by permission of WILEY-VCH. The crown ethers are: 12-crown-4 (12C4), 15-crown-5 (15C5), 18-crown-6 (18C6) and dibenzo-24-crown-8 (DB24C8). PEO is poly(ethylene oxide) of 10^5 molecular weight

and hectorite, in this case from acetonitrile solutions, giving novel solid electrolytes. The oxyethylene units of PEO could operate in a way similar to crown ethers, acting as a complexing agent for M^{n+} interlayer cations. For different PEO/M^{n+}-clay intercalated materials ($M^{n+} = Li^+$, Na^+, K^+, Ba^{2+}, $CH_3CH_2CH_2NH_3^+$), well ordered 00l reflections in the X-ray diffraction (XRD) patterns are observed. The corresponding increase in the interlayer distance (Δd_L) is close to 0.8 nm [58, 59]. This Δd_L value is compatible with two possible arrangements of the polymer chains in the interlayer space: (i) as a bilayer of PEO chains in a planar zig-zag disposition or (ii) as a single-layer intercalated polymer adopting a helical conformation. The latter intracrystalline PEO arrangement (Figure 6) is consistent with IR, ^{13}C and ^{23}Na NMR spectroscopic data indicating that interlayer cations remain associated with the PEO oxygen atoms [58, 59]. Wu and Lerner [60] have reported two phases with d_L values of 1.36 and 1.77 nm, assigned respectively to mono- and bi-layers respectively, for PEO intercalation in Na-montmorillonite. The swelling of Na-montmorillonite aqueous systems favors the rapid adsorption of polymers and could determine the formation of multilayers.

Other preparation methods for PEO–clay nanocomposites (Table 1) could be carried out under dry media experimental conditions. Thus, mixtures of clay and PEO heated above the melting point of the polymer (> 350 K) leads to nanocomposites characterized by the formation of a single phase with basal spacing of 1.77 nm [61]. An alternative procedure consists of the microwave irradiation of clay–PEO mixtures for short periods of time (~ 10 min). The optimized process requires the control of the water content in the clay–polymer mixture because the microwave irradiation acts on the water molecules heating the system to provoke melting and the subsequent penetration of the PEO in the smectite interlayer space [62]. The latter two procedures give nanocomposites comprising a blend of intercalated PEO–clay particles and non-crystalline PEO (non XRD detectable), ensuring a good connectivity between the organo–inorganic microparticles (Figure 7). Such behavior is reflected in the isotropic character of the ionic conductivity, contrary to that of nanocomposites prepared from PEO solutions.

In general, PEO–clay nanocomposites show higher ionic conductivities than the parent silicate, increasing with temperature until a maximum at around 600 K. For higher temperatures, the conductivity strongly decreases owing to PEO elimination by a pyrolysis process in which oxyethylene units are progressively lost. In the case of Li^+- and Na^+-montmorillonite–PEO complexes, the maximum conductivity values measured by the a.c. impedance technique in the direction parallel to the (a,b) plane of the silicate (layer plane) are in the 10^{-5}–10^{-4} S/cm range [58]. Such values are similar to those of conventional PEO–salt complexes, but in this case measured at room temperature. Nanocomposites prepared from PEO by the melt intercalation procedure show isotropic and practically constant ionic conductivity (10^{-6} S/cm) from room temperature to 350 K, as reported by Vaia and coworkers [61]. A particular case is the ammonium-exchanged silicates, where stronger PEO–cation interactions, which involve hydrogen bonding of the ammonium ions and the

Table 1 Microstructure and behavior of clay polymer nanocomposites

Methods of preparation	Experimental procedure	Structural arrangement	Conductivity	References
Adsorption of PEO in aqueous media	PEG adsorption on Ca-mont. at r.t.	d_L from 1.3 to 2.2 nm (various phases)	[a]	[55, 56]
	PEO adsorption on swollen Na-mont.	Two phases: $d_L = 1.36$ and 1.77 nm (mono- and bi-layer)	$\sigma(max) \approx 10^{-5}$ S/cm (520 K) σ is anisotropic	[60]
Adsorption of PEO from organic solvents (acetonitrile)	PEO and PEG adsorption in M^{n+}-montm. and hectorite from acetonitrile at r.t.	Li^+, Na^+, $PrNH_3^+$, Ba^{2+}: one phase $d_L \approx 1.7$ nm (helical arrangement) NH_4^+, Al^{3+}, Cr^{3+}: two phases $d_L \approx 1.75$ and ≈ 1.0 nm Ca^{2+}: one phase $d_L \approx 1.5$ nm	σ is anisotropic $\sigma_{=}$Na-mont. $\approx 10^{-4}$ S/cm (570 K) $\sigma_{\perp}$Na-mont. $\approx 10^{-8}$ S/cm (570 K) $\sigma_{=}$Li-mont. $\approx 10^{-5}$ S/cm (550 K) $\sigma_{\perp}$Li-mont. $\approx 10^{-8}$ S/cm (560 K) $\sigma_{=}NH_4^+$-mont. $\approx 10^{-7}$ S/cm (480 K) $\sigma_{=}NH_4^+$-hectorite $\approx 10^{-7}$ S/cm (450 K)	[15, 58, 59, 63]
Polymer melt intercalation	PEO-Na- and PEO-Li-montm. heating the polymer–clay mixture at 453 K, 6 h	One phase $d_L = 1.77$ nm	σ is isotropic and constant from r.t. to 350 K σ (Li-mont) $\approx 10^{-6}$ S/cm	[61]
Microwave assisted blending-intercalation	PEO-Na- and PEO-Li-montm. 10 min MW irradiation of polymer–clay mixtures (with controlled water content)	Mixture of intercalated and non-intercalated phases varying the clay/PEO ratio	σ is isotropic	[63]
		Full intercalated phase at $d_L \approx 1.5$ nm (helical conformation)	σ_{max}(Li-mont) $\approx 10^{-5}$ S/cm (500 K)	

[a] Non reported conductivity. Samples were prepared to control the water content in clay systems.
Abbreviations: mont. montmorillonite; r.t. room temperature; MW: microwave irradiation.

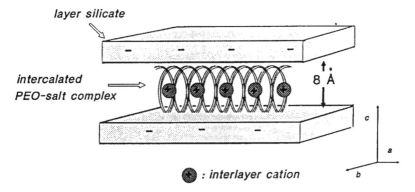

Figure 6 Structural model representing the intercalation of PEO in a homoionic smectite, after Ruiz-Hitzky and Aranda [15]. Reproduced by permission of WILEY-VCH

oxyethylene groups, can be responsible for the lower conductivities found (10^{-7} S/cm at 480 K) [63].

A significant difference between conventional and PEO–clay systems is that in these last materials only the cations are able to move, avoiding the typical problems of ion pair formation that occur in PEO–salt compounds. Therefore, the transport number, which was supposed to be equal to one (i.e. $t_+ = 1$), has recently been

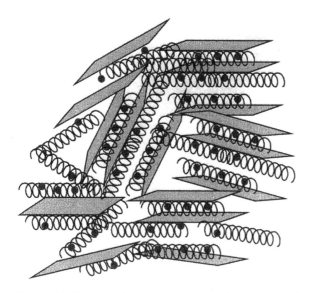

Figure 7 Schematic representation of an isotropic PEO–clay nanocomposite material prepared by melt intercalation of PEO, induced by conventional heating [61] or by microwave irradiation [62]

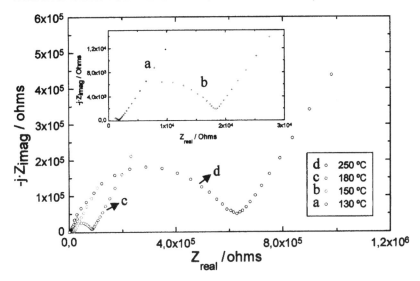

Figure 8   Evolution of Nyquist plots as a function of temperature for a PEO–Li-montmorillonite/LiCF$_3$SO$_3$ sample

determined using solid-state dielectric polarization techniques [64]. This behavior is of great interest for potential applications in electrochemical devices, as, for instance, in solid-state batteries [3, 4]. The addition of certain salts of alkaline metals (NaSCN, LiI, etc.) to clays enhances their ionic conductivity from 10^{-6} to 10^{-4} S/cm at 473 K [65]. This approach has also been applied in an attempt to increase the number of charge carriers in PEO–clay nanocomposites and therefore their conductivity. Wu and Lerner [60] observed that the addition of NaClO$_4$ to PEO–Na-montmorillonite produces a decrease in the conductivity at low temperature. However, isotropic PEO–clay nanocomposites prepared by microwave-assisted blending intercalation incorporating alkaline salts (e.g. LiCF$_3$SO$_3$ or NaCl) present slightly higher conductivities than the former nanocomposites. This effect is more significant at low temperatures (300–400 K) where the conductivity is several orders of magnitude greater in the sample containing excess salt [62], as deduced from impedance diagrams (Figure 8).

PEG–clay nanocomposites based on laponite† [66] and kaolinite [67] have been successfully attempted. For kaolinite, preliminary expansion of silicate layers by intercalation of DMSO is required prior to the insertion of the melted polymer. Kaolinite is a natural silicate belonging to the family of kandites (1 : 1 silica/ alumina, non-charged phyllosilicates, without interlayer cations) with a basal spacing of about 0.7 nm, which increases to about 1.1 nm after PEG intercalation.

---

† Laponite is a synthetic hectorite (2 : 1 trioctahedral magnesium charged phyllosilicate with OH replaced by fluorine) produced by Laporte, UK

The lack of charge carriers precludes the use of such nanocomposites as solid-state electrolytes, but the addition of alkaline salts, as suggested by Tunney and Detellier [67], could be an attractive way of preparing a second generation of electroactive nanocomposites with anisotropic behavior.

The intercalative polymerization of ethylene oxide is apparently an alternative way of preparing PEO–clay nanocomposites [68]. Nevertheless, only homoionic montmorillonites exchanged with cations exhibiting Lewis/Brönsted acidity are able to catalyze the polymerization. Therefore, this route is limited to certain homoionic clays, which precludes the preparation of nanocomposites containing charge carriers such as Li^+ or Na^+ which are of particular interest as solid-state electrolytes.

4 ELECTRONICALLY CONDUCTING POLYMER INTERCALATION IN CLAYS

Alkyl- and ary-lamines, as well as other nitrogenated bases, show an ability to intercalate smectite clays, frequently giving protonated species on account of the acidic character of the interlayer space of such phyllosilicates [8]. Among this type of molecule, aniline and pyrrole spontaneously intercalate different clays, giving in certain conditions PANI and PPy, respectively. Following Kanatzidis, such general procedure of interaction was named *in situ intercalative polymerization* [48]. The nature of the intracrystalline environment, i.e. the acidity, the oxidant character, the hydration state and other characteristics of the interlayer cations, are determining factors inducing the polymerization.

Three general procedures have been reported for preparing PANI–clay nanocomposites (Figure 9). The first one consists of the use of transition metal ions, e.g. Cu^{2+}, introduced as exchangeable cations into homoionic smectites to induce direct topotactic PANI formation. The second procedure is based on the previous exchange of interlayer cations (i.e. Na^+) by anilinium species which are subsequently polymerized by means of an oxidizing agent such as ammonium peroxodisulphate or by an electrochemical process. Finally, the third method involves the addition of PANI to the reaction media in which the clay is hydrothermally synthesized.

Cloos and coworkers [69, 70] studied the interlayer adsorption of aniline into Cu^{2+}- and Fe^{3+}-montmorillonites from aqueous solutions. For Cu^{2+}-exchanged silicates they observed the formation of black-colored intercalation materials. This feature was attributed to the generation of polymeric species when high initial aniline concentrations were employed. Using EPR and IR spectroscopy, they concluded that aniline adsorbed in Cu^{2+}-montmorillonite is oxidized to radical species, giving PANI as Cu^{2+} ions are reduced to Cu^+ ions, which are rapidly reoxidized as initiators for the growth of the polymer chain. Resonance Raman spectroscopy showed that aniline adsorption from the vapor phase at about 400 K forms benzidine dications, whereas the adsorption from liquid aniline gives PANI [71]. Interlayer

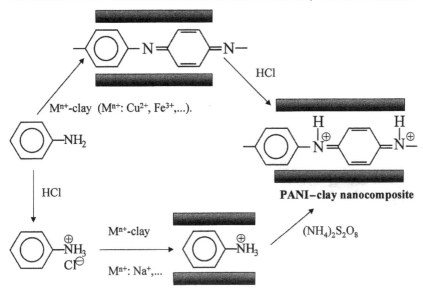

Figure 9 Schematic representation of two routes to the preparation of PANI–clay nanocomposites

PANI produced under those conditions exhibits Raman spectra consistent with a conventional polyaniline electrochemically synthesized in acidic solutions.

Mehrotra and Giannelis [72] carried out the intercalative polymerization of aniline using a Cu^{2+}-exchanged synthetic hectorite (*fluorohectorite*). The electronic absorption spectrum of the resulting nanocomposites showed the characteristic insulating PANI form (i.e. the emeraldine base). To obtain conducting materials it is necessary to dope the nanocomposite by exposure to HCl vapors. As a consequence of this treatment, PANI emeraldine salt is formed in the interlayer region. A significant increase in the electrical conductivity is observed, which reaches high in-plane d.c. conductivity values (5×10^{-2} S/cm) with an anisotropic ratio close to 10^5 [72].

In a second general procedure, PANI can also be formed in the interlayer space of smectites by a previous ion-exchange process allowing the introduction of anilinium cations by treatment with aniline hydrochloride aqueous solutions [73]. The resulting homoionic $C_6H_5–NH_3^+$ smectites, when oxidized with ammonium per-oxodisulfate, give intracrystalline PANI in its emeraldine salt form as deduced from the electronic absorption spectra [73]. Alternatively, the oxidation of aniline (in acidic media, i.e. the protonated form) could be induced by electrochemical procedures [74]. The electrical conductivity value of PANI–clay materials prepared from anilinium–montmorillonite is about 10^{-6} S/cm [73], which is lower than that of the corresponding nanocomposites obtained as described by Mehrotra and Giannelis [72]. Such discrepancies could be tentatively ascribed to the high anisotropy behavior of conductivity in this type of material. Besides, the nature of the PANI chains (length, oxidation and protonation extent, etc.) may also be different from one system to the other, thereby affecting their conductivity. In this context, it should also be taken into account that the maximum conductivity of PANI–clays depends on the aggregation state of PANI chains and that, for the intercalated materials containing isolated molecular wires, the conductivity is lower than in bulk PANI.

The basal spacing of PANI–smectite clay nanocomposites varies from 1.3 to 1.5 nm, depending on the experimental conditions adopted in the polymerization of aniline/anilinium species [72, 73, 75]. It should be assumed that monolayers of PANI chains form with the rings in a pseudo-planar or tilted arrangement with respect to the (a,b) silicate plane. However, Porter and coworkers [75] claim the possible formation of mono- and bilayers of PANI in the interlayer regions of clays.

The third general procedure, recently reported by Carrado and Xu [76], strives to synthesize PANI–hectorite nanocomposites by *in situ* hydrothermal crystallization of an aqueous mixture of a gel of SiO_2, $Mg(OH)_2$, LiF and PANI which is refluxed for 2 days. Interestingly, the resulting PANI nanocomposites may contain controllable amounts of PANI that strongly influence the structural arrangement and the properties of the resulting materials. The fact that delaminated clays are formed as inorganic host lattices could affect the connectivity of PANI chains, enhancing the electrical conductivity of the system. Such studies could be of interest for future research on this subject, in particular taking into account the ability of PANI–clay systems to make chemical sensors to detect gases and vapors, such as ethanol, hexane, etc., in agreement with previous results reported by Porter and coworkers [75, 77–79].

Smectite clays exchanged with transition metal ions adsorb pyrrole, inducing immediately its polymerization and giving PPy–clay nanocomposites as reported for the first time by Vande Poel in Cu^{2+}-montmorillonite [80]. More recently, Mehrotra and Giannelis [81] describe the formation of this type of composite using a Cu^{2+}-exchanged fluorohectorite. The increase in the interlayer distances deduced from XRD patterns ranges from 1.41 to 1.51 nm, depending on the experimental conditions in the preparation [80, 81], as with PANI–clay formation. The variation in the basal spacing (Δd_L) is consistent with the molecular thickness (0.45–0.50 nm) of polymer chains with the heterocyclic rings oriented parallel to the (a,b) plane of the silicate. From IR results [80] it could be inferred that PPy is directly formed in its conductive state. According to the work by Mehrotra and Giannelis [81], an IR broad band at 4000–2000 cm^{-1} is indicative of the presence of polarons/bipolarons in conducting PPy, whose formation into Cu^{2+}-hectorite was confirmed by Raman spectroscopy. Doping of PPy–hectorite with iodine vapours produces a strong increase in the d.c. conductivity from 2×10^{-5} (undoped) to 1.2×10^{-2} S/cm (I_2 doped). It is noteworthy that the electrical conductivity of the latter systems is highly anisotropic, with a $\sigma_=/\sigma_\perp$ ratio of 4×10^3 [81]. The electrical conductivity increases with temperature in the 100–400 K range, as illustrated in Figure 10. Above 400 K a dramatic decrease in the conductivity is observed, which could be interpreted as

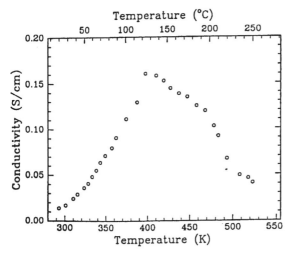

Figure 10 Evolution of the conductivity in the (*a,b*) plane versus temperature for a PPy–fluorohectorite nanocomposite. Reprinted from Mehrotra and Giannelis [81] Copyright 1992, with permission from Elsevier Science

strong oxidation of PPy. In fact, according to Vande Poel, metallic copper could be produced in agreement with XPS results [80].

Smectites exchanged with transition metal ions (e.g. Cu^{2+}) have also been intercalated with other aromatic molecules, such as benzene, biphenyl, thiophene, etc., showing the formation of 'pi' and 'sigma' complexes, as first reported by Mortland and Pinnavaia [82]. This chemistry was used to induce the formation of different polymers of potential interest for their electrical properties, such as poly(*p*-phenylene) [83]. Polymerization of thiophene and methylthiophenes can also be attained after intercalation of the corresponding monomers into smectites exchanged with transition metal ions such as Cu^{2+}, Fe^{3+} and VO^{2+} [84–87]. Alternatively, Oriakhi and Lerner [88] report a route for incorporating previously formed PTh or PPy polymers in colloidal dispersions of montmorillonite. This procedure does not require the presence of transition metal ions in the smectite to induce a further polymerization of the species.

Current work is being devoted to applications of conducting polymers associated with inorganic solids including clays. Such classes of polymers present remarkable electrochemical behavior, attracting interest in applications that include modified electrodes, biosensors, solid-state batteries, smart windows and other electrochemical devices [21, 89–91]. Thus, for instance, PPy–clay nanocomposites have been envisaged for the development of modified electrodes used as sensors [92–94] or as devices for electrocatalysis [95]. Ramachadran and Lerner [96] report that PPy–montmorillonite nanocomposites exhibit a redox chemistry approximately 1 V more negative than chemically synthesized PPy. Although the capacities of the materials are similar, the charge/ discharge plateau is also shifted approximately 1 V. These observations could be related to the close interaction of the polymer with the anionic clay layers, stabilizing the PPy cationic form. Changes in the reaction mechanism could also be invoked. Modified electrodes based on PPy–clay nanocomposites are in general more stable than either the clay or the polymer electrode. Combining the ability of clays to inmobilize electroactive species (i.e. a redox couple) by ion exchange with the electronic conductivity ensured by the PPy could be useful for electroanalytical purposes and also for the immobilization of electrocatalysts [92]. In this context, Besombes and coworkers [94] have shown that modified electrodes based on PPy–laponite nanocomposites incorporating polyphenol oxidase or cholesterol oxidase present higher stability than the corresponding electrodes made without clay.

5 CONDUCTING POLYMER NANOCOMPOSITES BASED ON CLAY RELATED SOLIDS

5.1 INORGANIC LAYERED SOLIDS

Similarly to clays, other two-dimensional inorganic solids can act as host lattices for conducting polymer intercalations [4]. Among such solids are transition-metal oxides (e.g. V_2O_5 xerogel), halides (e.g. α-$RuCl_3$) and oxyhalides (e.g. FeOCl), chalcogenides (e.g. MoS_2), phosphates [e.g. $Cu(UO_2PO_4)_2.10H_2O$] and trichalcogenide phosphates (e.g. $CdPS_3$). Also, layered double hydroxides (LDHs) and graphite oxide have been used to prepare a large variety of electroactive polymer nanocomposites (Table 2).

On the basis of the nature of both the two-dimensional solid and the electroactive polymer, various synthesis procedures can be adopted:

(a) adsorptive polymer intercalation from solutions,
(b) *in situ* intercalative polymerization,
(c) delamination and entrapping–restacking.

The first method consists of the spontaneous penetration of the polymer in the two-dimensional interlayer space, the driving force mainly related with entropic changes. Complexing of interlayer cations, water replacement, 'crawling' of polymers such as PEO to the interior of solids, etc., are factors that could be invoked to explain the intercalation mechanisms. The cohesion between the stacked layers of the host solids has to be broken to facilitate the polymer penetration. A strategy to assist such processes includes the deliberate introduction of certain cations or the controlled addition of water to the solvent. Examples of nanocomposites formed by these procedures are PEO-V_2O_5 [29] and PEO–$CdPS_3$ [97–99]. The lack of large layer expansions avoids the loss of stacking order in the resulting two-dimensional nanocomposites.

Table 2

2D inorganic host lattice	Intercalated conducting polymer	Preparation and other features	Electrical conductivity (S/cm)	References
FeOCL	PPy, PANI, PTh polyfurane	The intercalative polymerization firstly reported (PPy) Evidence of Fe(III) reduction	0.1 (PPy); 10^{-3}–10^{-1} (PANI) 5 (PTh)	[48, 116–121]
α-RuCl$_3$ Li$_{0.2}$RuCl$_3$	PANI, PPy PEO, PVP	In situ polymerization and/or delamination—entrapping–restacking	0.1 (PANI); 23 (PPy) $4.5\cdot10^{-3}$ (PEO); 10^{-8} (PVP)	[101, 122]
M$_x$MoO$_3$ (M = Li$^+$, Na$^+$).	PPV, PANI, PVP PEO, PEG	PPV formed through cationic intermediates Non-ionic polymers intercalate via delamination–entrapping–restacking	0.5 (PPV); 10^{-3} (PANI); 10^{-5}–10^{-4}(PEO)	[123–129]
MO$_2$ (M = Co, Ni)	PANI	Intermediate lithiated phases treated with (NH$_4$)$_2$S$_2$O$_8$ and anilinium chloride	—	[130]
V$_2$O$_5$ (xerogel and aerogel)	PANI, PPy, PTh PEO, PEG	Intercalative polymerization delamination– entrapping–restacking Adsorptive intercalation from solutions	10^{-2}–10 (PANI); 10^{-4} (PPy); 10^{-4}–10^{-3} (PEO)	[29, 51, 103–105, 131–136]
LDHs (Cu$_{1-x}$Cr$_x$(OH)$_2$·nH$_2$O)	PANI	Intercalative polymerization of aniline into pre-intercalated LDHs	—	[137]
C (graphite) oxide	PEO, PANI	Delamination–entrapping–restacking	—	[138–141]
MX$_2$ (M = Mo, Ti, Zr, Ta, Nb..) (X = S, Se)	PEO, PVP PANI, PPy	Delamination–entrapping–restacking using intermediate lithiated phases	10^{-3}–10^{-1} (PEO); 10 (PANI); 100–250 (PEO, PVP-NbSe$_2$)	[33, 142–153]
MPS$_3$(M = Mn, Cd, Zn)	PEO PANI	Intermediate Li- or Na-phases followed by adsorptive intercalation from solutions	$<10^{-7}$	[97–99, 154, 155]
Layered phosphates	PPy, PANI	In situ intercalative polymerization	$<10^{-10}$	[100, 156–163]

In some cases (α-VOPO$_4$) external + interlayer polymerization. Evidence of V(V) reduction

In situ polymerization is a procedure first reported by Kanatzidis and coworkers [48] for preparing PPy–FeOCl nanocomposites (Figure 11). This method consists in the insertion of molecules or ions acting as monomers which could be induced to polymerize within the intracrystalline region of the two-dimensional solid. The presence of transition metal ions either as exchangeable cations or included in the structure of the layered solids is used to obtain composites containing intercalated PPy, PANI, PTh, etc. It has been clearly shown through spectroscopic techniques that the role of the oxidizing transition metal ions (such as Fe^{3+} in FeOCl or V^{4+} in α-VOPO$_4$) determines the conducting form of the polymer and, therefore, the electrical behavior of the resulting nanocomposite [48, 100].

The third general preparative method is delamination followed by entrapping–restacking (*encapsulative precipitation* after Kanatzidis [101]). This process implies two consecutive steps:

(a) expansion along the *c* axis of the solid layers provoked by the entry of molecules of solvent into the intracrystalline region, giving a colloidal phase, previous treatment being necessary in some cases to produce expandable intermediate phases (e.g. $MoS_2 + BuLi \rightarrow Li_xMoS_2$);

(b) entrapping of the polymer during the restacking step of the dispersed layers, giving rise to the intercalated material. Alternatively, precursors of polymers could be entrapped and subsequently polymerized ('*in situ* polymerization coupled with encapsulative precipitation', after Kanatzidis [101]).

The last procedure is advantageous for obtaining a large variety of nanocomposites, the only requirement being an affinity and compatibility of solvents towards both the guest species and the host solids. However, the resulting nanocomposites frequently show lower crystalline order than intercalation compounds obtained by the above described procedures. Poorly stacked and interstratified materials are often obtained following this last preparative method.

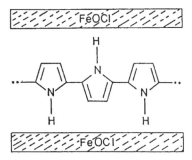

Figure 11 Schematic representation of a PPy–FeOCl intercalated nanocomposite. Reprinted with permission from Kanatzidis *et al.* [48]. Copyright 1987 American Chemical Society

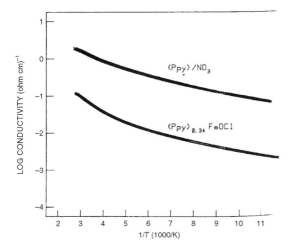

Figure 12 Conductivity versus temperature for a bulk doped (NO_3^-) PPy polymer and a PPy–FeOCl nanocomposite. Reprinted with permission from Kanatzkidis *et al.* [48]. Copyright 1987 American Chemical Society

The electrical conductivity of polymer–two-dimensional solids depends on the nature of both the host inorganic solid and the guest polymeric species. As a general rule, the intercalation of conducting polymers always produces less conductive nanocomposite materials compared with bulk polymers (Figure 12), as indicated elsewhere [4]. Nevertheless, the synergetic or complementary behavior between the polymer and the two-dimensional inorganic solid is interesting for electrochemical devices applications. Applications of such nanocomposites obtained by intercalation of electroactive polymers into transition metal oxides (e.g. V_2O_5 xerogel) and chalcogenides (e.g. MoS_2) are promising as modified positive electrodes for rechargeable batteries based on lithium insertion into those host solids. In fact, intercalated polymers such as PEO can facilitate the mobility of Li^+ ions, and consequently the efficiency of these systems could be enhanced. Intercalation of electronically conducting polymers such as PPy and PANI, and, more recently, polymers containing polythiadiazole units, ensures a greater electronic conductivity of the material, which is also a requirement for the design of efficient battery electrodes [102–109].

5.2 SILICA AND ZEOLITES

Table 3 shows the main features of nanocomposites prepared from different conducting polymers and silicic materials with three-dimensional networks. These

Table 3 Silica and 3D as host lattice for conducting polymer insertion

3D inorganic host lattice	Inserted conducting polymer	Preparation and behavior	Electrical conductivity (S/cm)	References
SiO$_2$xerogel	PPy, PANI, PPV	Adsorption in colloidal silica *In situ* PPy polymerization on Cu-SiO$_2$ matrices generated by solgel. I$_2$ doping. PANI, PPV entrapping into SiO$_2$ matrices generated by solgel.	10^{-3} (PPy)	[164–167]
Mesoporous silica (MCM41)	PANI	*In situ* polymerization into silica nanopores.	10^{-3}	[111]
Porous Vycor glass	PPy, PANI	*In situ* oxidative polymerization of PPy into Vycor glass pores doped with Cu^{2+}	10^{-7} (PPy)	[168–170]
Zeolites	PPy, PANI, PTh polyfurfural	*In situ* polymerization into micropores of the zeolites induced by (i) the presence of redox cationic pairs, (ii) oxidant agents as ammonium peroxodisulphate or (iii) electrochemical oxidation	$<10^{-7}$ 7×10^{-6} (polyfurfural)	[73, 171–178]

materials contain cavities with variable dimensions and topology, as well as different crystalline organization. Except for polymers entrapped by sol–gel during the generation of the silica network, the general procedure is based on *in situ* polymerization. Zeolites are the most studied host lattices [110]. They resemble clays, having similar composition and ion-exchange properties. Zeolites only differ from clays in the shape and size of the intracrystalline cavities that control the growth of individual polymer chains isolated one from another by the silicate matrix.

The electrical conductivity of these nanocomposites is of the order of that of clays, or even lower than that of phyllosilicate-based nanocomposites. The growth of polymers into large pore materials such as MCM41 mesoporous silica (about 3 nm in pore diameter) gives nanowires with higher conductivities due to the polymer chain aggregations [111]. Larger-pore materials (inorganic porous membranes) allow for the formation of thicker polymer wires and nanotubes within the solid matrix which can reach conductivity values even greater than that of the bulk polymers [112–114]. Porous alumina membranes obtained by electrochemical anodization of aluminum and microporous mica membranes prepared by 'track-etch' methods can be used as templates for the synthesis of nanowires and nanotubules of conducting polymers such as PANI and PPy. For instance, PPy synthesized in the pores of a microporous mica membrane (Figure 13) presents conductivity values greater than 100 S/cm, although the connectivity among polymers in different pores results in heterogeneous behavior of the resulting nanocomposite (P. Aranda and C.R. Martin, unpublished results).

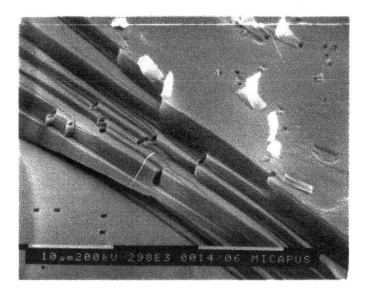

Figure 13   SEM image of PPy microtubules included into the track-etched pores in a mica (after P. Aranda and C.R. Martin, unpublished results)

6 CONCLUSIONS

Clay minerals were the first solids to intercalate electroactive polymers or to induce the polymerization in their interlayer space, giving electroactive organic–inorganic systems. Smectites and other charged phyllosilicates are versatile materials particularly well adapted to produce a large variety of nanocomposites because of their unique ion-exchange, expandability and colloidal properties. The ability to insert neutral polar molecules into interlayers containing complexing cations is the basis for preparing PEO–clay solid electrolytes. The ability to introduce transition metal ions, such as Cu^{2+}, together with the affinity of clays towards intercalation of nitrogenous molecules, such as pyrrole, makes it possible to induce intracrystalline polymerization, giving electroactive materials after appropriate doping. Therefore, clays are excellent models for preparing anisotropic functional nanocomposites that show peculiar electrical properties. Note, for instance, that PEO–clay materials are pure cationic conductors.

The electrical conductivity of electroactive polymer–clay nanocomposites is always lower than that of the corresponding parent guest polymer without the clay. In fact, these composites result from the assembly of a conducting polymer separated by individual silicate layers ('ceramic') of insulating character. Moreover, electrical transport through short and narrow intercalated polymer chains is strongly disfavoured. The constrained polymer in the intracrystalline region precludes interchain interactions and can impose chain conformation changes that also contribute to a decrease in the conductivity of the system. Besides, the microparticulate nature of clays is another factor affecting the connectivity needed to procure good electrical conductivity.

Electroactive polymer–clay nanocomposites and related material can be regarded as a novel class of materials exhibiting singular behavior. In spite of their low conductivity, polymer–clay nanocomposites offer complementary or synergetic effects that improve and open a way to various applications. Stabilization and enhancement of the mechanical and thermal properties of the conducting polymers are examples that illustrate how these nanocomposites can be useful as modified electrodes, electrocatalysts and other future applications.

7 REFERENCES

1. *Appl. Clay. Sci.* (special issue on Clay Mineral–Polymer Nanocomposites), **15** (1999).
2. Ruiz-Hitzky, E. *Mol. Cryst. Liq. Cryst.*, **161**, 433–452. (1988).
3. Ozin, G. A. *Adv. Mater.*, **4**, 612–649 (1992).
4. Ruiz-Hitzky, E. *Adv. Mater.*, **5**, 334–340 (1993).
5. Ruiz-Hitzky, E., Aranda, P., Casal, B., and Galván, J. C. *Adv. Mater.*, **7**, 180–184 (1995).
6. Giannelis, E. P. *Adv. Mater.*, **8**, 29–35 (1996).
7. Ruiz-Hitzky, E. and Aranda, P. *An. Quím. Int. Ed.*, **93** 197–212 (1997).

8. Rausell-Colom, J. A. and Serratosa, J. M. in Newman, A. C. D. (Ed.), *Chemistry of Clays, and Clay Minerals*, London, 1987, pp. 371–422.
9. Ogawa, M. and Kuroda, K. *Bull. Chem. Soc. Jpn*, **70**, 2593–2618 (1997).
10. Fripiat, J. J., Jelli, A., Poncelet, G. and André, J. *J. Phys. Chem.*, **69**, 2185–2197 (1965).
11. Calvet, R. and Mamy, J. *C.R. Acad. Sci. Paris*, **273**, 1251–1253 (1971).
12. Hoffman, U., Endell, K. and Will, D. *Z. Krist.*, **86**, 340–348 (1933).
13. Ruiz-Hitzky, E. and Casal, B. *Nature*, **276**, 596–597 (1978).
14. Aranda, P., Galván, J. C., Casal, B. and Ruiz-Hitzky, E. *Electrochim. Acta*, **37**, 1573–1577 (1992).
15. Ruiz-Hitzky, E. and Aranda, P. *Adv. Mater.*, **2**, 545–547 (1990).
16. Allcock, H. R. *Science*, **255**, 1106–1112 (1992).
17. Armand, M. B. *Adv. Mater.*, **2**, 278–286 (1990).
18. MacCallum, J. R. and Vincent, C. A. (Eds.), *Polymer Electrolyte Reviews*, Elsevier Applied Science, London, Vols 1 and 2, 1987, and 1989.
19. Vincent, C. A. *Chem. Br.*, 391–395 (1989).
20. Ratner, M. A. and Shriver, D. F. *Chem. Rev.*, **88**, 109–124 (1988).
21. Salaneck, W. R., Clark, D. T. and Samuelsen, E. J. (Eds.), *Science and Application of Conducting Polymers*, Adam Hilger, Bristol, 1991.
22. Syed, A. A. and Dinesan, M. K. *Talanta*, **38**, 815–837 (1991).
23. Pedersen, C. J. *J. Am. Chem. Soc.*, **89**, 7017–7036 (1967).
24. Izatt, R. M. and Christensen J. J. (Eds.), *Progress in Macrocyclic Chemistry*, John Wiley & Sons, New York, Vols 1 and 2, 1979, and 1981.
25. Ruiz-Hitzky, E. and Casal, B. in Setton, R. (Ed.), *Chemical Reactions in Organic and Inorganic Constrained Systems*, D. Reidel, Dordrecht, NATO-ASI Series C, Vol. 165, 1986, pp. 179–189.
26. Casal, B., Ruiz-Hitzky, E., Van Vaeck, L. and Adams, F. C. *J. Incl. Phenom.*, **6**, 107–118 (1988).
27. Casal, B., Aranda, P., Sanz, J. and Ruiz-Hitzky, E. *Clay Miner.*, **29**, 191–203 (1994).
28. Aranda, P., Casal, B., Fripiat, J. J. and Ruiz-Hitzky, E. *Langmuir*, **10**, 1207–1212 (1994).
29. Ruiz-Hitzky, E., Aranda, P. and Casal, B. *J. Mater. Chem.*, **2**, 581–582 (1992).
30. Glueck, D. S., Brough, A. R., Mountford, P. and Green, M. L. H. *Inorg. Chem.*, **32**, 1893–1902 (1993).
31. Herber, R. H. and Casell, R. A. *J. Chem. Phys.*, **75**, 4669–4678 (1981).
32. Guerrero, A. R., Peinado-García, J. and López-González, J. de *An. Quím.*, **78**, 270–273 (1982).
33. Lara, M. and Ruiz-Hitzky, E. *Braz. Chem. Soc.*, **7**, 193–197 (1996).
34. Armand, M. B. *Solid State Ionics*, **9–10**, 745–754 (1983).
35. Plocharski, J. and Wieczorek, W. *Solid State Ionics*, **28–30**, 979–982 (1988).
36. Wieczorek, W., Such, K., Wycislik H. and Plocharski, J. *Solid State Ionics*, **36**, 225–257 (1989).
37. Capuano, F., Croce, F. and Scrosati, B. *J. Electrochem. Soc.*, **138**, 1918–1922 (1991).
38. Shirakawa, H., Louis, E. J., MacDiarmid, A. G., Chiang, C. K. and Heeger, A. J. *J. Chem. Soc., Chem. Commun.*, 578–580 (1977).
39. Skotheim T. A. (Ed.), *Handbook of Conducting Polymers*, Dekker, New York, Vols 1 and 2, 1986.
40. Baseku, N., Liu, Z. X., Moses, D., Heeger, A. J., Naarmann, H. and Teophilou, H. *Nature*, **327**, 403–405 (1987).
41. Naarmann, H. and Teophilou, H. *Synth. Metals*, **22**, 1–8 (1987).
42. Roth, S. *Synth. Metals*, **34**, 617–622 (1989).
43. Roth S. in Salaneck, W. R., Clark, D. T. and Samuelsen E. J. (Eds.), *Science and Applications of Conducting Polymers*, Adam Hilger, Bristol, 1991, pp. 129–139.

44. MacDiarmid, A. G., Chiang, J. C., Halpern, M., Huang, W. S., Mu, S. L., Somasiri, N. L. D., Wu, W. and Yaniger, S. I. *Mol. Cryst. Liq. Cryst.*, **121**, 173–180 (1985).
45. Syed, A. A. and Dinesan, M. K. *Talanta*, **38**, 815–837 (1991).
46. Diaz, A. F., Kanazawa, K. K. and Gardini, J. P. *J. Chem. Soc., Chem. Commun.*, 635–636 (1979).
47. Tourillon, G. and Garnier, F. *J. Electroanal. Chem.*, **135**, 173–178 (1982).
48. Kanatzidis, M. G., Tonge, L. M., Marks, T. J., Marcy, H. O. and Kannewurf, C. R. *J. Am. Chem. Soc.*, **109**, 3797–3799 (1986).
49. Casal, B. and Ruiz-Hitzky, E. *Opt. Pur. Apl.*, **18**, 49–58 (1985).
50. Casal, B. and Ruiz-Hitzky, E. *Clay Miner.*, **21**, 1–7 (1986).
51. Aranda, P., Casal, B., Galván, J. G. and Ruiz-Hitzky E. in Bernier, P., Fischer, J. E., Roth, S. and Solin, S. A. (Eds.), *Chemical Physics of Intercalation II*, Plenum Press, New York, NATO-ASI Series B, Vol. 305, 1993, pp. 397–400.
52. Aranda, P., Galván, J. C., Casal, B. and Ruiz-Hitzky, E., *Colloid Polym. Sci.*, **272**, 712–720 (1994).
53. Slade, R. C. T., Barker, J., Hirst, P. R., and Hilstead, T. K. *Solid State Ionics*, **24**, 289–295 (1987).
54. Aranda, P. PhD Thesis, Universidad Complutense, Madrid, 1991.
55. Parfitt, R. L. and Greenland, D. J. *Clay Miner.*, **8**, 305–315 (1970).
56. Parfitt, R. L. and Greenland, D. J. *Clay Miner.*, **8**, 317–323 (1970).
57. Theng, B. K. G. *Formation, and Properties of Clay-Polymer Complexes*, Elsevier Science, New York, 1979, p. 82.
58. Aranda, P. and Ruiz-Hitzky, E. *Chem. Mater.*, **4**, 1395–1403 (1992).
59. Aranda, P. and Ruiz-Hitzky, E. *Acta Polymer.*, **45**, 59–67 (1994).
60. Wu, J. and Lerner, M. M. *Chem. Mater.*, **5**, 835–838 (1993).
61. Vaia, R. A., Vasudevan, S., Krawiec, W., Scanlon, L. G. and Giannelis, E. P. *Adv. Mater.*, **7**, 154–156. (1995)
62. Aranda, P., Galván, J. C. and Ruiz-Hitzky, E. in Laine, R. M., Sanchez, C., Brinker, C. J. and Giannelis E. (Eds.), *Organic/Inorganic Hybrid Materials*, MRS Symposium Proceedings, Warrendale, 1998, Vol. 519, pp. 375–380.
63. Aranda, P. and Ruiz-Hitzky, E. *Appl. Clay Sci.*, **15**, 119–135 (1999).
64. Ruiz-Hitzky, E., Aranda, P., Perez-Cappe, E., Mosquedo-Laffita, Y. and Villanueva, A. *Rev. Cubana Quim.*, **12**, 58–63 (2000).
65. Kawada, T., Tokokawa, H. and Dokiya, M. *Solid State Ionics*, **28–30**, 210–213 (1988).
66. Doef, M. M. and Reed, J. S. *Solid State Ionics*, **115**, 109–115 (1998).
67. Tunney, J. J. and Detellier, C. *Chem. Mater.*, **8**, 927–935 (1996).
68. Pusino, A., Gennari, M., Premoli, A. and Gessa, C. *Clay, Clays Miner.*, **38**, 213–215 (1990).
69. Cloos, P., Moreale, A., Braers, C. and Badot, C. *Clay Miner.*, **14**, 307–321 (1979).
70. Moreale, A., Cloos, P. and Badot, C. *Clay Miner.*, **20**, 29–37 (1985).
71. Soma, Y. and Soma, M. *Clay Miner.*, **23**, 1–12 (1988).
72. Mehrota, V. and Giannelis, E. P. *Solid State Commun.*, **77**, 155–158 (1991).
73. Chang, T.-C., Ho, S.-Y. and Chao, K.-J. *J. Chin. Chem. Soc.*, **39**, 209–212.(1992)
74. Ioune, H. and Yoneyama, H. *J. Electroanal. Chem.*, **233**, 291–294 (1987).
75. Porter, T. L., Thompson, D., Bradley, M., Eastman, M. P., Hagerman, M. E., Attuso, J. L., Votava, A. E. and Bain, E. D. *J. Vac. Sci. Technol. A*, **15**, 500–504 (1997).
76. Carrado, K. A. and Xu, L. *Chem. Mater.*, **10**, 1440–1445 (1998).
77. Eastman, M. P., Hatgerman, M. E., Attuso, J. L., Bain, E. D. and Porter, T. L. *Clays Clay Miner.*, **44**, 769–773 (1996).

78. Porter, T. L., Eastman, M. P., Zhang, D. Y. and Hagerman, M. E. *J. Phys. Chem. B*, **101**, 11 106–11 111 (1997).
79. Porter, T. L., Manygoats, K., Bradley, M., Eastman, M. P., Reynolds, B. P., Votava, A. E. and Hagerman, M. E. *J. Vac. Sci. Technol. A*, **16**, 926–931 (1998).
80. Vande D. Poel, PhD Thesis, Catholic University of Louvain, Leuven, 1975.
81. Mehrota, V. and Giannelis, E. P. *Solid State Ionics*, **51**, 115–122 (1992).
82. Mortland, M. M. and Pinnavaia, T. J. *Nature*, **229**, 75–77 (1971).
83. Soma, Y., Soma, M. and Harada, I. *J. Phys. Chem.*, **88**, 3034–3038 (1984).
84. Camerlynck, J. P. Mémoire Grade d'Ingénieur Chimiste et des Industries Agricoles, Catholic University of Louvain, Leuven, 1972.
85. Cloos, P. Vande Poel, D. and Camerlynck, J. P. *Nature, Phys. Sci.*, **243**, 54–55 (1973).
86. Pinnavaia, T. J., Hall, P. T., Cady, S. S. and Mortland, M. M. *J. Phys. Chem.*, **78**, 994–999 (1974).
87. Soma, Y., Soma, M., Furukawa, Y. and Harada, I. *Clays Clay Miner.*, **35**, 53–59 (1987).
88. Oriakhi, C. O. and Lerner, M. M. *Mater. Res. Bull.*, **30**, 723–729 (1995).
89. Brédas, J. L. and Chance, R. R. (Eds), Conjugated Polymeric Materials: Opportunities in Electronics, Optoelectronics and Molecular electronics, Kluwer Academic Publishers, Dordrecht, NATO-ASI Series E, Vol. 182, 1990.
90. Linford, R. G. (Ed.), *Electrochemical Science and Technology of Polymers*, Elsevier Applied Science, Essex, Vols and 2, 1987 and 1990.
91. Lyons, M. E. G. (Ed.), *Electroactive Polymer Electrochemistry*. Part 2. *Methods and Applications*, New York, 1996.
92. Rudzinski, W. E., Figueroa, C., Hoppe, C., Kuromoto, T. Y. and Root, D. *J. Electroanal. Chem.*, **243**, 367–378 (1988).
93. Fan, F.-R. F. and Bard, A. J. *J. Electrochem. Soc.*, **133**, 301–304 (1986).
94. Besombes, J. L., Cosnier, S. and Labbé, P. *Talanta*, **44**, 2209–2215 (1997).
95. Faguy, P. W., Ma, W., Lowe, J. A., Pan, W. and Brown, T. *J. Mater. Chem.*, **4**, 771–772 (1994).
96. Ramachadran, K. and Lerner, M. M. *J. Electrochem. Soc.*, **144**, 3739–3743 (1997).
97. Lagadic, I., Léaustic, A. and Clément, R. *J. Chem. Soc., Chem. Commun.*, 1396–1397 (1992).
98. Jeevananadm, P. and Vasudevan, S. *Chem. Mater.*, **10**, 1276–1285 (1998).
99. Jeevananadm, P. and Vasudevan, S. *J. Chem. Phys.*, **109**, 8109–8117 (1998).
100. de Estefanis, A., Foglia, S. and Tomlinson, A. A. G. *J. Mater. Chem.*, **5**, 475–483 (1995).
101. Wang, L., Brazis, P., Rocci, M., Kannewurf, C. R. and Kanatzidis, M. G. in Laine, R. M., Sanchez, C., Brinker, C. J. and Giannelis E. (Eds), *Organic/Inorganic Hybrid Materials, MRS Symposium Proceedings*, Warrendale, 1998, Vol. 519, pp. 257–264.
102. Kloster, G. M., Thomas, J. A., Brazis, P. W., Kannewurf, C. R. and Shriver, D. F. *Chem. Mater.*, **8**, 2418–2420 (1996).
103. Leroux, F., Koene, B. E. and Nazar, L. F. *J. Electrochem. Soc.*, **143**, L181–L183 (1996).
104. Leroux, F., Goward, G., Power, W. P. and Nazar, L. F. *J. Electrochem. Soc.*, **144**, 3886–3895 (1997).
105. Goward, G., Leroux, F., Power, W. P. and Nazar, L. F. *Electrochim. Acta*, **43**, 1307–1313 (1998).
106. Lira-Cantú, M. and Gómez-Romero. P. *J. Electrochem. Soc.*, **146**, 2029–2033 (1999).
107. Lira-Cantú, M. and Gómez-Romero. P. *J. New Mater. Electrochem. Syst.*, **2**, 142–144 (1999).
108. Lira-Cantú, M. and Gómez-Romero. P. *Int. J. Inorg. Mater.*, **1**, 111–116 (1999).
109. Shouji, E. and Buttry, D. A. *Langmuir*, **15**, 669–673 (1999).
110. Ozin, G. A., Kuperman, A. and Stein, A. *Angew. Chem. Int. Ed. Engl.*, **28**, 359–376 (1989).

111. Wu, C.-G. and Bein, T. *Science*, **264**, 1757–1759 (1994).
112. Martin, C. R. *Science*, **266**, 1961–1966 (1994).
113. Martin, C. R. *Chem. Mater.*, **8**, 1739–1746 (1996).
114. Hulteen, J. C. and Martin, C. R. *J. Mater. Chem.*, **7**, 1075–1087 (1997).
115. Chatani, Y. Y. and Okamura S. *Polymer*, **28**, 1815–1820 (1987).
116. Kanatzidis, M. G., Wu, C.-G., Marcy, M. O., DeGroot, D. C., Kannewurf, C. R., Kostikas, A. and Papaefthymiou, V. *Adv. Mater.*, **2**, 364–366 (1990).
117. Kanatzidis, M. G., Hubbard, M., Tonge, L. M., Marks, T. J., Marcy, M. O. and Kannewurf, C. R. *Synth. Metals t*, **28**, C89–C95 (1989).
118. Kanatzidis, M. G., Marcy, M. O., McCarthy, W. J., Kannewurf, C. R. and Marks, T. J. *Solid State Ionics*, **32-33**, 594–608 (1989).
119. Wu, C.-G., Marcy, M. O., DeGroot, D. C., Schindler, J. L., Kannewurf, C. R., Leung, W.-Y., Benz, M., LeGoff, E. and Kanatzidis, M. G. *Synth. Metals*, **41–43**, 797–803 (1991).
120. Kanatzidis, M. G., Wu, C.-G., Marcy, M. O., DeGroot, D. C., Schindler, J. L., Kannewurf, C. R., Benz, M. and LeGoff, E. in Bein, T. (Ed.), *Supramolecular Architecture: Synthetic Control in Thin Films and Solids*, ACS Symp. Series, Vol., **499**, 1992, pp. 194–219.
121. Wu, C.-G., DeGroot, D. C., Marcy, H. O., Schindler, J. L., Kannewurf, C. R., Bakas, T., Papaefthymiou, V., Hirpo, W., Yesinowski, J. P., Liu, Y.-J. and Kanatzidis, M. G. *J. Am. Chem. Soc.*, **117**, 9229–9242 (1995).
122. Wang, L., Brazis, P., Rocci, M., Kannewurf, C. R. and Kanatzidis, M. G. *Chem. Mater.*, **10**, 3298–3300 (1998).
123. Nazar, L. F., Zhang, Z. and Zinkweg, D. *J. Am. Chem. Soc.*, **114**, 6239–6240 (1992).
124. Nazar, L. F., Wu, H. and Power, W. P. *J. Mater. Chem.*, **5**, 1985–1993 (1995).
125. Kerr, T. A., Wu, H. and Nazar, L. F. *Chem. Mater.*, **8**, 2005–2015 (1996).
126. Goward, G. R., Kerr, R. A., Power, W. P. and Nazar, L. F. *Adv. Mater.*, **10**, 449–452 (1998).
127. Bissessur, R., de Groot, D. C., Schindler, J. L., Kannewurf, C. R. and Kanatzidis, M. G. *J. Chem. Soc., Chem. Commun.*, 687–689 (1993).
128. Wang, L., Kannewurf, C. R. and Kanatzidis, M. G. *J. Mater. Chem.*, **7**, 1277–1283 (1997).
129. de Farias, R. F., de Souza, J. M., de Melo, J. V. and Airoldi, C. *J. Colloid Interface Sci.*, **212**, 123–129 (1999).
130. Ramachandran, K., Oriakhi, C. O., Lerner, M. M. and Koch, V. R. *Mater. Res. Bull.*, **31**, 767–772 (1996).
131. Kanatzidis, M. G., Wu, C. G., Marcy, H. O. and Kannewurf, C. R. *J. Am. Chem. Soc.*, **111**, 4139–4141 (1989).
132. Liu, Y.-J., DeGroot, D. C., Schindler, J. L., Kannewurf, C. R. and Kanatzidis, M. G. *Chem. Mater.*, **3**, 992–994 (1991).
133. Liu, Y.-J., Schindler, J. L., DeGroot, D. C., Kannewurf, C. R., Hirpo, W. and Kanatzidis, M. G. *Chem. Mater.*, **8**, 525–534 (1996).
134. Wong, H. P., Dave, B. C., Leroux, F., Dunn, B. and Nazar, L. F. *J. Mater. Chem.*, **8**, 1019–1027 (1998).
135. Harreld, J., Wong, H. P., Dave, B. C., Dunn, B. and Nazar, L. F. *J. Non-Cryst. Solids*, **225**, 319–324 (1998).
136. Casal, B. and Ruiz-Hitzky: E. *Communication to 3rd National Congress of French Society of Chemistry*, SCF 88, Nice, France, 1988, Abstracts, p. 8P26.
137. Challier, T. and Slade, R. C. T. *J. Mater. Chem.*, **4**, 367–371 (1994).
138. Matsuyo, Y., Tahara, K. and Sugie, Y. *Carbon*, **34**, 672–674 (1996).
139. Matsuo, Y., Tahara, K. and Sugie, Y. *Carbon*, **35**, 113–120 (1997).
140. Higashika, S., Kimura, K., Matsuo, Y. and Sugie, Y. *Carbon*, **37**, 351–358 (1999).
141. Kotov, N. A., Dékány, I. and Fendler, J. H. *Adv. Mater.*, **8**, 637–641 (1996).

142. Ruiz-Hitzky, E., Jiménez, R., Casal, B., Manríquez, V., Santa Ana, A. and González, G. *Adv. Mater.*, **5**, 738–741 (1993).
143. Bissessur, R., Kanatzidis, M. G., Schindler, J. L. and Kannewurf, C. R. *J. Chem. Soc., Chem. Commun.*, 1582–1585 (1993).
144. Kanatzidis, M. G., Bissessur, R., DeGroot, D. C., Schindler, J. L. and Kannewurf, C. R. *Chem. Mater.*, **5**, 595–596
145. Bissessur, R., Schindler, J. L., Kannewurf, C. R. and Kanatzidis, M. G. *Mol. Cryst. Liq. Cryst.*, **245**, 249–254 (1994).
146. Wang, L., Schindler, J., Thomas, J. A., Kannewurf, C. R. and Kanatzidis, M. G. *Chem. Mater.*, **7**, 1753–1755 (1995).
147. Tsai, H.-L., Schindler, J. L., Kannewurf, C. R. and Kanatzidis, M. G. *Chem. Mater.*, **9**, 875–878 (1997).
148. Lemmon, J. P. and Lerner, M. M. *Chem. Mater.*, **6**, 207–210 (1994).
149. Lemmon, J. P. and Lerner, M. M. *Solid State Commun.*, **94**, 534–537 (1995).
150. Wypych, F., Seefeld, N. and Denicoló, I. *Quím. Nova*, **20**, 356–360 (1997).
151. Wypych, F., Adad, L. B. and Grothe, M. C. *Quím. Nova*, **21**, 687–692 (1998).
152. González, G., Santa Ana, M. A. and Benavente, E. *J. Phys. Chem. Solids*, **58**, 1457–1460 (1997).
153. Hernán, L., Morales, J. and Santos, J. *J. Solid State Chem.*, **141**, 323–329 (1998).
154. Clément, R., Lagadic, I., Léaustic, A., Audière, J. P. and Lomas, L. in Bernier, P., Fischer, J. E., Roth, S. and Solin, S. A. (Eds), *Chemical Physics of Intercalation II*, Plenum Press, New York, NATO-ASI Series B, Vol. 305, 1993, pp. 315–324.
155. Manríquez, V., Galdámez, A., Ponce, J., Brito, I. and Kasaneva, J. *Mater. Res. Bull.*, **34**, 123–130 (1999).
156. Martínez-Lara, M., Barea-Aranda, J. A., Moreno-Real, L. and Bruque, S. *J. Incl. Phenom. Mol. Recogn. Chem.*, **9**, 287–299 (1990).
157. Jones, D. J., El Mejjad, R. and Rozière, J. in Bein, T. (Ed.), *Supramolecular Architecture: Synthetic Control in Thin Films and Solids*, ACS Symp. Series, Vol. 499, 1992, pp. 220–230.
158. Bonnet, B., El Mejjad, R., Herzog, M.-H., Jones, D. J. and Rozière, J. *Mater. Sci. Forum.*, **91–93**, 177–182 (1992).
159. Liu, Y.-J. and Kanatzidis, M. G. *Inorg. Chem.*, **32**, 2989–2991 (1993).
160. Chao, K. J., Chang, T. C. and Ho, S. Y. *J. Mater. Chem.*, **3**, 427–428 (1993).
161. Matsubayashi, G. E. and Nakajima, H. *Chem. Lett.*, 31–34 (1993).
162. Nakajima, H. and Matsubayashi, G. E. *J. Mater. Chem.*, **5**, 105–108 (1995).
163. Torres-Maireles, P., Olivera-Pastor, P., Rodríguez-Castellón, E., Jiménez-López, A. and Tomlinson A. A. G. *J. Incl. Phenom. Mol. Recogn. Chem.*, **14**, 327–337 (1992).
164. Maeda, S. and Armes, S. P. *J. Colloid Interface. Sci.*, **159**, 257–259 (1993).
165. Mehrotra, V., Keddie, J. L., Miller, J. M. and Giannelis, E. P. *J. Non-Cryst. Solids*, **136**, 97–102 (1991).
166. Mattes, B. R., Knobbe, E. T., Fugura, P. D., Nishida, F., Chang, E.-W. and Pierce, B. M. *Synth. Metals*, **41–43**, 3183–3187 (1991).
167. Lee, K.-S., Kim, H.-M., Wung, C. J. and Prasad, P. N. *Synth. Metals*, **55–57**, 3992–3977 (1993).
168. Maia, D. J., Zarbin, A. J. G., Alves, O. L. and de Paoli, M.-A. *Adv. Mater.*, **7**, 792–794 (1995).
169. Zarbin, A. J. G., de Paoli, M. A. and Alves, O. L. *Synth. Metals*, **9**, 227–235 (1999).
170. Zarbin, A. J. G., de Paoli, M.-A. and Alves, O. L. *Synth. Metals*, **84**, 107–108 (1997).
171. Chao, T. H. and Erf, H. A. *J. Catal.*, **100**, 492–499 (1986).
172. Bein, T. and Enzel, P. *Synth. Metals*, **29**, E163–E168 (1989).
173. Bein, T. and Enzel, P. *Angew. Chem. Int. Ed. Engl.*, **28**, 1692–1694 (1989).

174. Bein, T. and Enzel, P. *Angew. Chem.*, **101**, 1737–1738 (1989).
175. Enzel, P. and Bein T. *J. Chem. Soc., Chem. Commun.*, 1326–1327 (1989).
176. Roque, R., de Oñate, J., Reguera, E. and Navarro, E. *J. Mater. Sci.*, **28**, 2321–2323 (1993).
177. Roque-Malherbe, R., de Oñate-Martinez, J. and Navarro, E. *J. Mater. Sci. Lett.*, **12**, 1037–1038 (1993).
178. Phani, K. L. N., Pitchumani, S. and Ravichandran, S. *Langmuir*, **9**, 2455–2459 (1993).

3

Polymer–Clay Nanocomposites Derived from Polymer–Silicate Gels

K. A. CARRADO,[†] L. XU,[†] S. SEIFERT,[†] R. CSENCSITS,[‡] AND C. A. A. BLOOMQUIST[†]

[†] Chemistry Division and [‡] Materials Science Division, Argonne National Laboratory, Argonne, IL, USA

1 INTRODUCTION

In this chapter, we show that polymer-containing silicate gels can be hydrothermally crystallized to form polymer–clay nanocomposite systems based on the magnesium silicate clay hectorite. Gels consist of silica sol, magnesium hydroxide sol, lithium fluoride, and a water-soluble polymer. Dilute solutions of gel in water are refluxed, then isolated via centrifugation, washed, and air dried. Polyaniline (PANI)–, polyacrylonitrile (PACN)–, polyvinylpyrrolidone (PVP)–, and hydroxypropylmethyl cellulose (HPMC)–clays have been successfully prepared. Final polymer loadings are determined by thermal analysis. Delamination is evident for PACN– and PANI–clay nanocomposites from X-ray diffraction (XRD), small angle X-ray scattering (SAXS), and transmission electron microscopy (TEM) results. The highest polymer loadings are observed for the PACN–clays at 88 % and the PANI–clays at 60 wt %. For the cationic polymer polydimethyldiallyl ammonium chloride, the loading could not be increased beyond about 20 % owing to charge compensation limitations. Polyethylene oxide and, to a lesser extent, polyvinyl alcohol and polyacrylamide do not yield nanocomposites owing to complexation of ions (especially Li) from the starting gel, which in turn prevents clay crystallization. The synthetic nanocomposites can be re-dispersed in water without loss of either organic material or structure.

Polymer–clay nanocomposites Edited by T. J. Pinnavaia and G. W. Beall
© 2000 John Wiley & Sons Ltd

There are two end-members which define the realm of structures possible in polymer–clay nanocomposites [1]. At one end are the well-ordered and stacked multilayers that result from intercalated polymer chains within host silicate clay layers. At the other end are exfoliated materials in which the host layers have lost their entire registry and are well dispersed in a continuous polymer matrix. These two cases are shown schematically in Figures 1a and c. Exfoliated (also called delaminated) polymer–clay nanocomposites display acceptable stiffness, strength, and barrier properties with far less ceramic content than is used in conventional particulate-filled polymer composites. Generally, the larger the degree of exfoliation in polymer–clay nanocomposites, the greater is the enhancement of these properties. Researchers at Toyota have shown that exfoliated nylon–clay nanocomposites are improved in strength and modulus compared with both pure nylon and conventional composites, without a sacrifice in impact resistance [2].

The synthesis of polymer–clay nanocomposites is normally carried out by the following methods. Typically, intercalation of a suitable monomer promotes delamination and dispersion of the host clay layers (usually a natural montmorillonite). This is followed by polymerization of the monomer to yield either linear or crosslinked polymer matrices. Often, however, to obtain the most homogeneous and stable products, the clay must first be dispersed using a preliminary preswelling step of long-chain quaternary ammonium intercalation. Alternatively, direct polymer melt intercalation into a host silicate lattice is done. The materials made via these two processes have been summarized [3–5]. Other unique methods for blending polymers directly with clays have employed microwaves [6] and a latex–colloid interaction [7]. More sophisticated pretreatments of the clay to render them organophilic in designed ways have also been reported recently [8–12].

Our new method of polymer–clay formation, which involves their hydrothermal crystallization from silicate sol–gels, has added to the repertoire of synthetic techniques. Results from XRD and small angle neutron scattering (SANS) experi-

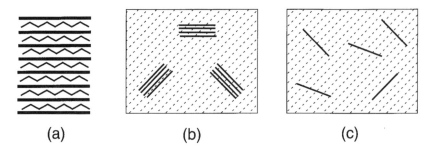

(a)   (b)   (c)

Figure 1 Possible polymer–clay nanocomposite structures; (a) well-stacked intercalated clay layers with polymer chains; (b) small stacks of polymer-intercalated clay layers dispersed within a polymeric matrix (indicated by dashed lines); (c) completely exfoliated clay layers randomly dispersed within a polymeric matrix

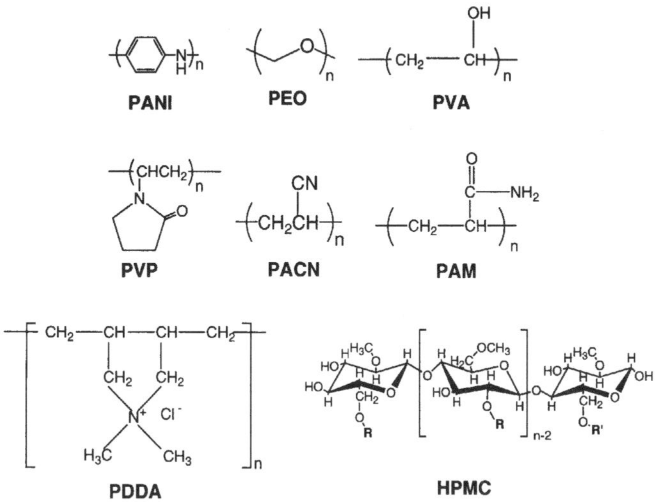

Figure 2 Structures of the polymers used in this study: polyaniline (PANI), polyethylene oxide (PEO), polyvinyl alcohol (PVA) polyvinylpyrolidone (PVP), polyacrylonitrile (PACN), polyacrylamide (PAM) and analogues, polydimethyl-diallylammonium (PDDA), and hydroxypropylmethylcellulose [HPMC, R = $CH_2CH(OH)CH_3$, R′ = CH_3]

ments [13] have led us to propose a structure for synthetic polymer–clays, displayed in Figure 1b, that can be thought of as intermediate between the two end-members in Figures 1a and c. In this scenario, small-stacks of polymer-intercalated clay crystallites are well dispersed within a continuous polymer matrix. This can be referred to as a 'semi-exfoliated' system. The synthetic polymer–clay nanocomposites made via the *in situ* technique will be summarized here, along with their characterization by XRD, SAXS, and TEM. Figure 2 displays the polymers that were utilized in this study.

2 EXPERIMENTAL

2.1 CLAY SYNTHESIS

The typical method for *in situ* hydrothermal crystallization of the polymer–hectorite clays is to create a 2 wt % gel of silica sol, magnesium hydroxide sol, lithium fluoride, and polymer in water, and to reflux for 2 days. Complete details can be found elsewhere [14]. Since the formation of hectorite normally occurs under mild conditions, this clay has become our silicate of choice for organic incorporation [15]. When the polymers are positively charged, their incorporation into the gallery space occurs electrostatically. Neutral polymers can also be incorporated by adding excess lithium fluoride to the mixture to account for charge compensation. All chemical reagents, including a 30 wt % silica sol Du Pont product, were purchased from Aldrich. The polyacrylonitrile (PACN), polyvinylpyrrolidone (PVP), and emeraldine salt polyaniline (PANI) were provided with average M_W values of 86 000, 10 000, and >15 000 K, respectively. The following polyacrylamides were utilized: poly-acrylamide (PAM) with molecular weight values of 1500 and 10 000 poly(acryla-mide-co-acrylic acid) (PAMA), M_W values of 200 000 and 5 000 000 and poly(acrylamide-co-diallyldimethylammonium chloride) (PAMD). Polydimethy-ldiallyl ammonium chloride (PDDA) was used as a 20 wt % solution in water, with a provided M_W of 100 000–200 000. Polyethylene oxide (PEO) had an M_W of 100 000. Hydroxypropylmethyl cellulose (HPMC) was obtained from Dow Chemi-cal Co. as Methocel 240S with a M_W of 1 500 000. For comparison purposes, some tests were done with Laponite RD, a synthetic hectorite obtained from Southern Clay Products, Gonzales, TX, and SHCa-1, a California Na^+-hectorite from the CMS Source Clay Minerals Repository, Columbia, MO. SHCa-1 was first purified by sedimentation techniques to remove the 50 wt % $CaCO_3$ and iron impurities.

Polymers are added initially at various wt % loadings of the total gel solids components. For example, when 2 g gel solids [SiO_2, $Mg(OH)_2$, LiF] is dispersed in 200 ml, 0.50 g polymer is added ($0.5/2.5 \times 100 = 20$ %). This mixture is refluxed for 48 h and a portion is removed for work-up and testing. Another portion of polymer is then added to the remaining slurry, and the mixture is heated for another 24 h. When only one extra addition was made, the second heating was also carried out at 100 °C. When up to four or five repetitions and polymer additions were carried out, the heatings after the first 48 h reflux were performed at only 60 °C. A convention of naming samples as 'polymer-*n* (gel wt %)' is adopted to denote the number of polymer loading sequences, n, at a particular loading (wt %) for each polymer–clay sample. For example, PVP-3 (20 %) is synthetic PVP–hectorite after three loadings at 20 wt % polymer as gel solids. For loading on to pre-existing clays, 1 g purified SHCa-1 and Laponite RD was dispersed in 100 ml water, 10 wt % water-soluble polymer was added, and the mixture was refluxed overnight followed by centrifugation, washing, and air drying.

2.2 CHARACTERIZATION

XRD analyses were carried out on a Rigaku Miniflex+ instrument (30 kV, 15 mA) using Cu K_α radiation, a NaI detector, variable slits, a 0.05 ° step size, and a 0.50 ° 2θ/min scan rate. The variable slit data are presented. Powders were loosely packed in horizontally held trays. Thermal gravimetric analysis (TGA) measurements were obtained on a SDT 2960 from TA Instruments. The samples were measured against an alumina standard in a 100 ml/min O_2 flow with a temperature ramp of 10 °C/min to 800 °C. Total polymer loadings were calculated by measuring the weight loss over the approximate temperature range of 200–600 °C.

Small angle X-ray scattering (SAXS) experiments were carried out on the SAXS instrument constructed at ANL and used on the Basic Energy Sciences Synchrotron Radiation Center CAT undulator beamline ID-12 at the Advanced Photon Source [16]. Clay samples were suspended in the beamline path using either scotch tape or 1.00 or 2.00 mm glass capillary tubes. SAXS data were collected in 5 min scans. Monochromatic X-rays at 10.0 keV were scattered off the sample and collected on a $19 \times 19\,cm^2$ position-sensitive two-dimensional gas detector. The scattered intensity has been corrected for absorption, blank scotch tape or glass capillary scattering, and instrument background. The differential scattering cross section can be expressed as

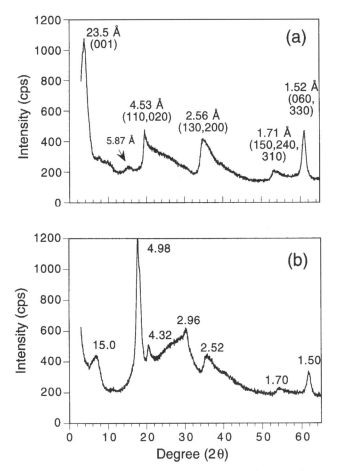

Figure 3 XRD patterns (variable slit data) of synthetic hectorites prepared with (a) PVP at the 30 wt% final polymer loading level [PVP-3 (20%)] and (b) PACN at the 78 wt% final polymer loading level [PACN-2 (80%)]. [Reproduced by permission of the American Chemical Society from *Chem. Mater.*, **5**, 1442 (1998)]

a function of the scattering vector q, which is defined as: $q = 4\pi(\sin\theta)/\lambda$, where λ is the wavelength of the X-rays and θ is the scattering half angle. The value of q is proportional to the inverse of the length scale (Å^{-1}) as $(2\pi/q) = d$, with the Bragg reflection, d, in Å. The instrument was operated at a sample-to-detector distance of 68.5 cm to obtain data at $0.04 < q < 0.7\ \text{Å}^{-1}$. Mylar windows were used because mylar does not have diffraction peaks in this q range. Samples were calibrated to the diffraction peaks of both a known clay (SWy-1, a natural Na(I)-montmorillonite from Wyoming) and silver behenate.

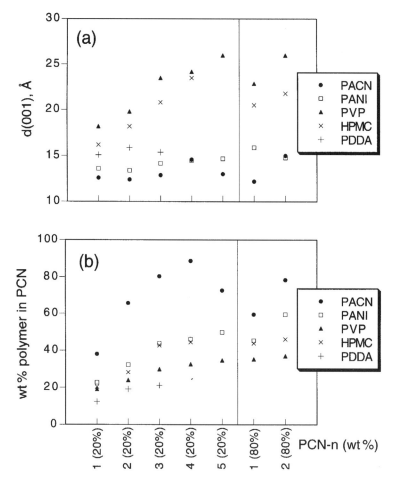

Figure 4   Plots of the physical data displayed in Table 1 for synthetic polymer–clay nanocomposites: (a) XRD basal spacings and (b) wt% polymer determined via TGA. The *x* axis represents the particular sample as polymer-*n* (gel wt%), where *n* is the sequence number of polymer loading; the vertical line demarcates the 20% and 80% loading series

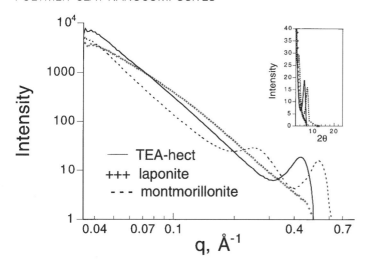

Figure 5　SAXS data for reference clay minerals including a montmorillonite (SWy-1), a laponite, and a synthetic TEA-hectorite (TEA = tetraethyl ammonium). The inset displays the data on the more familiar XRD scale of degrees 2θ

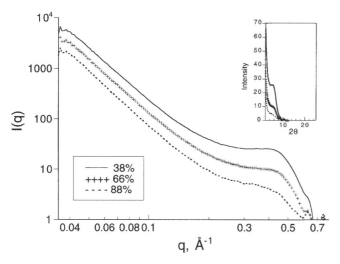

Figure 6　SAXS data for samples from the PACN–hectorite loading series from 38 to 88 wt% polymer as indicated in the legend. The inset displays the data on the more familiar XRD scale of degrees 2θ

Specimens for TEM were imaged in a JEOL 100CXII instrument operating at 100 kV. Powder samples of polymer–clay nanocomposites were dispersed in methanol and sonicated for 30 s. Copper grids with 'holey' carbon films were then dipped into the suspension and allowed to dry for 2 h in a vacuum oven at 100 °C. Bright field TEM images of polymer–clay nanocomposites are shown in Figures 7 and 8; selected area electron diffraction patterns are shown as insets.

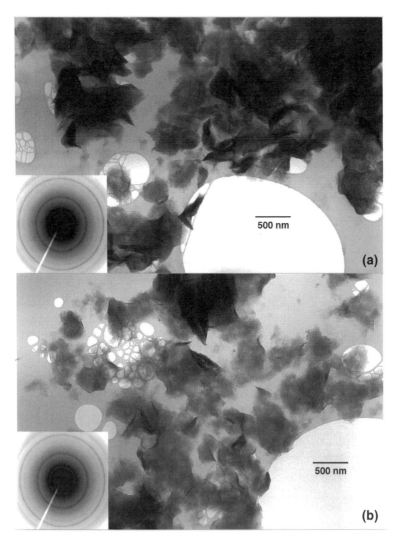

Figure 7 TEM images of (a) PANI-1 (20%) at 23 wt% polyaniline and (b) pani-5 (20%) AT 50 wt% polyaniline

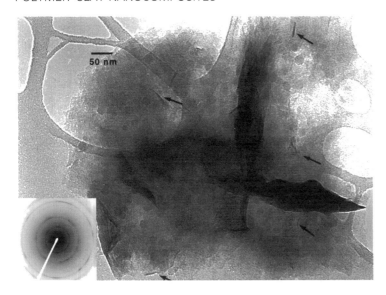

Figure 8 TEM image of PACN-4 (20%) at 89 wt% polyacrylonitrile; the arrows indicate dispersed clay as single layers or in stacks of a few layers

3 RESULTS

Dilute solutions of precursor hectorite gel and polymer are refluxed for various lengths of time, then isolated via centrifugation, washed, and air dried. The wt % loadings of polymer are increased synthetically by several successive reflux and addition treatments until the effect begins to taper off. Polymer loadings of up to 88 % are attained by adding more polymer to the solutions after 48 h reaction times, then reacting for another 24 h, and continuing this process prior to isolation. PANI– and PACN–clay samples contain up to 60 and 78 % polymer, respectively, after just one sequential addition at high polymer loadings. PANI–, PACN–, PVP– and HPMC–clay series have also been prepared by several sequential additions at lower polymer loadings to the silicate gel during crystallization (see Table 1). Along with the ability to tune the amount of organic present in the nanocomposite this way, in some cases (most notably for PACN) this is the preferred method for obtaining the highest loadings. In most cases, however, we found no advantage in doing several smaller sequential additions over just one or two additions at larger initial loading. See Figure 3 for plotted versions of the data in Table 1.

That said, using sequential treatments to load the polymer in stages is sometimes also a preferred method because of the concern that initially high polymer solution concentrations may negatively affect the crystallization of the clay layers. This has been observed when polyvinyl alcohol (PVA) is used [12]. In this case, the PVA is

Table 1 Basal spacings and polymer loading values for polymer–clay nanocomposite (PCN) reaction products

	Polymer gel loading sequence No. (wt %)						
PCN XRD and TGA	1 (20 %)	2 (20 %)	3 (20 %)	4 (20 %)	5 (20 %)	1 (80 %)	2 (80 %)
PACN, d(001) (Å)	12.6	12.4	12.9	14.6	13.0	12.2	15.0
PACN loading (wt %)	38.1	65.9	80.4	88.7	72.9	59.6	78.5
PANI, d(001) (Å)	13.6	13.4	14.2	14.5	14.7	15.9	14.8
PANI loading (wt %)	22.8	31.4	43.9	46.2	49.9	45.4	59.6
PVP, d(001) (Å)	18.2	19.8	23.5	24.2	26.0	22.9	26.0
PVP loading (wt %)	19.1	24.0	30.0	32.7	34.8	35.4	37.1
HPMC, d(001) (Å)	16.2	18.2	20.8	23.5		20.5	21.8
HPMC loading (wt %)	20.4	28.3	42.8	44.6		44.0	46.2
PDDA, d(001) (Å)	15.1	15.8	15.4				
PDDA loading (wt %)	12.5	19.2	21.2				

intercalated between the layers, but then the polymer chains apparently cap the edges of the growing silicate layers in some fashion because only very small particle sizes occur. This also appears to occur to some extent for the polyacrylamides, as demonstrated in Table 2. Unmodified PAMs do not foster significant loadings (<5 wt %). Copolymers of PAMs modified with polyacrylic acid (PAMA) or a cationic ammonium group (PAMD) are moderately better, but certainly not in the realm of interest to nanocomposite research. PEO polymers complex cations such as Mg(II) and Li(I) so well from the starting gel that clay does not form at all. Finally, polymer loadings cannot be increased beyond about 20 % for the cationic polymer PDDA (Table 1). This is due to electrostatic interactions that balance the negatively charged sites on the silicate lattice with those on the cationic polymer chain. Beyond

Table 2 Basal spacings and polymer loadings for synthetic polyacryla-mide–clay nanocomposite reaction products

Polymer M_w (wt %)[a]	d(001) (Å)	Polymer loading (wt %)
PAM 1500 (20 %)	15.5	2.4
PAM 1500 (50 %)	15.6	4.0
PAM 10 000 (20 %)	14.2	3.0
PAM 10 000 (50 %)	15.0	4.2
PAMA 200 000 (70 %)	15.5	15.4
PAMA 5 000 000 (20 %)	14.6	4.9
PAMD (20 %)	14.2	9.5

[a] Initial polymer content of the gel prior to reaction.

charge compensation there is no driving force for further incorporation. Charge compensation in the case of the neutral polymers is attained by interlayer Li(I) cations.

Figure 4 displays representative XRD patterns for the synthetic polymer–clays, in particular PVP– and PACN–hectorites with 30 and 78 % final polymer loadings, respectively. A synthetic hectorite with Li(I) exchangeable cations (no polymer) displays a pattern typical of hectorites, with a basal or $d(001)$ spacing of 14.3 Å ($6.18°\ 2\theta$). There is no sign of unreacted $Mg(OH)_2$, which would display peaks at 4.77 Å ($18.6°\ 2\theta$) and 2.37 Å ($38.0°\ 2\theta$) for the (001) and (101) reflections of the magnesium hydroxide mineral brucite. For the PVP–clay sample containing 30 % polymer shown in Figure 4a, the $d(001)$ is clearly evident at 23.5 Å and the remainder of the pattern is that of a typical hectorite. The pattern in Figure 4b for PACN–clay with a final polymer loading of 78 % is quite different, however. All of the expected clay peaks are present, but the pattern is dominated by new peaks near 5.0 and near 3.0 Å. These peaks match the pattern of pure PACN and are visible because of the high concentration of the organic present in this sample. The highest loadings are observed for the PACN–clays at up to 88 %.

As shown in Table 1, PACN and PANI provide the highest polymer loadings among those studied. The $d(001)$ values for these nanocomposites are very low at 12–15 Å and they are also of markedly lower intensity than is normally observed (see Figure 4). These polymers apparently have promoted a large degree of exfoliation of the hectorite layers. Loss of most of the layer registry will manifest itself in the loss of a basal spacing since the layers are no longer organized in stacks. All of the other (hkl) clay peaks are observed, which ensures that clay has indeed been made. To help ascertain the degree of exfoliation and rule out the existence of basal spacings larger than would be observable by XRD, small-angle X-ray scattering (SAXS) was performed since it can access a lower q region (proportional to 2θ).

Figure 5 displays SAXS plots for representative clays for comparison purposes. A natural montmorillonite (which is structurally analogous to hectorite) has two strong peaks in the basal region corresponding to 14.5 and 25.0 Å. The occurrence of two peaks instead of just one basal spacing is not unusual for this SWy-1 sample, and is due to ordered interstratification [17]. A well-crystallized synthetic hectorite made using tetraethylammonium cations (TEA-hectorite) also has a strong basal reflection corresponding to 14.5 Å. The final control mineral is laponite, a commercially available synthetic hectorite of extremely small particle size and well known to be exfoliated. As expected, no evidence of a peak is seen in the SAXS of laponite. Figure 6 shows curves from selected samples of the PACN series. All display a peak at about 0.43 Å^{-1} in q (14.6 Å), but it becomes significantly weaker in intensity as the percentage of polymer increases from 38 to 88 %, again indicating dispersion and at least some degree of exfoliation. Importantly, there is no evidence of a higher d spacing out to 0.04 Å^{-1} in q (which corresponds to 160 Å). The overall intensity of the entire SAXS curve decreases as the percentage of polymer increases, because of the progressively lower contribution of clay to the scattering in each case. The insets in both Figures 5 and 6 display the SAXS data on the scale of conventional XRD in terms of the often more familiar 2θ values.

Since exfoliation can only be inferred by the absence of a $d(001)$ peak in XRD and SAXS, direct observation becomes desirable using a technique such as TEM. Figure 7 shows TEM images for polyaniline–clays containing 23 and 50 wt% polymer [PANI-1 (20 %) and PANI-5 (20 %), respectively]. Readily noticeable from these images is the relative difference in contrast: As the polymer content increases there is less of the 'darker' clay phase present and the overall image appears 'brighter'. This phenomenon is observed for all of the polymer–clay nanocomposites. Figure 8 is a TEM image of a sample with the highest polymer loading attained by this technique, 89 wt % PACN–hectorite [PACN-4 (20 %)]. While larger particles of aggregated clay are apparent, also clearly evident are small dispersed clay particles (indicated by arrows in Figure 8) that are about 50 nm in length and approximately 2–5 nm in width. Note the random orientation of these particles as well. Electron diffraction patterns in all of the TEM images match those of crystalline hectorite clay, with the noted absence of a basal spacing.

Since these synthetic polymer–clay nanocomposites are made in an aqueous medium and the polymers are water soluble, a critical test of their stability is to redisperse them in water and check for structural degradation due to polymer loss. Table 3 shows the results of selected samples for this test. Samples were stirred in water at room temperature for 2–4 h, then centrifuged and air dried. Within experimental error, TGA results show that there is no loss of organic. There are also minimal changes in XRD data; for the PACN– and PANI–clays the $d(001)$ appears to be equilibrating at about 15–16 Å. One exception to this is the noticeable 29 % weight loss for the HPMC–clay sample, although the d spacing does not change significantly with this loss (20.5–21.8 Å). Overall, the lack of change is very

Table 3 Polymer loadings and basal spacings for PCN reaction products before and after resuspension in water

	Before		After	
Polymer–PCN	wt %^{a}	$d(001)$, (Å)	wt %^{a}	$d(001)$, (Å)
PVP-1 (10 %)^b	9.3	18.2	9.7	16.7
PVP-1 (30 %)^b	20.1	24.8	20.0	24.0
HPMC-1 (80 %)	44.0	20.5	31.2	21.8
PANI-2 (20 %)	31.4	13.4	36.0	16.2
PANI-1 (80 %)	45.4	15.9	46.5	16.2
PANI-2 (80 %)	59.6	14.8	62.2	15.6
PACN-2 (20 %)	65.9	12.4	70.6	15.5
PACN-3 (20 %)	80.4	12.9	82.2	14.8

^{a} 200–600 °C weight loss by TGA.
^{b} M_w of PVP was 29 000.

Table 4 Polymer–clay nanocomposites made with various hectorite samples[a]

PCN	Polymer gel loading (%)	Polymer MW	Polymer wt % in PCN	$d(001)$ (Å)
PVP-1	10	10 000	9.8	18.2
PVP–SHCa	10	10 000	6.9	16.2
PVP–laponite	10	10 000	—	—[b]
PVP-1	10	360 000	8.7	None
PVP–SHCa	10	360 000	10.8	15.6
PVP–laponite	10	360 000	38.1	24.1
PANI-1	20	15 000	22.8	13.6
PANI–SHCa	10	15 000	8.1	15.5
PANI–laponite	10	15 000	8.2	18.4
PACN-1	20	86 000	38.1	12.6
PACN–SHCa	10	86 000	13.5	15.5
PACN–laponite	10	86 000	39.9	15.1

[a] Synthetic PCNs are indicated as polymer-X, where X is the sequence No. (1 in this case); other samples are made by dispersing clay and polymer in water.
[b] sample was not isolable.

encouraging and demonstrates the stability of the synthetic polymer–clay nanocomposites (except perhaps for samples containing cellulose) to aqueous exposure.

Finally, a preliminary comparison has been made of the synthetic polymer–clay nanocomposites made with water-soluble polymers versus those made by simple mixing in water. For the latter experiments, a natural hectorite (SHCa-1) and a commercially available pre-existing synthetic hectorite (Laponite RD) were utilized. Table 4 lists TGA and XRD results for selected samples; initially, only low loadings have been studied. Note that, upon polymer incorporation, laponite, a completely exfoliated clay, becomes highly oriented with large and intense d spacings. It also incorporates large amounts of PACN and PVP (40 and 38 % respectively) at very low gel loadings (10 wt %). This reflects the rather low yield of products, which is due to the high dispersion of the very small laponite clay crystallites into solution. For an as yet undetermined reason, this does not occur for PANI. Future experiments will compare samples made at higher polymer loadings.

4 DISCUSSION

Comparisons can be made, in terms of preliminary characterization, between these synthetic polymer–clay nanocomposites and those made using natural clays and more conventional methods [3]. They will now be updated and summarized below.

There are a few previous studies involving the interaction of PVP with clays. One examines the adsorption of PVP (M_w values ranging from 5000 to 600 000) on the surface of the non-swelling clay mineral kaolinite [18]. The results focus on polymer structure (loops versus flattened coils) in terms of M_w and surface coverage. Kaolinite intercalated with PVP was prepared via the *in situ* polymerization route without any evidence of exfoliation [19]. More recently, PVP was directly intercalated into kaolinite using a refined guest displacement method [20]. Monolayer coverage was indicated in the layers (d spacing 12.4 Å), and once again this occurred without any evidence of exfoliation. Interestingly, this PVP was easily removed by washing with water, whereas our synthetic PVP–hectorites retain all of their polymer upon such a treatment. A PVP–tetrasilicic mica [21] (TSM) has a basal spacing of 23 Å, indicating bilayer coverage, along with a polymer loading of 42 % (PVP $M_w = 1000$). We are not able to achieve a loading this high by the *in situ* synthetic technique, and our samples with a 23 Å basal spacing correspond to only 28 % PVP. In another study, a PVP–montmorillonite [22] (PVP $M_w = 10\,000$) affords a basal spacing of 23.4 Å with 30 wt % PVP, which is more in line with our results. Hectorite and montmorillonite are in fact very similar clays with respect to structure and layer charge density, while TSM has a significantly larger layer charge density. There is only one report of clay exfoliation using PVP. Beall [22] discovered that, by mixing a large ratio of dry PVP to dry clay and then adding a certain amount of water, a spontaneous exothermic reaction caused exfoliation.

Polymer–clay nanocomposites containing polyaniline are of interest because of the conducting properties of the polymer. A conventionally made polyaniline–smectite that has been extensively studied by Giannelis [23] has a loading of only 20 wt % polymer. Even at this low loading, the fracture toughness of the hybrid was shown to be superior to the clay matrix alone [24]. In addition, the initially insulating nanocomposite became conductive upon exposure to HCl vapors, with an in-plane conductivity [25] of 0.05 S/cm. One other report of a polyaniline–montmorillonite used it to prepare an electrorheological fluid [26]. Both of these conventionally prepared polyaniline–clays are reported to be intercalated, while the synthetic polyaniline–clay made by our method is a semi-exfoliated material. The synthetic PANI–clays have much higher polymer loadings at up to 57 wt %. A comparison of their conducting and toughness properties would be relevant. The conductivity of the 57 wt % PANI–hectorite sample [27] was 1.6×10^{-5} S/cm; doping with protons would increase this value.

Polyacrylonitrile–clay studies to date have focused more on the thermal characteristics of PACN than on nanocomposite properties. There is a report that summarizes work on PACN–clays by Kato and coworkers [28], wherein sialon ceramic is made by heating PACN–montmorillonite to 1300 °C. The PACN–clay precursors are made via *in situ* polymerization from a montmorillonite preswelled with a long chain alkyl ammonium cation, resulting in a sample containing 65 wt % PACN [29]. The XRD pattern is reported as 'poorly-defined, with a rising baseline in the 30 Å region', while their TEM images show small stacks of platelets about 21 Å apart. Wei and coworkers [30] exploit the functionality of the nitrile group to convert polymer–silica hybrids to various polyacrylics. Finally, in a very recent study, a styrene–acrylonitrile copolymer is intercalated into montmorillonite via emulsion polymerization [31]. No evidence of exfoliation is observed since the XRD d spacing is 16 Å and TEM shows 'good order'. PACN behaves differently from all the other polymers examined for the *in situ* synthetic method. Firstly, the final polymer loadings are significantly higher at up to 89 wt %. Secondly, this maximum value is achieved at an intermediate step within the 20 % loading series. Sample PACN-3 (20 %) has 80 wt % polymer, PACN-4 (20 %) has 89 wt %, and then PACN-5 (20 %) actually decreases to 73 wt % polymer. It also demonstrates at least one case where several low loadings work better than one initially high loading of polymer [PACN-1 (80 %) = 60 wt %] in the growing silicate gel solution. Furthermore, the synthetic PACN–hectorite is of the semi-exfoliated structure shown in Figure 1b, as evidenced by XRD and TEM results.

For most naturally occurring smectite clays the aspect ratio of the platelets is fairly large [32], typically in the range 200–2000. This large aspect ratio may be partly responsible for the enhanced polymer–clay nanocomposite properties in general. Considering the low loading of the inorganic phase, however, it has been proposed that the dramatic change in properties is due more to induced ordering of the polymer chains near the polymer–clay interface [23]. For a delaminated system, only about 8 wt % clay is sufficient to induce complete ordering of polyimide chains, for example, whereas the amount of silicate required rises to 50 % if delamination is not complete (average of 10 layers stacked) [23]. The aspect ratios of our synthetic clays are generally one-third to one-half that of natural clays [13b]. These smaller sheets, which are similar to those in fine-grained clays, pack with less regularity in smaller stacks. A high dispersion of layers is then viable in a simple, one-step process. This disorder is inherently similar to the exfoliation or delamination processes that must be forced when using natural clays to make delaminated clay–polymer nanocomposites. This is the advantage of the synthetic alternative presented in this work, and could offer enhanced properties when compared with currently formed clay–polymer hybrids in material testing and processability.

5 CONCLUSIONS

Completely exfoliated and homogeneous dispersions of silicate layers have been achieved in only a relatively small number of cases in the literature, and primarily only when the polymers contain polar functional groups. This has been attributed to the compatibility of the polar hydroxy groups of the clay with such polymer functionality [33]. Other systems without such polar groups must be intensively modified, as is the case with polypropylene–clay hybrids where the clay must be pretreated with stearylammonium ions and melted with both the polymer and oligomers containing an appreciable number of polar groups [32]. The *in situ* hydrothermal crystallization method has the potential of promoting high dispersions of silicate layers in a one-step process. Any savings in the number of steps will improve the efficiency of materials preparation.

The polymer loadings have reached 89 wt % by this method thus far. Toyota's nylon–clay materials contain only 2–7 wt % clay in the matrix; beyond this amount the materials become more difficult to process by casting, extruding, etc. [2b]. We therefore continue to strive to increase the polymer loadings. A limiting factor to this method has been the requirement of polymer water solubility. However, it is quite interesting that (especially in the case of 89 wt % PACN–hectorite) so little clay is needed to precipitate and stabilize such a large amount of water-soluble polymer from an aqueous solution. Preliminary redispersion tests in water have been excellent, showing no loss of polymer back to solution after isolation. The ultimate test of the usefulness of these synthetic polymer–clay nanocomposites will be in their mechanical performance; measurements of tensile and elastic properties, impact strength, heat distortion properties, etc., are necessary.

6 ACKNOWLEDGEMENTS

Ms Melanie Lymon, a student sponsored by the ANL Division of Education Programs, performed the polyacrylamide studies. Drs R. E. Winans, M. Beno, K. Littrell, and J. Linton (all of ANL) are acknowledged for their help with various SAXS experimental details. D. Dees (ANL) performed the conductivity measurement. This work was performed under the auspices of the US Department of Energy, Office of Basic Energy Sciences, Divisions of Chemical Sciences (KAC, LX, SS, CAAB) and Materials Science (RC), under contract no. W-31-109-ENG-38, and benefited from the use of the Advanced Photon Source at ANL.

7 REFERENCES AND NOTES

1. Burnside, S. D. and Giannelis, E. P. *Chem. Mater.*, **7**, 1597–1600 (1995).
2. (a) Usuki, A., Kojima, Y., Kawasumi, M., Okada. A., Fukushima, Y., Kurauchi, T. and Kamigaito, O. *J. Mater. Res.*, **8**, 1179 (1993); (b) Kojima, Y., Usuki, A., Kawasumi, M., Okada, A., Fukushima, Y., Kurauchi, T. and Kamigaito, O. *J. Mater. Res.*, **8**, 1185 (1993); (c) Usuki, A., Kato, M., Okada, A. and Kurauchi, T. *J. Appl. Polym. Sci.*, **63**, 137 (1997).
3. Carrado, K. A. and Xu, L. *Chem. Mater.*, **10**, 1440 (1998).
4. Carrado, K. A. *Appl. Clay Sci.*, **17**, 1 (2000).
5. Aranda, P., Galvan, J. C. and Ruiz-Hitzky, E. *Mater. Res. Soc. Symp. Proc.*, **519**, 375 (1998).
6. Oriakhi, C. O. and Lerner, M. M. *Mater. Res. Bull.*, **30**, 723 (1995).
7. Fischer, H. R., Gielgens, L. H. and Koster, T. P. M. *Acta Polym.*, **50**, 122 (1999).
8. Laus, M., Camerani, M., Lelli, M. Sparnacci, K., Sandrolini, F. and Francescangeli, O. *J. Mater. Sci.*, **33**, 2883 (1998).
9. Vaia, R. A., Jandt, K. D., Kramer, E. J. and Giannelis, E. P. *Chem. Mater.*, **8**, 2628 (1996).
10. Tsipursky, S., Beall, G. W. and Vinakour, E. US Patent 5,849,830 (1998).
11. (a) Beall, G. W., Tsipursky, S., Sorokin, A. and Goldman, A., US Pat. 5,880,197 (1999); (b) Beall, G. W., Tsipursky, S., Sorokin, A. and Goldman, A. US Pat. 5,877,248 (1999).

12. Carrado, K. A., Thiyagarajan, P. and Elder, D. L. *Clays Clay Miner.*, **44**, 506 (1996).
13. Carrado, K. A., Thiyagarajan, P., Winans, R. E. and Botto, R. E. *Inorg. Chem.*, **30**, 794 (1991); (b) Carrado, K. A., Thiyagarajan, P. and Song, K. *Clay Miner.*, **32**, 29 (1997).
14. (a) Carrado, K. A., Forman, J. E., Botto, R. E. and Winans, R. E. *Chem. Mater.*, **5**, 472 (1993); (b) Carrado, K. A. *Ind. Eng. Chem. Res.*, **31**, 1654 (1992).
15. For a full description of the instrument, see http://www.bessrc.aps.anl.
16. Moore, D. M. and Hower, J. *Clays Clay Miner.*, **34**, 379 (1986).
17. Hild, A., Sequaris, J.-M., Narres, H.-D. and Schwuger, M. *Colloids Surf. A.*, **123/124**, 515 (1997).
18. Sugahara, Y., Sugiyama, T., Nagayama, T., Kuroda, K. and Kato, C. *J. Ceram. Soc. Jpn*, **100**, 413 (1992).
19. Komori, Y., Sugahara, Y. and Kuroda, K. *Chem. Mater.*, **11**, 3 (1999).
20. Ogawa, M., Inagaki, M., Kodama, N., Kuroda, K. and Kato, C. *J. Phys. Chem.*, **97**, 3819 (1993).
21. Miyaga, H., Sugahara, Y., Kuroda, K. and Kato, C. *J. Chem. Soc., Faradat Trans. I*, **83**, 1851 (1987).
22. Beall, G. W., Tsipursky, S., Sorokin, A. and Goldman, A., US Pat. 5,578,672 (1996).
23. Giannelis, E. P. *J. Miner. Metals Mater. Soc.*, **44**, 28 (1992).
24. Mehrotra, V. and Giannelis, E. P. *Solid State Comm.*, **77**, 155 (1991).
25. Kim, J. W., Kim, S. G., Choi, H. J. and Jhon, M. S. *Macromol. Rapid Commun.*, **20**, 450 (1999).
26. Four-probe measurement on a Schlumberger SI 1260 impedance FRA with the sample pellet sandwiched between shielding tape.
27. Sugahara, Y., Kuroda, K. and Kato, C. *J. Mater. Sci.*, **23**, 3572 (1988).
28. Bastow, T., Hardin, S. G. and Turney, T. W. *J. Mater. Sci.*, **26**, 1443 (1991).
29. (a) Wei, Y., Wang, W., Yang, D. and Tang, L. *Chem. Mater.*, **6**, 1737 (1994); (b) Wei, Y., Yang, D. and Tang, L. *Macromol. Chem., Rapid Commun.*, **14**, 273 (1993).
30. Noh, M. H., Jang, L. W. and Lee, D. C. *J. Appl. Polym. Sci.*, **74**, 179 (1999).
31. Normally, the $<2\,\mu$m fraction of natural clays is utilized, which gives rise to particles in the size range 200–2000 nm. Since one sheet is 1 nm thick, an estimate of the aspect ratio is therefore 200–2000.
32. Kawasumi, M., Hasegawa, N., Kato, M., Usuki, A. and Okada, A. *Macromolecules*, **30**, 6333 (1997).

4

Polymerization of Organic Monomers and Biomolecules on Hectorite

M. P. EASTMAN[†] AND T. L. PORTER[‡]
[†] Department of Chemistry and [‡] Department of Astronomy and Physics, Northern Arizona University, Flagstaff, AZ, USA

1 INTRODUCTION

This book is primarily devoted to describing the production, properties and applications of polymer–clay composites in which individual clay nanoparticles are uniformly dispersed in a polymer matrix. In most of the systems discussed, the amount of clay contained in the nanocomposite is on the order of a few percent by mass and the beneficial effect of the clay is attributed to the high aspect ratio of the clay particles and their effect on the polymer structure. The primary focus of our work and this chapter is on a very different type of composite material, formed when transition metal-exchanged clay films react with organic monomers or biomolecules, amino acids and nucleotides to form polymeric materials that are intimately connected with the clay matrix. In these systems, the surface properties of the clay particles, the ability of the clay to act as a lamellar inorganic host for the monomer and the chemical properties of the exchangeable interlayer transition metal ions combine to control the products and yield of the polymerization reaction. Typically, the composite formed by this type of reaction contains 15 % or less polymer and has mechanical properties similar to those of the clay matrix. Thus, the final product is most aptly termed a clay–polymer composite. The polymer that forms is distributed over a range of low molecular masses and has a structure that appears to depend on whether the polymer is formed on the surface of the clay or in the interlayer region. The final clay–polymer composite is unique in the sense that

Polymer–clay nanocomposites Edited by T. J. Pinnavaia and G. W. Beall

the clay matrix organizes and directs the formation of the polymer, with the properties of the final material being controlled to a large extent by the initial state of the clay.

The following section highlights those aspects of clay structure that are important for the synthesis of clay–polymer composites. Later sections in this chapter will look at specific types of reaction pathways that have been identified in the course of our work and discuss some of the properties of the composite materials synthesized. Recent work extending our research into the polymerization of biomolecules on clays and prebiotic chemistry will be described. The experimental results described in this chapter will depend extensively on techniques such as scanning force microscopy (SFM), electron spin resonance (ESR) and X-ray diffraction (XRD). Other experimental techniques making important contributions to our knowledge of clay–polymer composites include matrix assisted laser desorption/ionization time-of-flight mass spectrometry (MALDI) and attenuated total reflectance (ATR) spectroscopy.

2 IMPORTANT PROPERTIES OF THE CLAY MATRIX

The smectite clays such as montmorillonite and hectorite are the most important matrix systems in the field of clay–polymer composites; the ability of these clays to organize and direct polymer synthesis is rooted in a number of unique structural and chemical properties [1]. The smectite clays have an octahedral complex of a metal $[M(OH)_6]$ sandwiched between two sheets of linked $Si(O, OH)_4$ tetrahedra. Substitution defects, primarily located in the octahedral layers, give clay platelets a net negative charge; this charge is compensated for by exchangeable cations located in the interlayer region separating individual platelets. Figure 1 shows the structure of sodium-exchanged hectorite; in this smectite, Li^+ is substituted for Mg^{2+} in the octahedral layer [2].

The nominal composition of the clay in Figure 1 is $(Na_{0.33}[Mg_{2.67}Li_{0.33}]$ $Si_4O_{10}[OH]_2)$; however, hectorite formed in nature also contains some structural iron. A synthetic hectorite, Laponite (TM Laporte Industries Ltd), has the same composition as natural hectorite and essentially no structural iron. The d spacing for the system in Figure 1, when maintained at a relative humidity of 20 %, is 1.125 nm, and the distance between Na^+ ions is approximately 0.65 nm [2].

The properties of smectites that are important in their chemistry include the following: a substantial cation exchange capacity (approximately 100 meq./100 g for hectorite), reactive edges on individual clay platelets, an ability to interact with a wide range of natural and synthetic organic compounds, a large effective surface area for adsorption of organic compounds and strong, local electric fields in the interlayer region between platelets [1, 3, 4]. These electric fields have a significant effect on the electrochemical properties of the interlayer metal ions and play an important role in the chemistry described later in this chapter. Organic compounds

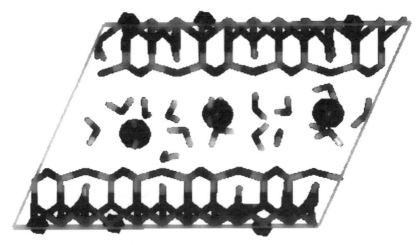

Figure 1 Schematic representation of the structure of hectorite. The (interlayer) intergallery region contains Cu^{2+} cations and water molecules. Reprinted with permission of the American Chemical Society

that interact with smectite clays have access to octahedral sites at platelet edges and tetrahedral sites at edges and on surfaces [1, 5]. Organic compounds also have access to metal ions in the interlayer region, the coordination of which may change with the extent and type of solvation. Uptake of a solvent by the clay gives the individual platelets mobility with respect to one another, and provides a mechanism for rearrangement of reactive sites into a variety of patterns. When an organic compound reacts with a clay, the clay may rearrange or swell to accommodate the reaction products.

It is possible to exchange a wide variety of metal ions into the interlayer region of hectorite and other smectite clays. It is sometimes possible to form thin films of these exchanged clays by allowing a small sample of an aqueous suspension to dry on a clean substrate such as a glass slide. It is particularly easy to form hectorite films containing Cu^{2+}, Fe^{3+}, Na^+ or Ca^{2+} in the interlayer [4, 6]. Films formed on interdigitated arrays are useful as chemiresistors and are effective in sensing compounds such as water that possess relatively large values of Hildebrand's 'solubility parameter' [7, 8].

3 SCANNING FORCE MICROSCOPY AND CLAY FILMS

Many clay minerals, when cast into thin films as described above, are suitable for high-resolution imaging using SFM. Variations in standard contact-mode SFM such as intermittent-contact (IC) or non-contact (NC) SFM have provided the highest-quality images, with the greatest level of reproducibility [9, 10]. In IC or NC SFM, the cantilever is driven into oscillation by a small piezoelectric element near one of its resonant frequencies (typically 50–350 kHz). The oscillating cantilever is then slowly brought near the sample surface to be imaged. In IC mode, the oscillating cantilever is brought near enough to the surface for the SFM tip on the end of the cantilever to make contact with the sample briefly during each downward excursion of the vibrating cantilever. This 'intermittent contact' results in a decrease in the cantilever vibration amplitude, the amount of which is sensed by the SFM instrument. A feedback loop adjusts the cantilever height above the sample surface to maintain a constant preset vibration amplitude as the tip is rastered over the surface. The vibrating cantilever thus follows the contours of the surface, generating an image. The IC mode works well on less rigid samples, producing little or no surface damage or modification during the imaging process [11].

In the NC mode, the vibrating cantilever is brought just near enough to the surface for the tip–cantilever assembly to begin to be affected by either the attractive Van der Waals force between the tip and the topmost surface atoms or by the repulsive force that comes into play nearer to the surface. The tip never makes actual contact with the surface. Within this 'force gradient', the cantilever vibration amplitude is decreased with respect to its amplitude in the far 'force-free' space. This amplitude decrease is sensed by the SFM instrument, and a feedback loop maintains the desired vibration amplitude as the tip is rastered across the surface. The tip thus follows the contours of the surface without ever touching it, generating a topographical image of the surface.

In Figure 2, SFM images of the surface of Cu(II)-exchanged hectorite (Figure 2a) and Na-montmorillonite (Figure 2b) thin films are shown [12]. These images were acquired in NC mode. The dimensions of the image in Figure 2a are 4500×4500 Å, while in Figure 2b the dimensions are 1×1 μm. The hectorite film displays large, flat regions (some extending several microns), the result of the individual hectorite platelets lying flat on the substrate during the film casting process. Although not all individual platelets are aligned parallel to the substrate in hectorite cast films, there can still be a large degree of disorder among the individual platelets. Using SFM, however, it is generally easy to locate regions on these cast films that exhibit order and flatness over the range of several microns. In many areas on the hectorite surface, step edges, cracks or faults are visible. These may be defects in individual platelets, or regions separating individual platelets lying on the surface. In some instances, tiny micropores are visible on otherwise flat layered silicate surfaces. These micropores are only a few tens of angstroms in dimension, and probably only extend a single layer in depth. The images of the films provide a strong indication that the clay layers are nearly completely ordered, with the plane of the interlayer parallel to the surface of the film. This is confirmed by ESR of the thin films which shows axial symmetry, g, and almost complete ordering of the Cu^{2+} sites. The ESR parameters that characterize the Cu^{2+} sites are: $g_{\parallel} = 2.353$, $g_{\perp} = 2.076$ and $A_{\parallel} = 151$ G [12]. In these ESR experiments, $g_{\parallel}$ is measured with the plane of the

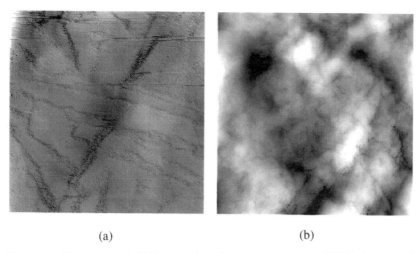

(a)   (b)

Figure 2 Non-contact (NC) scanning force microscope (SFM) images of the surface of (a) Cu^{2+}-exchanged hectorite thin film and (b) Na^+-exchanged montmorillonite thin film [4]. The dimensions of the images are 4500 × 4500 Å (a) and 1 × 1 μm (b). Reprinted with permission of the Clay Mineral Society

hectorite film vertical to the applied magnetic field and $g_\perp$ is measured with the film rotated by 90° with respect to the parallel orientation.

For the montmorillonite film, the structure is highly disordered, exhibiting few regions that are reasonably flat or parallel to the substrate. The particle size in the montmorillonite films is also much smaller, with the largest particles or platelets observed only a few tenths of a micron in dimension. We observed that the physical integrity of the montmorillonite cast films is much lower that of the hectorite films. Montmorillonite films were not used in any subsequent experiments.

Using one of the oscillating cantilever modes, one can also acquire information on the material properties of the surface being studied in addition to the surface topography. As the cantilever is driven into oscillation, the phase of the actual cantilever vibration with respect to the driving force can be monitored in real time. During IC or NC mode, the cantilever phase is altered slightly as the material properties of the underlying surface change during a scan. For example, a 'softer' region directly under the tip will generally result in a slight lag in phase, while 'harder' regions will result in little or no phase lag [13]. During a scan, a map of phase changes can thus be interpreted as changes in the material properties of the sample being imaged. This 'phase image' is acquired simultaneously with the topographical image and reflects material differences on the nanometer scale.

During phase imaging (PI), the cantilever phase is affected predominantly by the hardness of the material directly under the oscillating tip. Other properties of the sample may also affect the cantilever phase. Properties such as viscosity and local

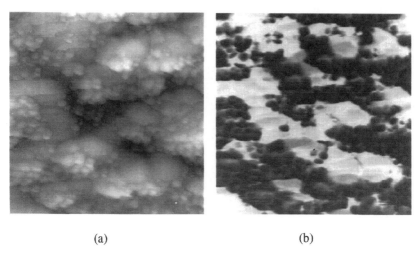

(a)   (b)

Figure 3 Intermittent-contact (IC) SFM images of the surface of a polymer (PVA)–carbon composite material. In the top image (a) the surface is presented in standard topographical form, while the image in (b) is a phase contrast (PC) representation of the same surface region. In this PC image, the carbon nanoparticles are represented by the small, dark spheres while the surrounding polymer is represented by the lighter regions. Reprinted from Scanning (2000, **ZZ**, 1–6) with permission from the Foundation for Advances in Medicine and Science, Inc.

rigidity may also affect the phase, so analysis of phase images must be performed carefully. In Figure 3, topographical and phase contrast images of a PVA film with carbon nanoparticles dispersed throughout the polymer are shown. In the topographical image (Figure 3a), the film features are clearly visible. In Figure 3b, the carbon particles (dark spheres) are clearly differentiated from the surrounding polymer matrix. Phase imaging can be especially useful in nanometer-scale analysis on two-component material systems.

4   REDOX REACTIONS AND POLYMERIZATION OF ORGANIC COMPOUNDS

Smectite clays are able to adsorb a wide variety of organic compounds; this property has been utilized for centuries to clarify wines and remove lanolin from wool [1]. Adsorption of organic compounds on transition metal-exchanged smectite clays often leads to interaction of the organic molecules with the metal ions in the interlayer of the clay. This interaction can be detected by a variety of spectroscopic techniques [14–19]. In some cases, the interaction of the organic compound with the transition metal ions in the interlayer is followed by a relatively slow electron transfer reaction which produces reactive organic free radicals. These radicals then serve to initiate polymerization of the organic compound [12, 15, 20]. This type of polymerization reaction has the potential to produce unique and interesting composite materials where the organic polymer component is ordered by the clay matrix in which it forms. This section describes the free radical-initiated polymerization of benzene, aniline, pyrrol and thiophene on hectorite and SFM investigations of the morphology of the resulting composites.

The reaction between benzene and transition metal-exchanged hectorite and montmorillonite can be carried out under a variety of conditions. Several workers have reported carrying out the reaction at room temperature by adsorbing benzene vapor on a clay film in a P_2O_5 desiccator [15, 20]. For example, Soma et al. [19] reacted Cu^{2+}- and Ru^{3+}-exchanged montmorillonite with benzene in this manner and studied the resulting material by resonance raman. A mechanism was presented in which poly(p-phenylene) polymer chains formed in the interlayer, with a portion of the aromatic groups of the polymer existing as cations. In our research we have carried out the reaction by placing vacuum-dried transition metal-exchanged hectorite films under vacuum in a sealed tube containing benzene and heating the system to temperatures between 65 and 93 °C, or by suspending dry hectorite films above refluxing benzene [4, 12]. It is noted that, at temperatures above about 93 °C, reactions between the metal ion in the interlayer and the clay may occur in addition to the desired polymerization. The resulting films are uniform dark brown or burgundy in color and are more flexible than those of the unreacted transition metal hectorite.

ESR provides an informative technique for following the progress of this reaction; it yields information on the environment and oxidation state of the paramagnetic metal ion as well as reporting on the presence of paramagnetic organic radicals. Initially, the ESR spectrum of thin films of transition metal-exchanged hectorite in benzene yields a spectrum characteristic of the metal ion in an axial site. As the reaction progresses, a single line ESR resonance characteristic of an organic free radical emerges and the resonance due to the transition metal decreases in intensity. Substitution of deuterobenzene for benzene in the reaction leads to a 30 % decrease in the width of the single line ESR resonance. This reduction in linewidth clearly indicates that the single line resonance arises from a radical derived from benzene and that the protons from benzene are coupled to the paramagnetic site. A careful analysis of the single line ESR spectrum shows that it is extremely unlikely that this radical is the benzene cation radical suggested by early workers, but rather is a large relatively immobile radical aligned by the film. These ESR results are consistent with the formation poly(p-phenylene) radical cations (as suggested by the resonance raman work of Soma et al.) or closely related species. If small amounts of water are added to the reaction mixture, the width of the single-line resonance increases and the shape of the line becomes gaussian in character. This observation leads to the conclusion that protons from the water are incorporated into the organic radical; such incorporation might reduce the conjugation in the system.

Figure 4 shows the SFM image of the surface of a Cu^{2+}-exchanged hectorite film after reaction with benzene. The surface is not flat like the surface of the unreacted

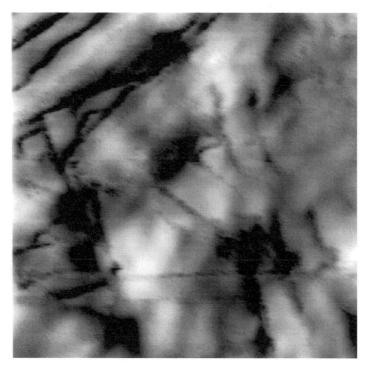

Figure 4 SFM image of Cu^{2+}-hectorite film exposed to benzene [4]. The inter-calated benzene has raised many of the platelets with respect to each other. Reprinted with permission of the Clay Mineral Society

film but is instead composed of many smaller regions with step-like transitions between them. The smallest step height measured is 0.7–1.0 nm and the other step heights are multiples of this separation; the largest step height measured is 7.8–8.3 nm. The image provides no evidence for polymerization of the benzene on the surface of the clay film. Benzene appears to polymerize only in the interlayer region of the hectorite where electric field and orientation effects are important. The polymerization in the interlayer 'raises' the clay platelets from the plane of the unreacted clay by a distance approximately equal to the 'width' of a benzene molecule (0.7 nm) or twice the 'thickness' of the benzene molecule (0.35 nm). The results suggest that the polymer in the interlayer is formed from benzene molecules oriented perpendicular to the surface of the film or in double layers parallel to the surface. The polymerization of benzene on hectorite films provides the only example we have observed where SFM shows little or no polymer forming on the surface of the clay film [12].

Aniline readily polymerizes on the surfaces and in the interlayer region of transition metal-exchanged smectite clays; furthermore, polymerization reactions involving aniline and inorganic hosts such as MoO_3 and $FeOCl$ have also been reported [10, 11, 21–23]. The redox reactions between transition metal-exchanged hectorite films and aniline occur much more readily than the redox reaction involving benzene. In typical experiments the reaction is essentially completed in a period of 24–48 h at room temperature. Furthermore, the extent of polymerization in hectorite/aniline systems is much greater than in the case of benzene. At the conclusion of the hectorite/aniline reaction, polymer is found both in the interlayer region and on the surface of the hectorite films [12]. Our research on hectorite polyaniline composites has focused on understanding the mechanism for the polymerization reaction and learning more about the factors that affect the morphology of the polymer synthesized [11, 12, 24]. In addition, we have investigated possible applications for hectorite/polyaniline composites in the area of chemical microsensors [24].

SFM images of a hectorite film exposed to aniline vapor provides important information on the mechanism of the polymerization reaction. Figure 5 shows both SFM topographical and phase contrast images, as a function of time, for a Cu^{2+}-exchanged hectorite film exposed to aniline vapor at room temperature and 1 atm pressure [11]. It is important to note that the inclusion of phase contrast imaging in these experiments allowed the simultaneous determination of the topography of the film surface and the material properties of the substances on the surface of the film. The images in Figure 5 were obtained by passing the aniline vapor over a Cu^{2+}-exchanged hectorite film enclosed in a 'microcell'. This microcell (manufactured by Park Scientific Instruments) provides a completely sealed environment for the exposure and subsequent *in situ* imaging of the film.

ESR provides chemical information concerning the polymerization reaction that is occurring on the film in Figure 5. The ESR parameters for Cu^{2+} in air-dried Cu^{2+}-exchanged hectorite films are given above; removal of water from the film broadens the resonance to the extent that identification of specific spectral features becomes difficult. Exposing a dehydrated film to aniline leads to the rapid emergence of a well-defined Cu^{2+} resonance with $g_{\parallel} = 2.369$, $g_{\perp} = 2.076$ and $A_{\parallel} = 144\,G$ [12]. This indicates that the aniline rapidly penetrates the interlayer and forms a complex with the Cu^{2+}. After a 24 h reaction time, the Cu^{2+} resonance disappears completely, polymer forms on the film and a free radical signal due to an organic free radical is observed. This signal is characterized by $g = 2.003$ and a linewidth of 13 G when the plane of the film is oriented perpendicular to the applied magnetic field; subsequently rotating the film by $90°$ increases the observed linewidth to 23 G.

Combining information from the SFM images in Figure 5 and the ESR results described in the preceding paragraph gives important information on the polymerization reaction. The images clearly show that polymer formation occurs on the surface by a modified island growth or modified Volmer–Weber growth mode. The ESR spectra indicate that aniline rapidly penetrates to the interlayer region and is spread throughout the hectorite film. This aniline undergoes a redox reaction with Cu^{2+}, with the resulting free radicals participating in the polymerization reaction.

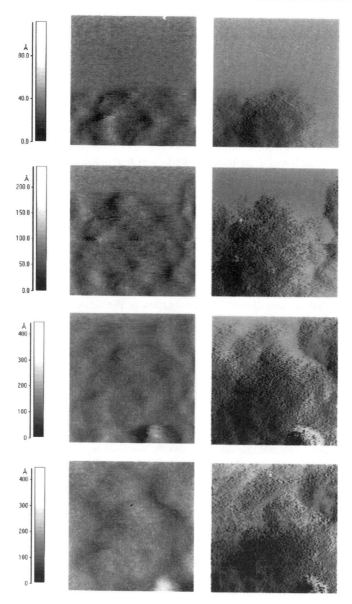

Figure 5 Series of 1 × 1 μm NC topographical (left-hand series) and PC (right-hand series) images of Cu^{2+}-exchanged thin film exposed to aniline vapor for 20 min (top), 1 h (second from top), 2 h (third from top) and 24 h (bottom) [11]. In the phase contrast images, darker regions correspond to the growing polyaniline surface layer, while the lighter regions correspond to the hectorite surface. Reprinted with permission from *J. Phys. Chem. B*, **101**, 11 106–11 111 (1997)

The anisotropy of the ESR signal from the organic radicals that are present in the system at the end of the reaction shows that the radicals are strongly oriented by the hectorite film.

A closer examination of the polymerization process is presented in the high-resolution topographical scan presented in Figure 6 [11]. In this topographical scan, light (more elevated) regions correspond to the surface polymer as opposed to the underlying hectorite film. Dozens of small nucleation sites occupied by polymer bundles are visible in this Figure, with the average nearest neighbor distance between bundles being of the order of 10 nm. The lateral size of these polymer bundles appears to range from less than 2 nm to about 10 nm, with coalescence of the bundles occurring when bundle diameters reach about 10 nm. At the top of the image in Figure 6 the underlying hectorite surface is evident. This surface is covered with small dark regions which correspond to individual micropores or faults in the hectorite surface structure. The lateral size and spacing of these micropores roughly correspond to the size and spacing of the observed nucleating polymer bundles in Figure 6. The depth of the micropores ranges from 1 to 2 nm, an indication that the faults extend into the top layer or two of the hectorite. It is expected that underlying layers would have similar fault structures offset from those on the top surface, and that a tortuous path through such faults would provide a path to all of the interlayer environments.

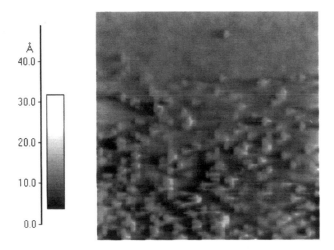

Figure 6 SFM 1200 × 1200 Å topographical scan of the hectorite surface near the edge of an expanding polyaniline island [11]. Numerous small polyaniline nucleation sites are visible, with an average distance separating them of about 100 Å. The size of the polyaniline bundles themselves ranges from about 20 Å to over 100 Å. Reprinted with permission from *J. Phys. Chem. B*, **101**, 11 106–11 111 (1997)

Both Figures 5 and 6 indicate that active surface regions (e.g. cracks, faults or pores) in the layered surface provide sites for the formation of polymer islands on the surface of the film. In addition, these faults and/or previously formed polyaniline must provide a 'gettering effect' for additional aniline arriving from the vapor phase or diffusing on the surface of the clay film. From Figure 6 it can be seen that the lateral growth of the islands apparently proceeds through the growth and subsequent coalescence of much smaller polymer bundles in the regions surrounding the islands. The images in Figure 5 indicate that, after 24 h exposure to aniline vapor at room temperature, the surface of the film is nearly completely covered by polyaniline.

In spite of the information presented above, it is still difficult to determine what precise surface, near-surface or interlayer effects are important in the polymerization reaction. Since Cu^{2+} will not initiate aniline polymerization by itself, the clay clearly plays an important role in the process. However, since Na^+-exchanged hectorite films show no sign of initiating aniline polymerization, one concludes that it is the unique properties of the transition metal-exchanged films that facilitate the polymerization reactions. As in the case of benzene, the role of electric fields within the hectorite interlayer would seem to be important in facilitating the initial redox reaction. In the case of aniline, unlike the case of benzene, the polymerization reaction can continue to proceed on the surface of the film, and coverage of the surface with polymer is possible.

Exposure of hectorite films to aniline vapor for a total of 48 h allows for polymer growth beyond that shown in Figure 5. An NC SFM topographical image of the polyaniline formed on a Cu^{2+}-exchanged hectorite film is shown in Figure 7 [11]. This Figure shows the aggregation of large nanometer-sized bundles of polyaniline; the morphology of this film is reminiscent of polyaniline films formed by standard chemical or electrochemical methods. XRD measurements of interlayer spacing on films like those imaged in Figure 7 yield an interlayer spacing of 1.51 nm; in comparison, measurements on air-dried Cu^{2+}-exchanged hectorite yield an interlayer spacing of 1.26 nm. It is interesting to note that, if the polymerization reaction is carried out on Cu^{2+}-exchanged hectorite films at elevated temperature (453 K), individual polymer bundles cannot be resolved in images taken at the scale of Figure 7 [12]. In images of $1.5 \times 1.5\,\mu m$ dimensions one can begin to resolve small, nearly spherical grains of polymer; thus, the morphology of the polymer deposited on the hectorite film reflects the temperature at which the polymerization reaction is carried out.

It is also possible to use SFM to image the polyaniline formed in the interlayer regions of the hectorite clay film. This is the material responsible for the 0.25 nm increase in interlayer spacing that occurs during the polymerization reaction and the material probably responsible for the strongly oriented free radical signal observed by ESR. This material can be exposed using standard lift-off techniques or by cleaving films, using a razor, to expose the inner layers. In Figure 8, $3 \times 3\,\mu m$ topographical (a) and phase contrast (b) images of an exposed intergallery section of the hectorite–polyaniline composite film is shown [24]. In Figure 8 a large,

Figure 7 NC SFM scan of Cu^{2+}-hectorite film surface after exposure to saturated aniline vapor for 48 h [11]. At this point, the initially blue clay surface is completely covered with a layer of black polyaniline. The large polyaniline bundles apparent in this image are typical of the morphology of polyaniline films formed using other techniques. Reprinted with permission from *J. Phys. Chem. B*, **101**, 11 106–11 111 (1997)

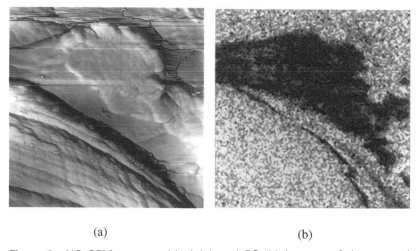

(a)   (b)

Figure 8 NC SFM topographical (a) and PC (b) images of the razor-cleaved intergallery region of a Cu^{2+}-polyaniline composite thin film [13]. The dimensions of these images are $3 \times 3 \mu m$. The intergallery polyaniline is represented by the dark regions in the PC image and is about 110 Å in thickness as measured by SFM. Reprinted by permission of the American Institute of Physics

polyaniline section runs diagonally through regions of hectorite. SFM shows that the thickness of the layer with respect to the neighboring hectorite is 110 nm. Approximately 30 layers of planar polyaniline would be required to obtain this total thickness. The polyaniline in the diagonal region shows planar regions and regions where there are some signs of the formation of the bundles that appear on the surface of fully reacted films. The planar regions are similar to those that appear on films where the interlayer has been exposed using lift off techniques and probably reflect ordering imposed by the clay layers. The sections where bundles are starting to form may reflect regions where there were defects or pores in the clay film. Polymerization in these regions would allow the formation of thicker layers of polymers and the beginning of structures resembling those of the bulk polymer. It is estimated that approximately 10 % of the overall polymer in the interlayer region is part of these thicker structures associated with defects. At present there is no information on the effect of temperature on the structure of the polymeric material in the interlayer.

The experimental information currently available allows the development of plausible chemical mechanisms for the formation of polyaniline on Cu^{2+}-exchanged hectorite films; such a mechanism is presented in Scheme 1 [11]. Here, one aniline radical molecule is initially oxidized by a Cu^{2+} (present in the interlayer region of the hectorite) to form an aniline radical cation and a Cu^+ ion. This radical cation then reacts with a neutral aniline molecule to form a resonance-stabilized radical cation dimer. The dimer then loses a proton to the environment; this step is followed by hydrogen abstraction of a *para* hydrogen by O_2 to form p-(phenylamino)aniline and an $HOO^.$ radical. This radical reacts with the p-(phenylamino)aniline to form H_2O_2 and a neutral dimer radical. The dimer radical reacts with neutral aniline monomer and steps (a), (b) and (c) in the mechanism are repeated with the subsequent formation of polyaniline in the leucoemeraldine form. The H_2O_2 that is generated may also serve to oxidize aniline, and Cu^{2+} may be regenerated from Cu^+ by oxidation reactions involving O_2 or H_2O_2. The mechanism for termination of the reaction is not well understood, although it would seem that polymer formation would eventually choke off the supply of aniline monomer to reactive sites. It may also be that the radicals observed by ESR in the later stages of the reaction represent paramagnetic 'defect sites' of lower reactivity than those described in Scheme 1. Accumulation of such defects could serve to terminate the reaction.

Both polyaniline and hectorite clay have been deposited on interdigitated arrays and used as the basis for useful chemiresistors [7, 8, 24]. Microsensor applications of the hectorite–polyaniline composite could capitalize on the ease of thin film production, the role of clay in maintaining film porosity and the fact that the composite should have higher thermal stability than the polyaniline alone. In addition, the electrical properties of the hectorite–polyaniline composite should be interesting because the oriented material in the interlayer region may have electrical properties differing from bulk polyaniline. Chemiresistors fabricated using Na^+-

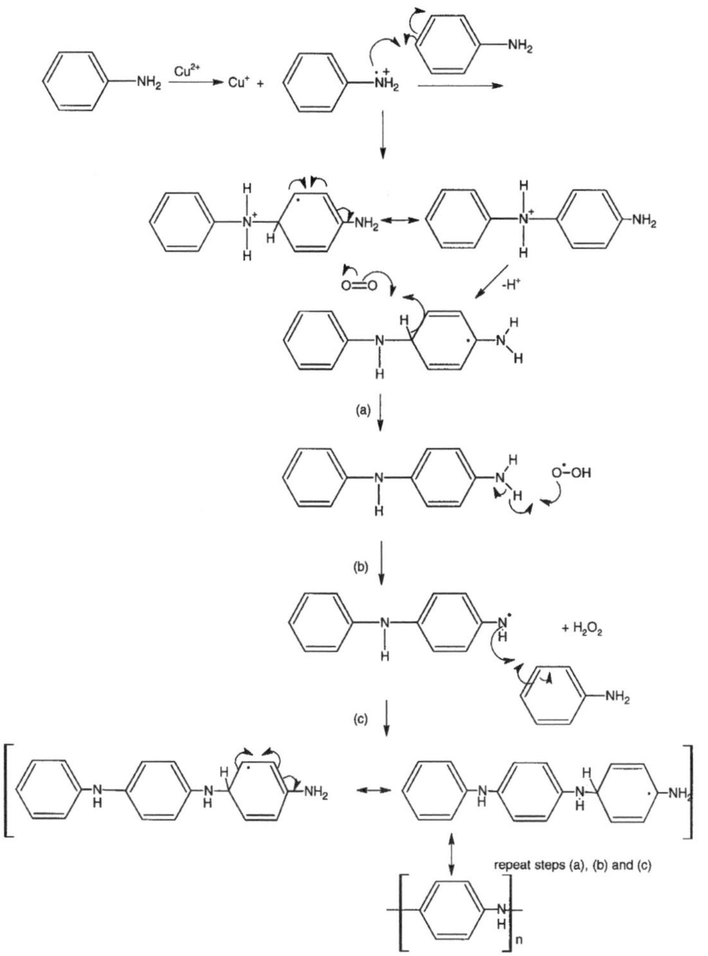

Scheme 1 Chemical mechanism for the formation of polyaniline on Cu^{2+}-exchanged hectorite films [11]. Reprinted with permission from *J. Phys. Chem. B*, **101**, 11 106–11 111 (1997)

hectorite films and hectorite films cast from suspensions of Na^+-hectorite in a gold sol have a.c. impedances that are very sensitive to the presence of water vapor and somewhat less sensitive to alcohols such as ethanol and methanol [7, 8]. The impedance of these films does not change in the presence of alkanes and other solvents having a low value of Hildebrand's solubility parameter. In a series of experiments we have shown that the a.c. impedance of films of hectorite–polyaniline composite changes significantly in the presence of hexane [24]. Unfortunately, these films delaminate from the surface of the interdigitated arrays after a few hours exposure to hexane; thus, in their present form they are of little practical use. Potential methods for binding the films to the surface of the array are currently being developed.

Other easily oxidized compounds such as thiophene and pyrrole also react with Cu^{2+}-exchanged hectorite films [9, 10]. In both cases the polymer reaction follows the pattern described above for aniline, with polymerization occurring both on the surface and in the interlayer region. In the case of pyrrole, the coverage of the surface by polymer is incomplete when the polymerization reaction is carried out at room temperature. Micron-sized bundles of polypyrrole are scattered over the surface of the clay film at the end of the reaction. Thiophene reacts with Cu^{2+}-exchanged hectorite to a much greater extent and the films that are formed are dark black, which suggests considerable conjugation in the thiophene polymer. At present there are no detailed SFM or ESR studies of the polymerization of thiophene or pyrrole on hectorite and no information about the suitability of the hectorite–polythiophene or hectorite–polypyrrole composites for microsensor applications.

5 ACID INITIATED POLYMERIZATION

Organic compounds adsorbed on transition metal-exchanged smectite clays may also undergo polymerization by mechanisms that do not involve electron transfer and the reduction of the metal ions in the interlayer. This type of polymerization, which occurs for methyl methacrylate and styrene, is less well studied than the free radical polymerizations described in the preceding section [6, 25]. This type of polymerization reaction occurs relatively slowly compared with the redox reactions previously described, and the polymer that forms seems, in most cases, to be primarily located on the surface of the clay film or in defects in the clay film.

Hectorite films exchanged with Cu^{2+} or Fe^{3+} react with methyl methacrylate monomer in a solventless process to form poly(methyl methacrylate). The reaction is slow and occurs over a period of 3 days to a week at room temperature [6]. A similar exposure of Ca^{2+}-exchanged hectorite to methyl methacrylate produces little evidence of reaction. SFM shows that PMMA forms on the surface of both Cu^{2+}- and Fe^{3+}-exchanged films, with the morphology of the resulting polymer reflecting both the mode of monomer delivery and the type of transition metal ion in the interlayer. Figure 9 shows SFM images of the surface of Cu^{2+}-hectorite exposed to liquid methyl methacrylate (a) and methyl methacrylate in the vapor phase (b) [6]. In both images the clay surfaces are completely covered by thick layers of PMMA. The two Figures differ in that vapor delivery of monomer seems to produce large polymer structures with dimensions of the order of 1–2 μm, while exposure of the films to liquid methyl methacrylate produces uniform, small bundles with dimensions of the order of 25 nm. Figure 10 shows an SFM image of Fe^{3+}-hectorite exposed to methyl methacrylate vapor [6]. Here one observes a dense packed polymer structure with polymer bundles exceeding 1 μm in size; exposure to liquid

(a)

(b)

Figure 9 SFM topographical scans of the surface of Cu^{2+}-hectorite exposed to (a) liquid methyl methacrylate and (b) methyl methacrylate in the vapor phase [6]. The dimensions of these images are $3 \times 3\,\mu m$ (a), and $7.5 \times 7.5\,\mu m$ (b). Reprinted from Eastman, M. P. *et al.*, *The Formation of Poly(methyl-methacrylate) on Transition Metal-Exchanged Hectorite*, Copyright 1999, pp. 173–185, with permission from Elsevier Science

methyl methacrylate produces a polymer with bundles of the order of 100 nm. Exposure of Fe^{3+}-exchanged hectorite to liquid methyl methacrylate swells the interlayer spacing from 1.55 ± 0.03 nm to 1.69 ± 0.04 nm; a similar exposure of Cu^{2+}-exchanged hectorite produces a relatively small change in the spacing from 1.24 ± 0.01 nm to 1.29 ± 0.04 nm. The mass change accompanying exposure to methyl methacrylate liquid is about 11 % for the Fe^{3+}-exchanged films but only about 4 % for the Cu^{2+}-exchanged films. The XRD and mass change measurements provide strong evidence that the Fe^{3+}-exchanged hectorite films produce polymer throughout the clay structure and at a higher yield than the Cu^{2+}-exchanged films. This conclusion is confirmed by ATR spectroscopy on the resulting clay–polymer composites. The spectrum for the polymer on the Fe^{3+}-exchanged film yields a strong carbonyl band at $1720\,cm^{-1}$, while evidence for a similar band from the polymer on the Cu^{2+}-exchanged films is weak [26]. Extraction of the polymer from hectorite–poly(methyl methacrylate) composite formed using Fe^{3+}-exchanged hectorite by acetone and subsequent analysis of the resulting material by MALDI show the formation of polymers containing up to 20 monomer units [26].

ESR shows no evidence for the formation of organic free radicals during or after the polymerization reaction, and no evidence for reduction of the transition metal ions in the interlayer region of the clay. We have presented a possible cationic mechanism for the polymerization of methyl methacrylate on transition metal-

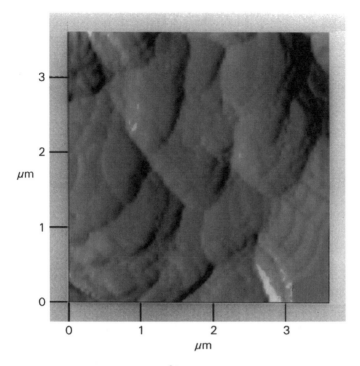

Figure 10 SFM image of Fe^{3+}-hectorite exposed to methyl-methacrylate vapor [6]. The dimensions of this image are 3 × 3 μm. Polymer bundles exceeding 1 μm in size are present in this image. Reprinted from Eastman, M. P. *et al.*, *The Formation of Poly(methly-methacrylate) on Transition Metal-Exchanged Hectorite*, Copyright 1999, pp. 173–185, with permission from Elsevier Science

exchanged hectorite (see Scheme 2) [6]. This mechanism relies on the enhanced Brönsted acidity of transition metals exchanged into the interlayer region of clay films and the potential for the stabilization of cationic species by the negatively charged silicate surface of the hectorite. While cationic polymerization of methyl methacrylate may be unusual, it may not be unprecedented. Gosh and Driscoll found that both styrene and methyl methacrylate polymerize in the presence of SO_2 [27]. This species can act as a Lewis acid and in the presence of small amounts of water as a Brönsted acid. The rate of the reaction with SO_2 is very similar to that observed on clays (3 days to a week).

Liquid styrene, when exposed to Cu^{2+}-exchanged hectorite films, reacts in a manner similar to methyl methacrylate [25]. The polymerization reaction requires about a week to complete at room temperature; longer reaction periods produce little evidence for the continued formation of polymer. ESR of the Cu^{2+} in the hectorite films shows only minor changes as the reaction with styrene proceeds. The Cu^{2+}

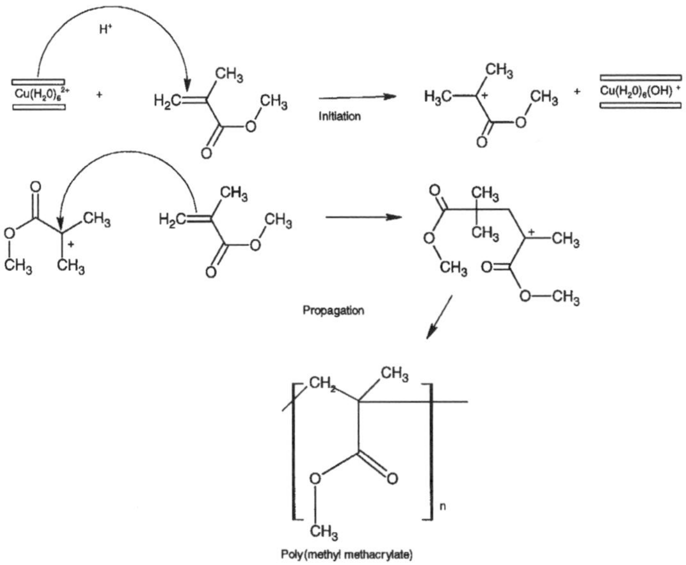

Scheme 2 Possible cationic mechanism for the polymerization of methyl methacrylate on transition metal-exchanged hectorite [6]. Reprinted from Eastman, M. P. et al., The Formation of Poly(methly-methacrylate) on Transition Metal-Exchanged Hectorite, Copyright 1999, pp. 173–185, with permission from Elsevier Science

resonance shows a slight broadening, but the integrated intensity is unchanged within experimental error ($\sim 10\,\%$) and there is no ESR evidence for the formation of organic free radicals. XRD of Cu^{2+}-exchanged hectorite before and after reaction with styrene shows essentially no change. Thus, ESR and XRD provide no evidence of the formation of polymer in the interlayer region. The similarity of the polymerization reaction rates, ESR data and XRD results for methyl methacrylate and styrene strongly suggest that both reactions proceed by a similar mechanism. In our original work on styrene we did not suggest a detailed mechanism for the polymerization but did allude to the possibility of redox reactions playing a role in the polymerization. Additional ESR work and our work on methyl methacrylate lead us now to believe that a redox mechanism for the styrene polymerization is very unlikely.

Figure 11a shows an $8 \times 8\,\mu m$ SFM image of a hectorite–polystyrene composite following reaction [25]. Amorphous polymer covers a large portion of the surface, with small areas where no apparent polymerization has occurred being evident in the left- and lower right-hand portion of the Figure. Analysis of this and several other large area scans suggest that approximately 95 % of the surface of the clay film is covered with polymer. SFM scans at higher resolution reveal an underlying polymer structure. Figure 11b shows a $1.2 \times 1.2\,\mu m$ SFM scan of the composite, with the

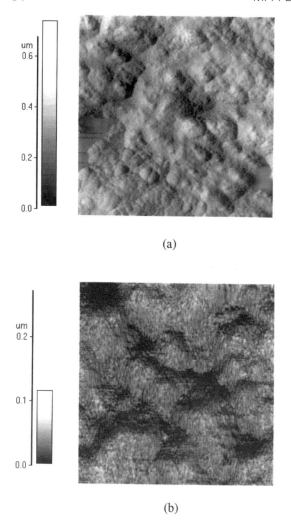

(a)

(b)

Figure 11 (a) SFM topographical scan on the surface of a Cu^{2+}-exchanged hectorite–polystyrene composite thin film [25]. The dimensions of this image are $8 \times 8\,\mu m$. The hectorite surface is about 95% covered by the topmost polystyrene layer. (b) SFM $1.2 \times 1.2\,\mu M$ scan of the same surface as in (a). At this resolution, we can see that the individual polymer strands have grouped together, forming a 'brush-like' structure. Reproduced by permission of John Wiley & Sons

image showing strands of vertically oriented polymer aggregates or bundles varying from about 10 to 50 nm in length. These bundles give the overall structure a 'brush-like' appearance [25].

While ESR and XRD data give no evidence for intercalation of the styrene monomer into the intergallery regions and subsequent polymer formation, it is important to note that hectorite–styrene composite films, after undergoing 'razor cleaving' or tape lift-off, yield SFM images that show the presence of polymer within the clay film. This polymeric material appears to form in defects in the clay film and to cover 20–30 % of an exposed inner surface. The defects in which polymer forms may be planar vacancies in the layered structure, or pores and other types of three dimensional faults. Figure 12 shows topographical (a) and phase contrast (b) SFM images of such a defect region [25]. Scan (b) shows the presence of three distinct types of material, one of which is the clay surface. The other two regions appear to be polymer with different mechanical properties: SFM indicates that one of these polymeric materials is similar to what forms on the surface of the clay while the other form is more rigid. These results provide additional evidence that the formation of polymers in the confines of a clay matrix can produce materials different from those that form in bulk polymerization.

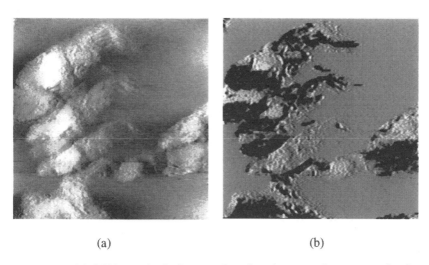

(a)   (b)

Figure 12   (a) SFM topological scan of an interlayer region exposed using the lift-off technique [25]. The dimensions of this image are $4 \times 4\,\mu m$. The height of the polymer regions above the surrounding clay layers ranges from $10\,\text{Å}$ to about $100\,\text{Å}$. For the sample as a whole, the polystyrene material occupies 20–30% of the exposed 'surface' area. (b) SFM phase contrast image of the same region. We can see from this image that the interlayer polymer material is composed of two phases, possibly polystyrene similar to that on the surface, along with a more rigid polystyrene form. Reproduced by permission of John Wiley & Sons

6 REACTIONS OF AMINO ACIDS AND ACTIVATED NUCLEOTIDES WITH HECTORITE

Biologically important organic molecules are known to form in many environments through a number of different natural means. These molecules have been found in such exotic environments as meteorites and comets, as well as numerous Earth-bound locations. Among the biological molecules needed for the emergence of life, amino acids and nucleic acid bases are of primary importance [28]. These molecules may be synthesized from more basic moieties through the action of electrical discharges, heat, UV light or other ionizing radiation. It is highly possible that these processes occurred not only on the prebiotic Earth but also extraterrestrially [29]. The pathways by which these molecules condensed into more complex biopolymers are currently the subject of much discussion and research [30–35]. About 50 years ago, Bernal [36] proposed that certain clay minerals might not only adsorb and concentrate these biomonomers but may subsequently have catalyzed the condensation reactions necessary to form the first polypeptides or polynucleotides. Currently, research is being undertaken with the goal of understanding how clay minerals may have participated in these important prebiotic reactions [37].

Experiments combining amino acids such as glycine, lysine or alanine with various clay minerals in bulk form have produced small peptides in simulated prebiotic reactions [33, 34, 38–42]. The clay minerals used in these studies are usually kaolinite, illite, montmorillonite or hectorite. In these studies, oligomers of up to six monomer units have been produced on the clays and identified using techniques such as HPLC, gas chromatography or mass spectrometry [10, 11]. Condensing agents such as polyphosphates or hydrogen cyanide derivatives have also been included in these reactions, as well as activated amino acids such as aminoacyl adenylates, producing molecular weights of up to 4000 for the organic molecules subsequently synthesized.

In addition to the research being conducted on the condensation of amino acids, other studies have focused on clay processed RNA-like molecules which may have formed on the primitive earth and subsequently condensed to form the first oligonucleotides [1, 30–32, 43–47]. It is felt by many that the origin of life on Earth began with the so-called RNA-world, in which the first viable, 'living' molecules were RNA-like. It has already been shown that certain clay minerals participate in reactions involving the condensation of nucleotide oligomers. For example, a mixture of an unblocked, activated monomer (phosphorimidazolide of adenosine, or ImpA) and montmorillonite produce oligomers of adenylic acid up to 11 mer in length [30, 31]. The extent of oligomer formation has been found to be highly dependent on the type of metal cation present in the clay. The oligomers produced generally contain a large majority of $3',5'$-phosphodiester bonds (over the $2',5'$-linkage). This allows them to participate in further chain elongation reactions by ribozymes or template-directed replication. The synthesis of polyadenylates up to 50 mer in length was recently accomplished through the elongation of the decanucleotide [32P]dA(pdA)8pA in the presence of Na^+-montmorillonite using periodic 'feedings' of the activated nucleotide ImpA [32].

The role of the clay minerals in the condensation reactions discussed above is currently under study and may contain many components. Clay surfaces or intergallery regions act to adsorb, intercalate and concentrate many organic species. Specific sites on or within the clay mineral (including sites where exchangeable metal cations are present) may initiate the reactions in which condensation of these molecules occurs. Properties of certain clay minerals related to their structure and local chemical or electrical properties may act to control the regiospecificity of the reaction, leading to more biologically viable molecules [28, 47, 48]. In our initial experiments involving amino acids, a solution of 30 mM glycine was applied to the surface of a Cu^{2+}-exchanged hectorite film prepared using standard techniques; this resulted in surface glycine coverage of approximately 0.05 monolayers for the clay films employed [12]. The samples were then subjected to 10 alternate cycles of heating to 90 °C for 24 h and rewetting with distilled water for 24 h. Identical cycling experiments using glycine exposed Na-hectorite films were also performed. Control films with identical glycine exposure but no cycling were also prepared, as were cycled clay films with no glycine exposure. Subsequent to scanning force microscopy (SFM) analysis, the organic material was extracted from the clay mineral with a 0.1 M calcium chloride solution. HPLC analysis was then used for chemical analysis of the reaction products.

In the experiments involving activated nucleotides, an ImpA solution (15 mM) was prepared in 0.2 M NaCl and 0.075 M $MgCl_2$ and subsequently adjusted to pH 8. Hectorite films were exposed to the solution until saturated, and then covered. At 24 h intervals, the samples were resaturated with the ImpA solution. After three cycles, the films were allowed to dry completely. Small sections of intact, reacted clay were mounted for SFM imaging, XRD and ESR analysis. The remainder of the clay samples were mixed with 0.1 M ammonium acetate, vortexed and allowed to stand for 24 h. This mixture was centrifuged, and the supernatant removed for commercial MALDI analysis.

Glycine polymerization was analyzed by HPLC on a BioCAD Sprint system using a HAIsil C18 column (ODS, 5 μm/250 × 4.6 mm; Higgins Analytical, Inc.). Glycine oligomers were detected by HPLC analysis using an absorbance of 195 nm. They were identified by their retention times relative to glycine, glycine anhydride (2,5-piperazinedione, DKP) and glycine ($n = 2$–6) oligomer standards. This method clearly indicated that the simulated prebiotic reactions formed glycine oligomers up to 6 units in length [28]. SFM images of both cycled and non-cycled glycine exposed clay films were then obtained using an Auto-Probe instrument from Park Scientific. All SFM images were acquired in non-contact mode under low (approximately 10 %) humidity conditions.

In Figure 13a, an SFM image of a 500 × 500 Å region on the surface of a film exposed to the glycine solution but not subjected to prebiotic cycles is presented. A single step in the otherwise generally smooth surface is seen. The height of this step

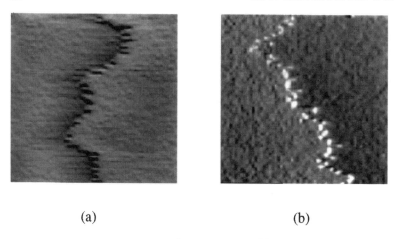

(a)   (b)

Figure 13   NC SFM 500 × 500 Å images of Cu^{2+}-exchanged hectorite films [28]. The image in (a) was obtained from a glycine exposed, non-cycled film. The image in (b) was taken from a glycine exposed film subjected to 10 simulated prebiotic cycles of alternate heating and wetting. The step edge in (a) is measured to be 14 ± 2 Å in height. In (b), a step of height 7 ± 2 Å is seen to be decorated with small glycine oligomers. Reprinted by permission of Springer-Verlag GmbH & Co. KG

is measured to be 14 ± 2 Å. This image is typical for the surface of hectorite thin films prepared in the manner described earlier, containing many flat regions separated by cracks, edges or faults.

In Figure 13b, a region on a film subjected to 10 alternate heating and wetting cycles is shown. The dimensions of this image are 500 × 500 Å. A single step of height 7 ± 2 Å traverses this section of the film surface from top to bottom. Many glycine oligomers appear to be attached or adsorbed to this site. The dimensions of the individual oligomers as measured by SFM range from 8 to 20 Å. SFM imaging of many areas on the cycled sample surface indicates that about 20 % of the surface visible step edges are similarly decorated with glycine oligomers. In general, flat surface regions (with no steps or cracks) are devoid of any trace of peptides, except in the case of the presence of micropore sites, which are discussed below. The presence of glycine oligomers at or near step edges suggests that activation of the biomonomer is occurring by complex formation with the gallery metal cations. These gallery cations are not on the flat clay surface, but instead at surface step edges, cracks or faults. The function of the surface edges is thus proposed to be twofold: to adsorb and constrain the reaction constituents, and to provide the necessary cation species to facilitate monomer activation and subsequent polymerization.

Prebiotic cycling experiments were also performed using Na^+-hectorite. For these experiments, no glycine polymerization occurred, and no oligomers were imaged at any sites on the surface. We conclude that the presence of Cu(II) cations at surface sites on the clay is imperative for the observed reactions to occur.

SFM was also used to examine the surface of hectorite films reacted with the activated nucleotide ImpA. In Figure 14, a non-contact $0.55 \times 0.55\,\mu m$ SFM image of the hectorite surface reacted with the ImpA solution is shown. These SFM clay samples were washed with ultrapure water prior to imaging in order to eliminate the presence of any salts on the surface. Most of the topmost surface regions imaged using SFM appeared to be free of any nucleotide oligomers, including step edges and micropore sites. This result is in contrast to the glycine reactions described above. Many areas on the clay surface, however, contain numerous apparent nucleotide oligomers. These surface areas are characterized by the presence of large concentrations of surface defects such as cracks and faults. In Figure 14 two large fault regions running the length of the image from top to bottom are seen near the left-hand side of the image.

In this image (and other images on this surface) the polyadenylate oligomers do not appear to be attached or adsorbed to the fault edges, in contrast to peptide oligomers imaged earlier. This is the case for virtually all of the adenylic acid

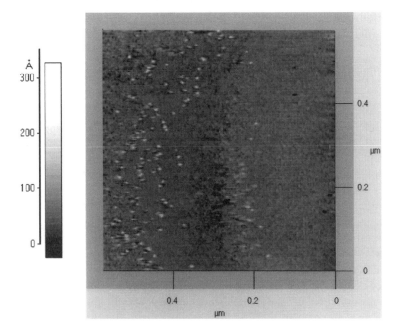

Figure 14 Non-contact SFM image of the hectorite surface reacted with ImpA [46]. Numerous small oligomers of adenylic acid are visible in this surface region. The dimensions of this image are $0.55 \times 0.55\,\mu m$. Reprinted with permission of the American Institute of Physics

oligomers that we observed on this clay. In one proposed mechanism for the formation of oligonucleotides in the presence of clay minerals, the nitrogen-containing activation groups of the monomer must contain either a positively charged nitrogen attached to the phosphorous or a nitrogen that is sufficiently basic to be protonated by acidic sites on the clay. The positive charge binds the activated nucleotide to local negative sites on the clay. Protonated ImpA does not bind as strongly to negative sites on the clay as other positively charged nitrogen-containing activating groups [46]. Also, the exchangeable Cu(II) cations present in hectorite are firmly located in the intergallery regions of the clay layers. Only a small fraction of these cations may be present on the topmost clay surface. The function of the surface cracks, faults or steps is then to present the interlayer cations to the biomolecules on the surface.

In another possible mechanism, polymerization of the activated nucleotide is taking place within the intergallery regions of the clay. Washing of the clay during simulated (or real) prebiotic cycles expels the oligomers to the surface. Evidence for this second mechanism using montmorillonites has been obtained by 'blocking' clay edge sites [43]. In this study, the blocked step edges had a minimal effect on the formation of subsequent RNA oligomers. X-ray diffraction (XRD) and electron spin resonance (ESR) provide additional evidence for intercalation of the monomer followed by intergallery polymerization. For dry, unexposed Cu(II) hectorite films, we measure a basal spacing of 12.6 ± 0.4 Å. This compares with a measured basal spacing of 14.9 ± 0.4 Å for films exposed to the ImpA monomer. In both sets of measurements, the relative humidity was held at 20–40%.

For the film not exposed to ImpA, the intergallery regions contain only the Cu(II) cations and water molecules. The water molecules present cause the interlayer spacing to increase from 9.4–9.5 Å to the XRD value quoted above. When the ImpA molecules are intercalated, some of the water molecules are displaced, resulting in an increase in basal spacing that is somewhat smaller than if no water were removed. Taking into account the removal of some water from the gallery regions, we calculate an available spacing of 5.5 Å to accommodate the intercalated ImpA molecules. Comparison of the dimensions of an ImpA molecule with these spacings suggests that the ring structures of the ImpA may be oriented in a variety of conformations within the clay intergallery regions, excluding only an orientation where the molecule is perpendicular to the gallery silica sheets.

We have also used ESR to study the intercalation of ImpA in Cu(II)-exchanged hectorite. For these experiments, Cu^{2+}-exchanged hectorite was exposed to buffered ImpA and allowed to air dry. ESR of these films at 77 K yields axially symmetric magnetic parameters for Cu^{2+}: $A_\parallel = 196$ G, $g_\parallel = 2.24$ and $g_\perp = 2.077$. Axial symmetry was also observed in the ESR of untreated Cu^{2+}-exchanged hectorite, but the magnetic parameters ($A_\parallel = 164$, G, $g_\parallel = 2.337$ and $g_\perp = 2.076$) differed substantially from the treated material. Both sets of spectral parameters are consistent with octahedrally coordinated Cu^{2+}. Cu^{2+}-exchanged hectorite treated

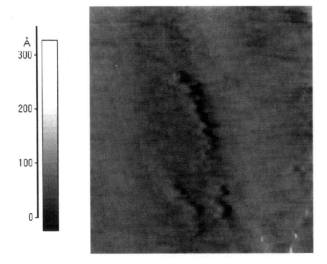

Figure 15 SFM $110 \times 110\,\text{Å}$ image containing three adenylic acid oligomers [46]. The largest oligomer measures approximately $35\,\text{Å}$ in length, while the smaller two are on the order of $10\text{--}12\,\text{Å}$ in dimension. The length of the largest oligomer suggests that it consists of at least eight monomer units. Reprinted with permission of the American Institute of Physics

with imidazole, adenine and adenine in an HEPES buffer solution yields substantially different spectra than those films treated with the full ImpA solution.

Adenylic acid oligomers on the Cu(II)-exchanged hectorite surface are found to exist in a predominantly globular, or folded, morphology. A small number of these oligomers, however, exhibit a more elongated morphology. In Figure 15, a $110 \times 110\,\text{Å}$ SFM image containing three oligomers is shown. The largest oligomer measures approximately $35\,\text{Å}$ in length, while the smaller two are of the order of $10\text{--}12\,\text{Å}$ in dimension.

7 CONCLUSIONS

The work described in this chapter is directed at understanding the chemistry accompanying the spontaneous polymerization of organic compounds and biomolecules on the surface of Cu^{2+}- or Fe^{3+}-exchanged hectorite and the factors influencing the morphology of the resulting polymer. In this work, hectorite was chosen for study because of its ability to form relatively uniform and well-oriented films suitable for investigation by SFM, ESR, XRD and ATR-FTIR. Organic molecules polymerize on hectorite clay by at least two distinct mechanisms. The polymerization of benzene, aniline, pyrrole and thiophene on hectorite occurs primarily by means of a redox mechanism, while an acid-based cationic reaction is the most plausible mechanism for the polymerization of methyl methacrylate and styrene. Both mechanisms depend on the special properties of transition metal ions in the interlayer region of hectorite and the unique structural aspects of clay. Important factors determining the morphology of the polymer that forms on hectorite include reaction temperature, mode of monomer delivery to the surface of the clay and the properties of the exchangeable metal ion in the clay interlayer. SFM shows that the morphology of polyaniline and polystyrene formed on the surface of hectorite differs from that of the polymer formed in defects and in the interlayer region. The polymerization reactions so far investigated stop or drastically slow down as the polymer loading on the hectorite increases, and the polymers produced appear to be of relatively low molecular mass. The termination or slowing down of the polymerization reaction may occur because polymer formation chokes off the delivery of monomer to reactive sites or because side reactions eliminate active sites.

The polymerization of amino acids and activated nucleotides also occurs on Cu^{2+}- or Fe^{3+}-exchanged hectorite. As in the case of organic monomers, the special properties of transition metal ions in the interlayer region of hectorite and the unique structural aspects of clay appear to be important in the polymerization reactions. SFM of Cu^{2+}-exchanged hectorite films following exposure to glycine under simulated prebiotic conditions shows peptide formation at surface step edges and micropore sites on the clay, while polymers formed from the activated nucleotide Imp(A) appear to wash out of the interlayer region and appear at large faults or cracks in the hectorite film. These results suggest that clays may not only facilitate the polymerization of biomolecules under prebiotic conditions but also template the reaction products for more complicated subsequent reactions.

In the future, our research will be directed towards developing MALDI techniques for characterizing the organic and biopolymers formed on hectorite and other clays, learning to build robust clay–polymer based chemical sensors, characterizing new polymeric materials formed in the confines of the clay interlayer and exploring templating reactions for biomolecules on clay surfaces. Present indications are that such research will significantly increase the scientific and practical significance of clay–polymer composites.

8 REFERENCES

1. Theng, B. K. G. *The Chemistry of Clay–Organic Reactions*, John Wiley & Sons, New York, 1974.
2. Hartzell, C. J., Cygan, R. T. and Nagy, K. L. *J. Phys. Chem.*, **102**, 6722–6729 (1998).
3. Farmer V. C. and Russell, J. D. *Trans. Faraday Soc.*, **67**, 2737–2749 (1971).
4. Eastman, M. P., Patterson, D. E. and Pannell, K. H. *Clays Clay Miner.*, **32**, 327–333 (1984).
5. Ingber, D. E. *Scient. Am.*, (1998).
6. Eastman, M. P., Bain, E., Porter T. L., *et al.*, *Appl. Clay Sci.*, **15**, 173–185 (1999).

7. Venedam, R., Eastman, M. P., Wheeler B. L. *et al.*, *J. Electrochem. Soc.*, **138**(6), 1709–1712 (1991).
8. Eastman, M. P., Hughes, R. C., Yelton G. *et al.*, *J. Electrochem. Soc.*, **146**, 3907–3913 (1999).
9. Porter, T. L., Eastman, M. P., Hagerman M. E. *et al.*, *J. Vac. Sci. Technol. A*, **14**, 1488–1493 (1996).
10. Porter, T. L., Hagerman, M. E. and Eastman, M. P. *Recent Res. Dev. Polym. Sci.*, **1**, 1–17 (1997).
11. Porter, T. L., Eastman, M. P., Zhang D. Y. *et al.*, *J. Phys. Chem. B*, **101**, 11106–11111 (1997).
12. Eastman, M. P., Hagerman, M. E., Attuso J. L. *et al.*, *Clays Clay Mine.*, **446**(6), 769–773 (1996).
13. Porter, T. L., Manygoats, K., Bradley M. *et al.*, *J. Vac. Sci. Tech. A*, **16**(3), 926–931 (1998).
14. Ogawa O. and Kuroda, K. *Chem. Rev.*, **95**, 399–438 (1995).
15. Rupert, J. P. *J. Phys. Chem.*, **77**, 784–790 (1973).
16. Hartzell, C. J., Yang, S.-W. and Morris, D. E. *J. Phys. Chem.*, **99**, 4205 (1995).
17. Weissmahr, K. W., Haderlein, S. B. and Nuesch, R. *Environ. Sci. Technol.*, **31**, 240 (1997).
18. Walter, D., Saehr, D. and Wey, R. *Clay Miner.*, **25**, 343–354 (1990).
19. Soma, Y., Soma, M. and Harada, I. *Chem. Phys. Lett.*, **99**, 153–156 (1983).
20. Pinnavaia, T. J., Hall, P. L., Cady S. S. *et al.*, *J. Phys. Chem.*, **78**, 994–999 (1974).
21. Krishnamootri, R., Vaia, R. A. and Ginnelis, E. P. *Chem. Mater.*, **5**, 1728 (1996).
22. Kerr T. and Nazar L. F., *Chem. Mater.*, **8**, 2005 (1996).
23. DeGroot, D. C., Schindler, J. C., Kannewurf C. R. *et al.*, *J. Chem. Soc., Chem. Commun.*, 687 (1993).
24. Porter, T. L., Thompson, D., Bradley M. *et al.*, *J. Vac. Sci. Tech. A*, **15**(3), 500–504 (1997).
25. Porter, T. L., Hagerman, M. E., Reynolds B. P. *et al.*, *J. Polym. Sci., Part B, Polym. Phys.*, **36**, 673–679 (1998).
26. Porter T. L. and Eastman, M. P. Unpublished results (1999).
27. Gosh, P. and Driscoll, K. F. *J. Polym. Sci.*, **4**, 519–525 (1966).
28. Porter, T. L., Eastman, M. P., Hagerman M. E. *et al.*, *J. Mol. Evol.*, **47**, 373–377 (1998).
29. Miller, S. J. and Orgel, L. E. *The Origins of Life on Earth*, Prentice Hall, 1974.
30. Ferris, J. P., and Ertem, G. *Science*, **257**, 1387–1388 (1992).
31. Ferris, J. P. and Ertem, G. *J. Am. Chem. Soc.*, **115**, 12270–12275 (1993).
32. Ferris, J. P., Hill, A. R., Liu R. *et al.*, *Nature*, **381**, 59–61 (1996).
33. Lahav, N., White, D. and Chang, S. *Science*, **201**, 67–69 (1978).
34. Lawless, J. G. and Levi, N. *J. Mol. Evol.*, **13**, 281–286 (1979).
35. Paecht-Horowitz, M. and Katachalsky, A. *J. Mol. Evol.*, **2**, 91–98 (1973).
36. Bernal, J. D. *The Chemical Basis of Life*, Routledge and Kegan Paul, London, 1951.
37. Cairns-Smith, A. G. and Hartman, H. *Clay Minerals and the Origin of Life*, Cambridge University Press, Cambridge, 1986.
38. Bujda'k, J., Eder, A., Yongyos Y. *et al.*, *J. Inorg. Biochem.*, **61**, 69–78 (1996).
39. Bujda'k, J. and Rode, B. M. *J. Mol. Evol.*, **43**, 326–333 (1996).
40. Bujda'k, J., Son, H. L. and Rode, B. M. *J. Inorg. Biochem.*, **63**, 119–124 (1996).
41. Bujdak, J., Slosiarikova, H., Texler N. *et al.*, *Monatsh. Chemie*, **125**, 1033–1039 (1994).
42. Fu, L., Weckhuysen, B. M., Verberckmoes A. A. *et al.*, *Clay Miner.*, **31**, 491–500 (1996).
43. Ertem, G. and Ferris, J. P. *Nature*, **379**, 238–240 (1996).
44. Fang, Y. and Hoh, J. H. *J. Am. Chem. Soc.*, **120**(35), 8905–8909 (1998).
45. Kawamura, K. and Ferris, J. P. *J. Am. Chem. Soc.*, **116**, 7564–7572 (1994).
46. Porter, T. L., Whitehorse, R., Eastman M. P. *et al.*, *Appl. Phys. Lett.*, **75**, 2674–2676 (1999).
47. Rao, M. Odom, D. G. and Oro, J. *J. Mol. Evol.*, **15**, 317–331 (1980).
48. Odom, D. G., Rao, M., Lawless J. G. *et al.*, *J. Mol. Evol.*, **12**, 365–367 (1979).

PART II

Nanocomposite Synthesis and Properties

5

Polymer–Clay Nanocomposites

M. KATO AND A. USUKI

Toyota Central R&D Labs, Inc., Nagakute, Aichi, Japan

1 INTRODUCTION

Polymers have been successfully reinforced by glass fibers and other inorganic materials. In these reinforced composites, the polymer and additives are not homogeneously dispersed on a nanometer level. If nanometer dispersion could be achieved, the mechanical properties might be further improved and/or new unexpected features might appear [1]. A clay mineral is a potential nanoscale additive because it comprises silicate layers in which the fundamental unit is a 1 nm thick planar structure as shown in Figure 1. Also, it undergoes intercalation with various organic molecules such as toluene and aniline [2]. The intercalation causes an increase in the distance between silicate layers, which is dependent on the molecular size of the organic molecule.

The Toyota CRDL group has shown that the silicate layers of a clay mineral can be dispersed on a nanometer level in an engineering polymer. In this chapter, some polymer–clay nanocomposite materials are described according to the method used for their synthesis. Five methods for the synthesis of polymer–clay nanocomposites will be described:

1. Monomer intercalation method.
2. Monomer modification method.
3. Covulcanization method.
4. Common solvent method.
5. Polymer melt intercalation method.

Polymer–clay nanocomposites Edited by T. J. Pinnavaia and G. W. Beall
© 2000 John Wiley & Sons Ltd

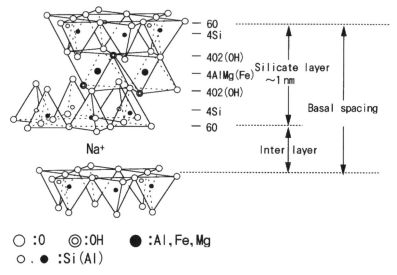

- 60
- 4Si

- 4O2(OH)
- 4AlMg(Fe) | Silicate layer ~1 nm
- 4O2(OH)

- 4Si
- 6O

Na+   Inter | layer

Basal spacing

○ :0   ◎ :OH   ● :Al, Fe, Mg
o . ● :Si (Al)

Figure 1   Structure of montmorillonite

2   SYNTHESIS METHODS

2.1   MONOMER INTERCALATION METHOD

Figure 2 illustrates the conceptual approach to the monomer intercalation method. Polymerization of a monomer occurs in the interlayer of the clay mineral, resulting in an expanded interlayer distance and silicate layers that are homogeneously dispersed on a nanometer level at the end of the polymerization.

Nylon 6 (polycaprolactam) has good mechanical properties and is a commonly used engineering polymer. The first polymer–clay nanocomposite was synthesized in

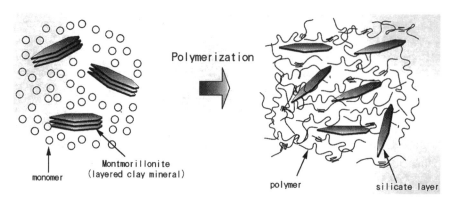

Polymerization

Montmorillonite
(layered clay mineral)
monomer

polymer   silicate layer

Figure 2   A conceptual picture of polymerization in the presence of montmorillonite (layered clay mineral)

a nylon 6 matrix using a monomer intercalation method by the Toyota CRDL group [3]. Because of the lack of affinity between the silicate layers of the clay mineral (montmorillonite) and the ε-caprolactam monomer of nylon 6, the insertion of ε-caprolactam into the clay gallery was of key importance. They found that organo-philic montmorillonite ion-exchanged with 12-aminododecanoic acid was swollen by molten ε-caprolactam [4].

Natural Na-montmorillonite is hydrophilic and not compatible with most organic molecules. Sodium cations in the interlayer space of montmorillonite can be exchanged with organic cations to yield organophilic montmorillonite. For the intercalation of ε-caprolactam, the Toyota CRDL group chose the ammonium cation of ω-amino acids since this acid catalyzes ring opening polymerization of ε-caprolactam. Figure 3 describes the basal spacing of montmorillonite cation-exchanged with various ω-amino acids of different carbon number, n, and the basal spacing of each 'n-montmorillonite' swollen by ε-caprolactam at 25 and 100 °C. The basal spacings were obtained from the peak positions in the X-ray diffraction pattern. The basal spacings of the 'n-montmorillonites' swollen by ε-caprolactam were equal at 25 and 100 °C when n was less than 8. They corresponded to the sum of the molecular length of the ω-amino acid and 1 nm silicate layer thickness at 25 °C. However, they exceeded this sum at 100 °C for longer n-montmorillonites. The schematic diagram is shown in Figure 4. For better swelling of ω-amino acids, n should be larger than 11. Therefore, they chose 12-aminododecanoic acid to prepare the first nylon 6–clay nanocomposite. They called this nanocomposite nylon 6–clay hybrid (NCH) [3].

The polymerization of nylon 6 occurs in the presence of n-montmorillonite after ε-caprolactam is intercalated into the gallery. Gradually, the silicate layers are uniformly dispersed in the nylon 6 matrix. A transmission electron micrograph (TEM) of a thin section of molded NCH is shown in Figure 5. The dark lines are the

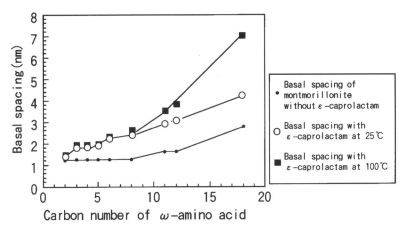

Figure 3 Basal spacing of n-montmorillonite with and without ε-caprolactam as a function of the amino acid carbon number

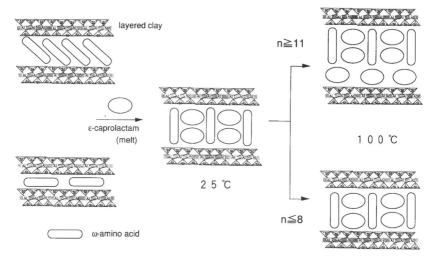

Figure 4   Swelling of *n*-montmorillonite by ε-caprolactam

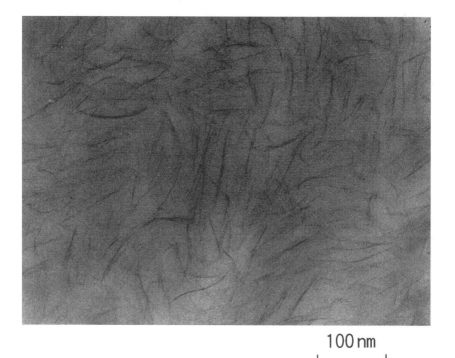

100 nm

Figure 5   Transmission electron micrograph of a thin section of an NCH at a montmorillonite content of 5.3 wt %

intersection of the sheet silicate of 1 nm thickness and the spaces between the dark lines are interlayer spaces. The silicate layers are dispersed on a nanometer level [4].

NCH materials provide significant improvements in mechanical, thermal, and gas barrier properties at loadings of only 2–5 wt % montmorillonite. The mechanical properties of NCH at about 5 wt % montmorillonite loadings, together with pristine nylon 6 and a conventional nylon 6–clay composite, are shown in Table 1 [5]. The conventional nylon 6–clay composite (NCC) in Table 1 was prepared by blending commercial nylon 6 and montmorillonite in a twin extruder. The tensile strength and tensile modulus of NCH were superior to those of both nylon 6 and NCC. The impact strength of NCH was identical to that of nylon 6. The most prominent effect was observed in the heat distortion temperature (HDT). The HDT of NCH containing about 5 wt % of montmorillonite was 152 °C, which was 87 °C higher than that of nylon 6. The resistance to water was also improved. The rate of water absorption in NCH was lowered by 40 % compared with that of nylon 6 and NCC [6]. The silicate layers, when dispersed on a nanometer level, are barriers to the transmission of water molecules. The molded specimen was found to be anisotropic. The coefficient of thermal expansion of NCH in the flow direction was less than half that in the perpendicular direction. Nylon 6 was isotropic and NCC was intermediate. TEM studies reveal that the sheets of silicate were parallel to the flow direction of the mold. The nylon 6 molecules in NCH were also oriented in the same direction. It seems that anisotropy of the thermal expansion results from the orientations of silicate and polymer chains.

The timing belt covers of automotive engines are usually made of glass fiber reinforced nylon or polypropylene. NCH shows a high modulus and a high distortion temperature, as mentioned above, and an attempt was made to make the timing belt covers from NCH by injection molding. The timing belt covers, which showed good rigidity and excellent thermal stability, were used in Toyota's automotive engine parts (Figure 6a). Also, for timing belt covers, the weight saving reached was up to 25 %, owing to the low content of inorganics compared with glass fiber reinforced nylon and polypropylene. This was the first example of industrialized use of a polymer–clay nanocomposite.

Nylon 6 film is usually used for food packaging film. NCH has a high gas barrier property because of the nanometer level dispersion of the silicate layers, as mentioned above. As shown in Figure 6b, NCH films at 2 wt % montmorillonite

Table 1 Tensile properties and impact strength of NCH and related materials

Specimen Montmorillonite (wt %)	Tensile strength (MPa)	Tensile modulus (GPa)	Charpy impact strength (KJ/m^2)
NCH-5 (4.2)	107	2.1	2.8
NCC-5 (5.0)	61	1.0	2.2
Nylon 6 (0)	69	1.1	2.3

M. KATO AND A. USUKI

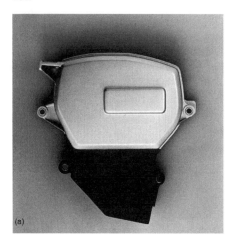

Figure 6 (a) Injection-molded timing belt cover formed with NCH for under-hood application and (b) food packaging film of NCH

loadings have half the oxygen permeability of straight nylon 6. The gas permeability decreased to about 20 % with a clay addition of 4.8 vol % (12 wt %) [7].

Another example of the monomer intercalation method was shown for an epoxy resin clay nanocomposite by Cornell University's group [8, 9]. In this case, the diglycidyl ether of bisphenol A (Epon-828, Shell) was used as an epoxide resin and poly-(ether amine) was used as a curing agent. Three types of organophilic montmorillonite were examined, which were prepared from Na-montmorillonite by cation-exchange reaction with $CH_3(CH_2)_{n-1}NH_3^+$, where $n = 7, 11, 18$. From the X-ray diffraction patterns of the mixture of the montmorillonite–epoxide–amine composites after curing, they claimed that the formation of exfoliated montmorillonite nanocomposites is dependent on the nature of the organophilic montmorillonites. Longer linear alkyl chains facilitate the formation of the polymer clay nanocomposites. Tensile properties were measured in the case of $CH_3(CH_2)_{17}NH_3^+$-montmorillonite. The tensile strength and modulus increased drastically under 2–20 wt % clay addition.

2.2 MONOMER MODIFICATION METHOD

Acrylic resins are useful polymers for coating materials and paints, which are synthesized by copolymerization of acrylic monomer. Acrylamide is one of the important monomers for water-borne acrylic resin paints because of the water solubility of the quaternary ammonium salt of acrylamide. A quaternary ammonium salt of N-[3-(dimethylamino)propyl]acrylamide (Q) is an exchangeable cation with the sodium cations of montmorillonite, and is ion-bonded to the silicate layers.

Table 2   Composition of acrylic resin and acrylic resin–clay nanocomposites

Acrylic resin–clay nanocomposite (aqueous suspension[a])	Acrylic resin composition			Montmorillonite content (wt %, based on acrylic resin)
	Ethyl acrylate (mol %)	Acrylic acid (mol %)	Acrylamide[b] (mol %)	
A	90.80	9.08	0.12	1.00
B	90.60	9.04	1.60	3.10
C	90.40	9.04	2.54	4.90
D	90.10	9.01	4.13	8.00

[a] Acrylic resin concentration is about 45 wt % in each suspension.
[b] N-[3-(Dimethylamino)propyl]acrylamide.

According to the composition as shown in Table 2, Q, ethyl acrylate (EA) and acrylic acid (Aa) are free radical copolymerized. Four kinds of acrylic resin–clay nanocomposite were prepared by dispersing montmorillonite in an aqueous suspension of those acrylic resins [10]. The clay content of the acrylic resin–clay nanocomposite is proportional to that of Q in acrylic resin, so the content is 1, 3, 5, and 8 wt % on the basis of solid acrylic resins. An aqueous suspension of acrylic resin–clay nanocomposite exhibited the properties of a pseudo-plastic fluid with a clay addition of above 3 wt %. The acrylic resin–clay nanocomposite films cross-linked by melamine are transparent, and the gas permeability of the films decreased to about 50 % with a montmorillonite addition of 3 wt %.

There are also some reports of acrylic resin–clay nanocomposites. A polymethyl methacrylate–clay nanocomposite was synthesized using modified organophilic clay in the same manner [11], and by emulsion polymerization [12].

2.3   COVULCANIZATION METHOD

Vulcanized rubbers are usually reinforced by carbon blacks to improve mechanical properties and also by inorganic minerals with some limitation. Carbon blacks are excellent in reinforcement owing to strong interaction with rubbers, but they often decrease the processability of rubber compounds at high volume loading. On the other hand, minerals have a variety of shapes suitable for reinforcement, such as needles and sheet, but they have only poor interaction with rubber. Therefore, it is interesting to disperse the silicate layers of the montmorillonite on a nanometer level. Rubbers are more hydrophobic than nylon 6, and it is consequently difficult to achieve dispersion of silicate layers in a rubber matrix with alkyl ammonium-treated montmorillonite.

Toyota's group prepared montmorillonite cation-exchanged with amine terminated butadiene–acrylonitrile (ATBN) oligomer in a solvent mixture of N,N'-dimethylsulfoxide, ethanol, and water. After this, organophilic montmorillonite was blended with nitrile–butadiene rubber (NBR) by roll milling, and the rubber was vulcanized with sulfur. According to TEM observations, the silicate layers were dispersed in the rubber matrix [13]. The tensile stress at 100 % of this rubber–clay nanocomposite, containing 10 parts per hundred of rubber (phr) of montmorillonite, is equal to that of the rubber containing 40 phr of carbon black. In this rubber–clay nanocomposite, the permeability of hydrogen and water decreased to 70 % by means of adding 3.9 vol % montmorillonite [14].

2.4 COMMON SOLVENT METHOD

Polyimides are used largely for microelectronics because of heat resistance, chemical stability, and superior electric properties. It would be desirable to reduce the coefficient of thermal expansion, amount of moisture adsorption, and dielectric constant, since those properties of polyimides are not sufficient for advanced electronics use.

The Toyota CRD group reported on the synthesis and properties of polyimide–clay nanocomposites [15]. In that report, it was found that, when a dodecyl ammonium ion was used as the intercalating reagent, montmorillonite was dispersed homogeneously in dimethylacetamide (DMAC). DMAC is a solvent for preparation of polyimides. The montmorillonite was dispersed in DMAC with polyamic acid (the precursor of polyimide). By removing the DMAC, a polyimide–montmorillonite nanocomposite was obtained.

For the synthesis of a polyimide–montmorillonite nanocomposite, the dispersibility of various organophilic montmorillonites in DMAC was examined. As the carbon number of the ammonium ion increases, the hydrophilicity of the organophilic montmorillonite decreases. Ammonium ions, with 10–12 carbon atoms, are suitable for organophilic montmorillonite to be dispersed in DMAC. Dodecylammonium-treated montmorillonite was especially capable of being dispersed homogeneously in DMAC. Longer carbon chains make the organophilic montmorillonite too hydrophobic.

Also, according to the X-ray diffraction patterns of the polyimide nanocomposite formed from onium ion treated montmorillonites, only in the case of dodecylammonium-treated montmorillonite was there no diffraction peak. These results indicated that the dispersibility of organophilic montmorillonite in DMAC is important for the synthesis of a polyimide–montmorillonite nanocomposite.

The permeability coefficient of water vapor decreased markedly with increasing content of montmorillonite in the polyimide nanocomposite. A montmorillonite addition of only 2 wt % brought the permeability coefficient of water vapor to a value less than half that of polyimide. However, the water absorption property of the polyimide nanocomposite was not improved compared with that of polyimide. The water adsorption at equilibrium of the polyimide nanocomposite was the same as that of polyimide. It was reported that the permeability of carbon dioxide was halved in the polyimide–clay nanocomposite [16].

2.5 POLYMER MELT INTERCALATION METHOD

Polypropylene (PP) is one of the most widely used polyolefin polymers. Therefore, it has been attractive to synthesize a PP–montmorillonite nanocomposite. However, two problems have been encountered in the case of PP nanocomposites. The first problem is the intercalation of a suitable monomer and then the exfoliating of the montmorillonite on a nanometer level by subsequent polymerization or polymer intercalation from a solution. It has been thought that polymer melt intercalation would be a promising new approach to fabricating PP nanocomposites using conventional polymer processing techniques. A PP nanocomposite would be formed by heating a mixture of PP and organophilic montmorillonite above the melting temperature of PP (about 160 °C). The second problem is that PP does not have any polar groups in its backbone and is one of the most hydrophobic polymers. Actually, the nanometer dispersion of silicate layers in a PP matrix was not realized even by using a montmorillonite treated with a dioctadecyldimethylammonium ion (DSDM-Mt), in which the polar surfaces of the clay mineral should be covered with one of the most hydrophobic alkylammonium ions. In the solution to this problem, a polyolefin oligomer that has polar groups (i.e. –OH, COOH) should be used as a compatibilizer.

In the approach to prepare a PP nanocomposite using a polyolefin oligomer as a compatibilizer, there seemed to be two important factors in terms of the structure of the oligomers. Firstly, the oligomer should include a certain amount of polar groups to intercalate between silicate layers through hydrogen bonding to the oxygen group of the silicate layers. Secondly, the oligomers should be miscible with PP. Since the content of polar functional groups in the oligomers will affect the miscibility with PP, there must be an optimum content of polar functional groups in the compatibilizer.

The Toyota CRD group examined two types of maleic anhydride modified PP oligomers containing different amounts of maleic anhydride groups (PP-MA-1001, acid value $= 26$ mg KOH/g, PP-MA-1010, acid value $= 52$ mg KOH/g) to prepare a PP nanocomposite [17]. For mixtures of octadecylammonium treated montmorillonite (C18-Mt) and PP-MAs formed by melt processing, the diffraction peaks of C18-Mt around $2\Theta = 4°$ are shifted to lower angles regardless of the kind of PP-MA. This clearly indicates the strong intercalation capabilities of PP-Mas in the silicate layers.

A mixture with the composition PP (70 wt %), PP-MA (22 wt %), and C18-Mt (8 wt %) was melt blended at 210 °C in a twin-screw extruder to obtain the PP nanocomposite. Figure 7 shows that the shapes of the XRD patterns of the obtained PP nanocomposite were dependent on the kind of PP-MAs used. A strong diffraction peak for montmorillonite was observed for the PP nanocomposite with PP-MA-

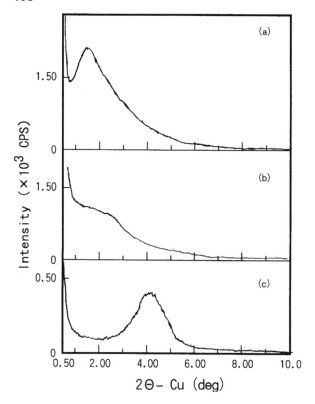

Figure 7 X-ray diffraction patterns of (a) PP nanocomposite with PP-MA-1010, (b) PP nanocomposite with PP-MA-1001, and (c) C18-Mt

1010. This XRD pattern indicates that the layers of the montmorillonite maintain relatively good stacking order in the PP nanocomposite with PP-Ma-1010. On the other hand, the XRD pattern of the PP nanocomposite with PP-MA-1001 exhibits a diffraction shoulder with a gradual increase in the diffraction intensity towards low angle. Completely dispersed silicate layers in nylon 6–clay hybrids exhibit no diffraction peak, but a gradual increase in the diffraction intensity occurs at lower angles. Therefore, the layers of montmorillonite in the PP nanocomposite with PP-MA-1001 should be more exfoliated and dispersed compared with the PP nanocomposite with PP-MA-1010.

The dispersibility of the clay layers is dependent on the miscibility of the PP-MA oligomers with the matrix PP polymer. The miscibility of PP and PP-MA-1001 is much better than that of PP and PP-MA-1010, as observed for mixtures of PP-MA and PP polymer formed in a twin extruder. These results show that the miscibility of the oligomers with the matrix polymer affects the dispersibility of the silicate layers in the nanocomposite.

Figure 8 illustrates the dispersion of C18-Mt in a PP matrix with the aid of PP-MAs. The driving force of the intercalation originates from the strong hydrogen bonding between the maleic anhydride group and the oxygen groups of the silicate layers. The interlayer spacing of the montmorillonite increases, and the interaction between the silicate layers is weakened. The montmorillonite intercalated by the PP-

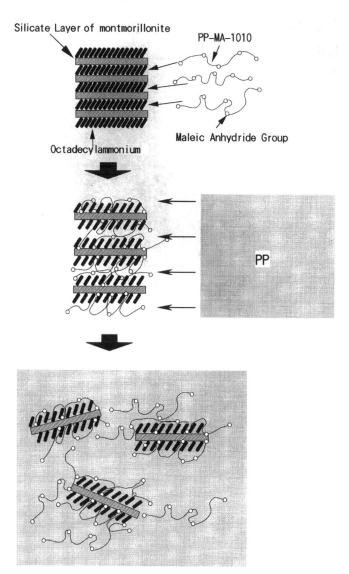

Figure 8 Schematic representation of the dispersion of C18-Mt in a PP matrix with the aid of PP-MAs

Table 3   Dynamic storage moduli of PP nanocomposites and related samples at various temperatures and their glass transition temperatures obtained from tan δ

Samples	Storage modulus^a (GPa)				Tg,^{b}°C
	−40 °C	20 °C	80 °C	140 °C	
PPCH-C18-Mt/1010	5.15 (1.31)	3.12 (1.58)	1.03 (1.59)	0.13 (0.60)	11
PPCH-C18-Mt/1001	5.26 (1.34)	3.09 (1.56)	1.10 (1.70)	0.21 (0.94)	8
PP/C18-Mt	4.50 (1.15)	2.36 (1.19)	0.82 (1.26)	0.28 (1.25)	9
PP/1010	3.92 (1.00)	1.99 (1.01)	0.60 (0.92)	0.15 (0.68)	13
PP/1001	4.04 (1.03)	2.02 (1.02)	0.55 (0.85)	0.14 (0.62)	10
PP	3.92	1.98	0.65	0.22	13

^{a} The values in parentheses are the relative values of the nanocomposites to those of PP.
^{b} The glass transition temperatures were measured at the peak tops of tan δ.

MA oligomers contact PP under a strong shear field. If the miscibility of PP-MA with PP occurs at the molecular level, the exfoliation of the intercalated montmorillonite should take place.

The dynamic storage moduli of the PP nanocomposites at −40, 20, 80, and 140 °C and the glass transition temperature, T_g, are listed in Table 3, along with those of PP, PP/PP-Mas, and PP/C18-Mt. The dynamic moduli of the PP nanocomposites were higher than that of PP below 80 °C. The lower moduli of the PP nanocomposites at 140 °C should be attributed to the lower softening temperature of the PP-Mas (PP-MA-1001, 154 °C, PP-MA-1010, 145 °C). The relative moduli of the PP nanocomposites compared with those of PP are 1.3–1.4 below T_g, and 1.6–1.7 above T_g, and those of PP/C18-Mt are slightly higher than those of PP. These results indicate that the enormous reinforcement effect observed in the PP nanocomposites should be attributed to the silicate layers on a nanometer level.

Recently, the polymer melt intercalation technique was used to prepare a polystyrene (PS) nanocomposite with styrene–methylvinyloxazoline copolymer (Psoz) as a compatibilizer [18]. Powdered polystyrene, Psoz, and montmorillonite treated with octadecyltrimethylammonium were melt blended at 180 °C using a twin-screw extruder. On the basis of TEM images of the obtained PS nanocomposites, it was found that the silicate layers were dispersed on a nanometer level. The dynamic storage moduli of the PS nanocomposites exhibited a reinforcement effect that was attributable to the presence of silicate layers on a nanometer level.

The polymer melt intercalation techniques in the absence of a compatibilizer have attracted industrial interest. Several groups of researchers have successfully prepared nylon 6 nanocomposites by intercalation directly into organophilic montmorillonites [19, 20, 21]. Other kinds of nylon-nanocomposites have also been synthesized, such as nylon 66 [20], nylon 11 [20], nylon 12 [20] and MXD 6 [21]-nanocomposites.

3 CONCLUSION

We have described the synthesis and properties of montmorillonite-based nanocomposites. Modification of the montmorillonite surface plays a principal role in this technology. Modified monomers and polymers or the use of a compatibilizer for organic materials and clay should be effective in opening a new methodology for montmorillonite-based nanocomposite production.

Many researchers have studied montmorillonite-based nanocomposites to obtain high-performance engineering plastics [22]. Future studies of the principles of montmorillonite dispersion and polymer performance should be done. From these further studies, new montmorillonite-based nanocomposite material may be produced. Montmorillonite reinforcement techniques should be developed as a universal method for obtaining high-performance materials.

4 REFERENCES

1. Giannelis, E. P. *J. Miner. Metals Mater. Soc.*, March, 28 (1992).
2. Jordan, J. M. *J. Phys. Colloid Chem.*, **53**, 245 (1950).
3. Usuki, A., Kawasumi, M., Kojima, Y., Fukushima, Y., Okada, A., Kurauchi, T. and Kamigaito O. *J. Mater. Res.*, **8**, 1179 (1993).
4. Usuki, A., Kojima, Y., Kawasumi, M., Okada, A., Kurauchi, T. and Kamigaito, O. *J. Mater. Res.*, **8**, 1174 (1993).
5. Kojima, Y., Usuki, A., Kawasumi, M., Fukushima, Y., Okada, A., Kurauchi, T. and Kamigaito, O. *J. Mater. Res.*, **8**, 1185 (1993).
6. Okada, A., Kawasumi, M., Kurauchi, T. and Kamigaito, O. *Polym. Prep.*, **28**, 447 (1987).
7. Messersmith, P. B. and Giannelis, E. P. *J. Polym. Sci. A, Polym. Chem.*, **33**, 1047 (1995).
8. Lan, T. and Pinnavaia, T. J. *Chem. Mater.*, **6**, 2216 (1994).
9. Wang, M. S. and Pinnavaia, T. J. *Chem. Mater.*, **6**, 468 (1994).
10. Usuki, A., Okamoto, K., Okada, A. and Kurauchi, T. *Kobunsi Ronbunshu*, **52**, 728 (1995).
11. Biasci, L., Aglietto, M., Ruggeri, G. and Ciardelli, F. *Polymer*, **35**, 3296 (1994).
12. Choo, D. and Jang, L. W. *J. Appl. Polym. Sci.*, **61**, 1117 (1996).
13. Fukumori, K., Usuki, A., Sato, N., Okada, A. and Kurauchi, T. in Proc. 2nd Japan International SAMPE Symp., 1991, p. 89.
14. Kojima, Y., Fukumori, K., Usuki, A., Okada, A. and Kurauchi, T. *J. Mater. Sci. Lett.*, **12**, 889 (1993).
15. Yano, K., Usuki, A., Okada, A. and Kurauchi, T. *J. Polym. Sci. A, Polym. Chem.*, **31**, 2493 (1993).
16. Lan, T., Kaviratna, P. D. and Pinnavaia, T. J. *Chem. Mater.*, **6**, 573 (1994).
17. Kawasumi, M., Hasegawa, N., Kato, M., Usuki, A. and Okada, A. *Macromolecules*, **30**, 6333 (1997).
18. Hasegawa, N. and Usuki, A. Proc. *Additives '99*, 1999.
19. Liu, L., Qi, Z. and Zhu, X. *J. Appl. Polym. Sci.*, **71**, 1133 (1996).
20. Kato, M., Okamoto, H., Hasegawa, N., Usuki, A. and Sato, N. in Proc. 6th Japan International SAMPE Symp., 1999, p. 693.
21. Lan, T., Liang, Y., Beall, G. W. and Kamena, K. Proc. *Additives '99*, 1999.
22. *Plastics Technol. Mag.*, **45**(6), 52 (1999).

6

In Situ Polymerization Route to Nylon 6–Clay Nanocomposites

K. YASUE, S. KATAHIRA, M. YOSHIKAWA
AND K. FUJIMOTO
Unitika, Ltd Research and Development Center, Kyoto, Japan

1 INTRODUCTION

Polymer materials are usually reinforced with fillers to improve mechanical properties. Such materials are widely used in diverse areas including transportation, construction, electronics and consumer products. One of the most common reinforcing materials is a fibrous filler in a randomly dispersed state. Glass fibers and carbon fibers are popular fibrous fillers. Theories by Cox [1] and Kelly [2] are commonly employed to estimate the reinforcing effect of fibrous fillers. According to these theories, the degree of reinforcement depends on the rigidity and aspect ratio of the filler itself, and the adhesive strength between the filler and the polymer matrix. In order to improve adhesive strength, the surface of glass or carbon fibers is organically treated. However, the size of such a filler is of micron order, which is large compared with the size of polymers which are of nanometer order. This significant difference in size between the two components often results in the degradation in properties of the neat polymer, such as decreased ductileness, poor moldability and poor surface smoothness in molded articles. Conventional polymer composites are not processible as films or fibers.

A new type of composite having a rigid rod-like polymer as a reinforcing filler was proposed in order to minimize the size difference between the filler and the matrix. This was a sort of polymer blend and was named a 'molecular composite' by Takayanagi [3], belonging to a category later defined as a 'nanocomposite' [4]. Molecular composites were not successful owing to a phase separation problem between the two component polymers, but the idea led to the synthesis of liquid

Polymer–clay nanocomposites Edited by T. J. Pinnavaia and G. W. Beall
© 2000 John Wiley & Sons Ltd

crystalline polymers, which overcame the problem of phase separation by copolymerizing rigid and flexible monomer components. A liquid crystalline polymer was initially called a 'self-reinforced polymer' at its earlier stage of development. However, generally a liquid crystalline polymer is not classified as a polymer composite because it is single component.

2   SWELLABLE CLAY MINERALS

A molecular composite having a rigid rod-like polymer as a reinforcing filler can be classified as a 'polymer–organic filler nanocomposite'. As mentioned above, this type of nanocomposite was not successful owing to the phase separation problem. On the other hand, conventional inorganic fillers such as talc and glass fibers are large in size compared with the matrix polymer. Even whisker fibers (e.g. potassium titanate fibers) are far larger than nanometer size. Because no inorganic filler with a nanometer size was known, polymer–inorganic filler nanocomposites were not suggested for a long time.

Swellable clay minerals are not nanometer-sized filler themselves but can produce a nanometer-sized filler. Clay minerals are called layered silicates because of the stacked structure of 1 nm thick silicate sheets with a variable basal distance. Silicate sheets can be regarded as rigid inorganic polymers consisting of mainly silicon and oxygen, and a small amount of aluminum, magnesium and other metal ions. Swellable clay minerals have alkali metals between the silicate sheets and can swell in polar solvents such as water. Montmorillonite is a typical example (Figure 1). Meanwhile, layered silicates such as talc and kaolin have no alkali metal between their layers and will not swell. They are consequently classified as non-swelling clays. Usually, the lattice energy of swellable clay minerals is so large that it is stable even at high temperature up to about $1000\,^{\circ}C$. However, in a polar solvent, the basal distance of silicate sheets expands by solvation of the polar solvent, and finally the silicate sheets come to exfoliate into individual sheets. When the concentration of the clay mineral in water is less than a few wt %, the clay mineral can form a colloidal solution.

3   NYLON 6–CLAY NANOCOMPOSITES

In 1970, Kato reported a 'polymer–clay complex' consisting of an organic polymer and a clay mineral [5]. He intercalated an acrylic acid monomer between the silicate sheets of montmorillonite and carried out polymerization of the acrylic acid monomer *in situ*, which resulted in an increased basal distance of montmorillonite from 9.6 to 17.4 Å. However, this polymer–clay complex could not be processed by conventional molding methods. In 1976, Fujiwara and Sakamoto [6] reported a 'nylon 6–clay nanocomposite' in two steps for the first time. They first obtained an

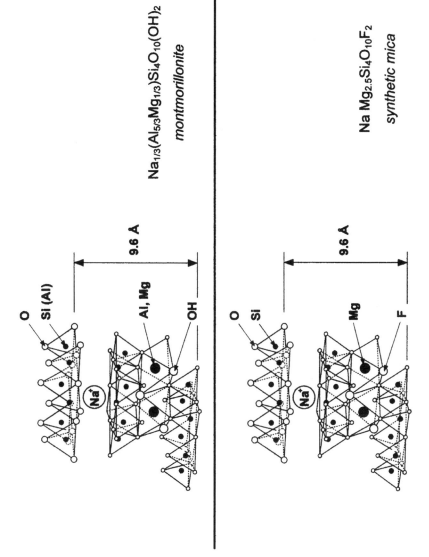

$Na_{1/3}(Al_{5/3}Mg_{1/3})Si_4O_{10}(OH)_2$
montmorillonite

$Na\ Mg_{2.5}Si_4O_{10}F_2$
synthetic mica

Figure 1 Chemical structure of montmorillonite and synthetic mica

aminocarboxylic acid–montmorillonite complex with an enlarged basal distance of 15.2 Å through an ion exchange reaction between the sodium ion in the interlayer of montmorillonite and a protonated aminocarboxylic acid. In the second step, ε-caprolactam was intercalated into the interlayer of the complex and then polymerized *in situ*. The nylon 6–montmorillonite nanocomposite thus obtained exhibited a large expanded basal distance (69.8 Å) [6]. This composite was processible by conventional molding methods. This procedure is the actual origin of the technology of nylon 6–clay nanocomposites. Usuki *et al.* developed a nylon 6–clay nanocomposite using a similar technology [7].

4 *IN SITU* POLYMERIZATION METHODS

As mentioned above, nylon 6–clay nanocomposites can be synthesized through a two-step method. However, this method is not economically suitable for the industrial production of polymer composites because of its high cost. The authors have tried to improve the production process and have finally accomplished an '*in situ* polymerization method' that directly produces a nanocomposite from only a specific layered silicate (a swellable synthetic mica) and ε-caprolactam in one step, without a step for the preparation of an 'aminocarboxylic acid–clay complex' [8]. This novel method has made possible the economic production of nanocomposites and has contributed to the rapid development into practical applications.

In situ polymerization includes the following processes which take place successively or simultaneously:

(a) formation of a filler of nanometer size (exfoliation of silicate sheets from a layered silicate),
(b) formation of a nylon 6 matrix (polymerization of ε-caprolactam)
(c) formation of a composite (mixing of nylon 6 and silicate sheets).

As is generally known, hydrolytic polymerization of ε-caprolactam to nylon 6 is divided into three elemental reactions (Scheme 1). The authors focused on the aminocaproic acid produced. The question was whether or not the sodium ion could be replaced by protonated aminocaproic acid during the polymerization process. If it could, the first step of the conventional two-step method for preparing an organic ion–clay complex would not be necessary. Starting from the above question, we found that a nylon 6–clay nanocomposite could be obtained by using a synthetic mica with an optimized ion exchange capacity and also by conducting hydrolytic polymerization under a specific condition.

Figure 2 shows a wide angle X-ray diffraction (WAXD) pattern of samples taken from the reactor during the polymerization process [9]. For synthetic mica as a raw material, diffraction peaks can be seen at 9.2 and 7.1 ° (2θ), which correspond to basal distances of 9.6 and 12.5 Å, respectively. The basal distance of 12.5 Å is attributed to hydration of the silicate sheets. For the samples taken from the reactor,

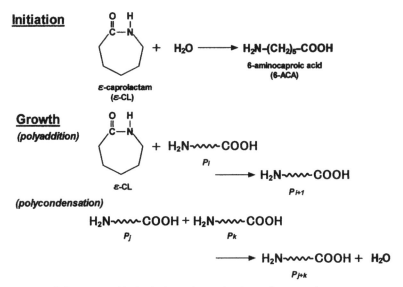

Scheme 1 Hydrolytic polymerization of ε-caprolactam.

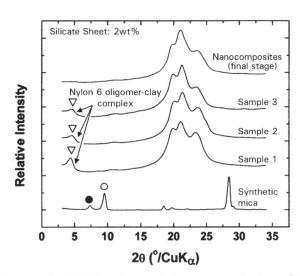

Figure 2 Changes in WAXD patterns during polymerization of the nanocomposite. Samples 1, 2 and 3 were taken from the reactor: ○ $d_{001} = 9.6\,\text{Å}$, ◐ $d_{001} = 12.5\,\text{Å}$, ● $d_{001} = 20.6\,\text{Å}$

the peaks at 20–24 ° (2θ) are attributed to the nylon 6 crystalline structure. A peak at 4.3 ° (2θ), corresponding to a basal distance of 20.6 Å, is also observed. The peak at 4.3 ° (2θ) was not seen for the raw materials, including the synthetic mica. The intensity of this peak decreased as polymerization proceeded, and disappeared when the polymerization was completed. We speculate that the peak corresponding to a 20.6 Å basal distance can be attributed to a complex of clay and intercalated nylon 6 oligomer (nylon 6 oligomer–clay complex).

At the end of polymerization, the relative viscosity of the nylon 6 matrix in the nanocomposite is 2.0–3.5 (in 96 % sulfuric acid at 25 °C), which indicates that the molecular weight of the matrix is high enough for use as an engineering plastic. Figure 3 shows a TEM micrograph of the cross-section of the nanocomposite. The silicate sheets are exfoliated and dispersed randomly into the nylon 6 matrix. The estimated size of the silicate sheet is ca 1 nm in thickness and ca 30–100 nm in length. This proves that the nanocomposite has already formed by the time polymerization is completed.

Table 1 shows additional data for a series of nanocomposite formation reactions. Table 1 provides a survey of properties for the reaction mixture at different stages of nanocomposite formation. From these data, the process of nanocomposite formation in *in situ* polymerization can be speculated to occur as follows.

At the first stage where ε-caprolactam, water and synthetic mica are mixed together and held at 80 °C, ε-caprolactam intercalates into the synthetic mica, and the basal distance of the synthetic mica expands to 15.3 Å. At this stage, the ion exchange reaction has not yet taken place, and hence the sodium ion is still entrapped in the interlayer of the mica.

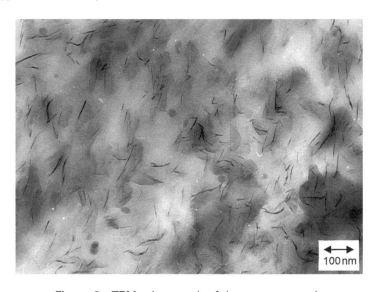

Figure 3 TEM micrograph of the nanocomposite

Table 1 Detailed data from a survey of nanocomposite formation

Stage	Temperature ($^\circ$C)	d_{001} (Å)	Degree of ion exchange (%)	Reaction degree of ε-caprolactam (%)
Synthetic mica	—	9.6	—	—
1st	80	15.3	0	0
2nd	220	20.6	75	50
3rd	260	No detection	95	90
Final	260	No detection	95	90

In the second stage, where the temperature of the mixed raw materials is raised to 220 $^\circ$C, ε-caprolactam is hydrolyzed to aminocaproic acid and some proportion is protonated. As soon as protonated aminocaproic acid is generated, it replaces the sodium ion between the layers through an ion exchange reaction. As a result, protonated aminocaproic acid is introduced into the interlayer of the synthetic mica, and the sodium ion is pushed out to the matrix phase. However, at this stage, the ion exchange reaction has not been completed. Residual ε-caprolactam in the reaction mixture subsequently intercalates into the layer, and a continuous addition reaction occurs in the protonated aminocaproic acid interlayers. Consequently, a nylon 6 oligomer–clay complex with a basal distance of 20.6 Å is formed. The degree of reaction of ε-caprolactam is approximately 50 % at this stage.

At the third stage, the addition reaction of ε-caprolactam proceeds to ca 90 %, which is the maximum value of this reaction type. The basal distance increases more and more by the further addition of ε-caprolactam, and the lattice energy of the synthetic mica decreases accordingly until it cannot endure the shearing stress of the reaction system any more. Thus, the layer structure is finally destroyed. No layer structure is observed by WAXD. That is, the dispersion of the silicate sheets was completed, and the ion exchange reaction reached equilibrium at this stage. The sodium ion extracted from between the layers by ion exchange is assumed to form salts with the carboxylic end group of the nylon polymer.

At the fourth and final stage, the degree of polymerization increases and the nanocomposite is finally produced. The total reaction time needed to synthesize the nanocomposite is almost equal to that needed to synthesize conventional hydrolytic polymerization of nylon 6. The nanocomposite thus obtained is purified with hot water and dried, as in usual nylon 6 production, and then subjected to further processing. During the purifying process, sodium existing in the nanocomposite as a salt with the carboxylic group is hydrolyzed and the resulting sodium ion is extracted from it.

Figure 4 shows a schematic illustration of the nanocomposite formation process. The *in situ* polymerization method has thus succeeded in utilizing not only the intercalation reaction of water and ε-caprolactam present at the earlier stage, but also

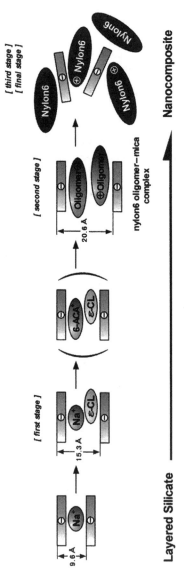

Figure 4 Schematic illustration of the nanocomposite formation process

the ion exchange reaction of protonated aminocaproic acid generated during the polymerization process. The solvation energy of such polar molecules in the above reactions participates in enlarging the basal distance of the layered silicate. Moreover, the heat generated during the addition reaction of ε-caprolactam and aminocaproic acid, which is exothermic, also contributes to expanding the basal distance.

5 PROPERTIES OF THE NYLON 6–CLAY NANOCOMPOSITE

The silicate sheet content in the nanocomposite is a few wt %, which is about one-tenth of the reinforcing filler content in conventional polymer composites. However, since the size of the dispersed silicate sheets is of nanometer order, the number of dispersed silicate sheets is so enormous that individual dispersed sheets in the matrix maintain a distance of several dozen nanometers from each other. Such a microstructure in the nanocomposite has not been accomplished in conventional polymer composites. Furthermore, because the functional groups of nylon 6 may interact with negatively charged silicate sheets, the affinity between the filler and the matrix is fairly good and aggregation of silicate sheets seldom occurs. The nanocomposite realized superior properties which conventional polymer composites never accomplished. Such properties may originate from the microstructure and the polar interaction, as well as the excellent mechanical properties of the silicate sheet, such as high strength and high modulus.

5.1 CRYSTALLIZATION OF THE NANOCOMPOSITE

Figure 5 shows a WAXD curve of the nanocomposite. Neat nylon has diffraction peaks at $20°$ (2θ) and $24°$ (2θ), both of which are attributed to the α-type crystalline structure of nylon 6, while the nanocomposite has an additional characteristic diffraction at $21°$ (2θ) which results from γ-type crystalline structure. The nanocomposite has protonated amino end groups, which are ionically bonded to a negatively charged silicate sheet. It can be speculated that the dominant γ-type crystal was caused by poor rearrangement of the polymer molecules which are tethered to the silicate sheets. Figure 6 shows a DSC curve of the nanocomposite. The nanocomposite has a main melting peak at 214 °C for the γ-type crystal and a small peak for the α-type crystal at 220 °C, while neat nylon 6 shows a single melting point at 220 °C which is of the α-type. In the cooling process of DSC measurement, the nanocomposite crystallizes in a narrower temperature range. This phenomenon indicates a high crystallization rate of the nanocomposite.

5.2 RHEOLOGICAL PROPERTIES

Figure 7 shows the dependence of melt viscosity on shear rate at some silicate sheet loadings. At low shear rates (less than 1), the influence of the silicate sheet content

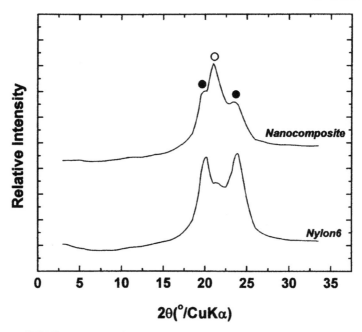

Figure 5 WAXD pattern of the nanocomposite and neat nylon 6: ● α-type crystalline structure, ○ γ-type crystalline structure

on melt viscosity is significant. At higher shear rates over 100, the silicate sheet loading influences melt viscosity, but only slightly. This rheological feature is quite suitable for injection molding and extrusion molding.

5.3 EFFECT OF SILICATE SHEET AS REINFORCING FILLER

Figure 8 shows the reinforcing effect of the silicate sheets in the nanocomposite in comparison with conventional fillers such as talc and glass fiber. It is obvious that the silicate sheets have a far better reinforcing effect than conventional fillers even at lower content. As mentioned earlier, the reinforcement effect depends on three factors of the reinforcing material: rigidity, aspect ratio and affinity with the matrix polymer. Silicate sheets of nanometer size satisfy these three factors.

5.4 TYPICAL PROPERTIES OF THE NANOCOMPOSITE

5.4.1 Flexural Modulus

Table 2 shows some practical properties of the nanocomposite. The relationship between the silicate sheet content and flexural modulus has already been shown in Figure 8. Silicate sheet has a larger aspect ratio than talc, while the rigidity of the two is assumed to be at the same level because of their similar basic structures as layered

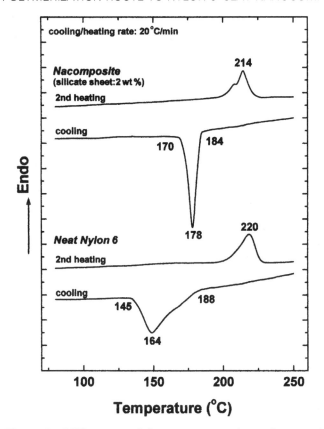

Figure 6 DSC curves of the nanocomposite and neat nylon 6

silicates. Compared with glass fiber, silicate sheet has almost the same aspect ratio, but superior rigidity. Thus, silicate sheet can be regarded as one of the best inorganic fillers.

Nylon 6 filled with talc exhibits a considerable decrease in flexural modulus at around the glass transition temperature, T_g, while for the nanocomposite the decrease is smaller. Furthermore, above T_g, the nanocomposite exhibits a better creep resistance than conventional composites filled with other inorganic fillers. This feature may be caused by the microstructure of the matrix polymer and the inorganic filler used in the nanocomposite.

5.4.2 Tensile Elongation

In general, tensile elongation of conventional polymer–inorganic filler composites is low. For example, talc-filled or glass fiber-filled nylon 6 breaks at only a few % elongation. On the other hand, the nanocomposite exhibits as large an elongation as

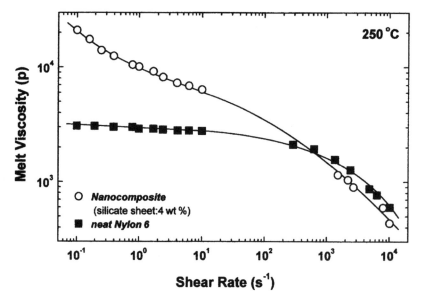

Figure 7   Shear rate dependence of melt viscosity

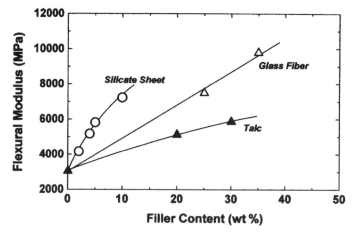

Figure 8   Reinforcing effect of silicate sheet on the nylon 6 matrix (nanocomposite), compared with other reinforcing fillers

Table 2 Typical properties of nylon 6–mica nanocomposite

Properties[a]	Nanocomposite		Conventional reinforced nylon 6		Neat nylon 6
Filler type	Silicate sheet		Talc		—
Filler content (wt %)	4	6	4	35	—
Specific gravity	1.15	1.17	1.15	1.42	1.14
Mechanical properties					
Elongation[b] (%)	4	4	4	4	100
Flexural strength (MPa)	158	176	125	137	108
Flexural modulus (GPa)	4.5	5.6	3.0	6.0	3.0
DTUL (at 1.8 MPa) (°C)	152	158	70	172	70

[a] Under dry condition.
[b] Thickness; 3.2mm.

over 100 % when its thickness is less than 1 mm. This property enables the nanocomposite to be converted to fibers and films.

5.4.3 Distortion Temperature under Load

The distortion temperature under load (DTUL) of the nanocomposite is shown in Table 2. Up to a silicate sheet content of 6 wt %, DTUL increases linearly. Over 6 wt %, the increase in DTUL becomes smaller.

5.4.4 Barrier Property

The nanocomposite has excellent barrier properties against oxygen, nitrogen, carbon dioxide, water vapor, gasoline, etc. Gas barrier properties for oxygen and water vapor are shown in Figure 9. The gas permeability of the nanocomposite is reduced to almost a half to one-third of that for the unfilled polymer. This effect is explained by the tortuous path principle. Dispersed platelets of the silicate sheet block the shortest path of gas molecules and force them to take a roundabout way. As a result, the permeation pathway is elongated.

5.4.5 Recycle Property

The nanocomposite maintains its excellent properties even when recycled. The glass fiber employed in conventional polymer composites easily breaks during injection molding or crushing processes, and so the mechanical properties of those composites usually deteriorate in recycling as the aspect ratio of the filler decreases. However, the silicate sheet used in the nanocomposite is ultrafine (nanometer order) and does not break during the recycling process. The nanocomposite shows no performance drop in repeated crushing and remelting.

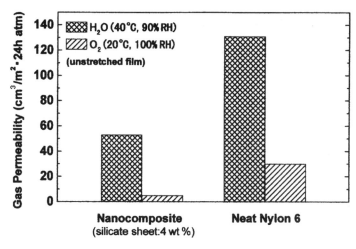

Figure 9   Gas permeability of the nanocomposite film (150 μm thickness)

6 APPLICATION OF THE NANOCOMPOSITE

6.1 INJECTION MOLDED ARTICLES

Though reinforced with inorganic material, the nanocomposite has a similar specific gravity and appearance as neat nylon 6. Both properties together make it difficult to distinguish one from the other. However, the mechanical properties of the nanocomposite are equal to those of conventional polymer composites. The nanocomposite has already been used widely in the fields of injection molding.

Examples of application include: engine cover (see Figure 10), timing belt cover, oil reservoir tank and fuel hose in the automobile field, floor adjuster and handrail in the construction field and various connectors in the electric field.

6.2 FILMS

The nanocomposite can be converted to films, while for conventional composites this was impossible. The T-die method and tubular methods are applicable. If necessary, oriented films can be made. The transparency of the nanocomposite film is generally similar to that of the neat polymer. Especially for unstretched film, the nanocomposite is superior to the neat polymer. Wrapping films and beverage containers are the main targets utilizing the gas barrier property or transparency.

6.3 FIBERS

The nanocomposite can be processed into fibers by conventional spinning methods. When being melt spun, it closely maintains the cross-sectional shape of the nozzle. Novel nanocomposite fibers are under development.

Figure 10   Nanocomposite engine cover

7   CONCLUSION

The nanocomposite presented here is a composite material reinforced with silicate sheet. Silicate sheet is an ultrafine filler of nanometer size, which is almost equal to the size of the matrix polymer. Although the content of the filler is as little as several wt %, individual filler particles exist at a distance as close as tens of nanometers from each other because of their ultrafine size. One end of the polymer is restrained strongly to the silicate sheet by polar interaction. Thus, the nanocomposite has a microstructure that has never been seen in conventional composites. The characteristic properties of the nanocomposite are derived from this very structure. Considering the properties, the nanocomposite may be, in a sense, an embodiment of the ideal polymer composite, or a completely novel composite. Silicate sheet can be regarded as a rigid inorganic polymer. In this sense, the nanocomposite realized is a molecular composite in which a silicate sheet is used instead of an organic rod-like polymer.

8   REFERENCES

1. Cox, H. L. *Br. J. Appl. Phys.*, **3**, 72 (1952).
2. Kelly, A. and Tyson, W. R. *High Strength Materials*, John Wiley & Sons, 1965, p. 578.
3. Takanyanagi, M. *Kobunshi*, **33**, 615 (1984).
4. Ziolo, R. F., Giannelis, E. P., Weinstein, B. A., O'Horo, M. P., Ganguly, B. N., Mehrotra, V., Russell, M. W. and Huffman, D. R. *Science*, **257**, 219 (1992).
5. Kato, C. *Kobunshi*, **19**, 758 (1970).
6. Fujiwara, S. and Sakamoto, T. Japanese Pat. JP-A-51-109998 (1976).

7. Usuki, A., Kojima, Y., Okada, M., Kurauchi, T., Kamigaito, O. and Deguchi, R. *Polym. Prepr. Jpn*, **39**, 2427 (1990).
8. Yasue, K., Tamura, T., Katahira, S. and Watanabe, M. Japanese Pat. JP-A-6-248176 (1994).
9. Katahira, S., Tamura, T. and Yasue, K. *Kobunshi Ronbunshu*, **55**, 83 (1998).

7

Epoxy–Clay Nanocomposites

Z. WANG, J. MASSAM AND T. J. PINNAVAIA
Department of Chemistry, Center for Fundamental Materials Research, and
Composite Materials and Structure Center, Michigan State University,
East Lansing, Michigan, USA

1 INTRODUCTION

1.1 NANOCOMPOSITE CONCEPT

Composite materials are formed when at least two distinctly dissimilar materials are mixed to form a monolith. For conventional composites, phase mixing typically occurs on a macroscopic (μm) length scale. In contrast, a nanocomposite is formed when phase mixing occurs on a nanometer length scale. The overall properties of a composite material are determined not only by the parent components but also by the composite phase morphology and interfacial properties. Nanocomposites usually exhibit improved performance properties compared with conventional composites owing to their unique phase morphology and improved interfacial properties. For these reasons, nanostructured organic–inorganic composites have attracted considerable attention from both a fundamental research and an applications point of view [1–4].

In this chapter we show that the exfoliation of smectite clays (e.g. montmorillonite) and layered silicic acid clays (e.g. magadiite) in epoxy matrixes greatly improves the tensile and compressive properties of the matrix through the reinforcement provided by the silicate nanolayers. The effects of nanolayer reinforcement are also manifested in terms of improved chemical resistance, barrier properties and dimensional stability, as well as reduced swelling by solvents. In addition to the enhanced strength and modulus, which can be more than an order of magnitude higher than those of the pristine polymer at 15 wt % loading, the toughness of the matrixes can be increased, particularly for systems with subambient glass transition temperatures. An overview of the methodology available for formulating epoxy–clay

Polymer–clay nanocomposites Edited by T. J. Pinnavaia and G. W. Beall
© 2000 John Wiley & Sons Ltd

nanocomposites and the potential of these materials for technological applications is provided.

1.2 SMECTITE AND LAYERED SILICATE CLAYS FOR NANOCOMPOSITE FORMATION

Layered materials are potentially well suited for the design of hybrid composites, because their lamellar elements have high in-plane strength and stiffness and a high aspect ratio [5]. Virtually all families of lamellar solids share these attributes, but the smectite clays (e.g. montmorillonite) and related layered silicates are the materials of choice for polymer nanocomposite design for two principal reasons. Firstly, they exhibit a very rich intercalation chemistry, which allows them to be chemically modified and made compatible with organic polymers for dispersal on a nanometer length scale. Secondly, they occur ubiquitously in nature and can be obtained in mineralogically pure form at low cost. Smectite clays can also be synthesized under hydrothermal condition, but purified natural clays normally offer a cost advantage over synthetic analogues.

The structure of a typical smectite clay nanolayer is represented in Figure 1. This same basic oxide lattice is also found in talc and mica. However, the smectite clays are distinguished from talc and mica by a unit cell layer charge (typically, in the range from -0.6 to -1.4 per O_{20} unit) that is intermediate between the layer charge

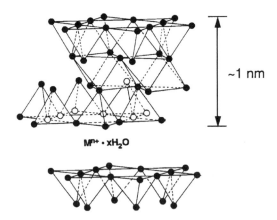

$M^{n+} \cdot xH_2O$

Figure 1 The oxygen framework (solid circles) of smectite clay nanolayers. Each nanolayer consists of two tetrahedral sheets (filled mainly by Si and occasionally Al) and a central octahedral sheet (occupied by Mg, Al, etc.). $M^{n+}.xH_2O$ represents the hydrated inorganic exchange cation (typically sodium and calcium) that occupies the gallery space between nanolayers in the naturally occurring mineral. Each silicate nanolayer has a lateral dimension of 200–2000 nm and a thickness of about 1 nm [5]. The stacking of nanolayers forms tactoids (disordered crystallines) that are typically 0.1–1 µm thick

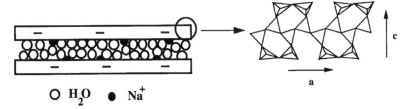

Figure 2 Proposed structure for magadiite according to Garces *et al.* [10] More recently, Schwieger has added to this structural model [11]. The bilayer of intragallery water molecules is based on the observed basal spacing of 15.6 Å for Na^+-magadiite

of talc (0.0 per O_{20} unit) and mica (-2.0 per O_{20} unit). This intermediate value of layer charge leads to cation exchange and gallery swelling properties that are not shared by talc and mica. As will be made apparent in the discussions below, the ion exchange property of smectite clays is an important aspect of their use in nanocomposite formation.

The fundamental chemistry of smectite nanolayer exfoliation has been extended to other layered clay systems, particularly layered silicic acids [6]. This latter family of layered silicates [7] includes five members, namely, kanemite ($NaHSi_2O_5.nH_2O$), makatite ($Na_2Si_4O_9.nH_2O$), ilerite ($Na_2Si_8O_{17}.nH_2O$), magadiite ($Na_2Si_{14}O_{29}.nH_2O$), and kenyaite ($Na_2Si_{20}O_{41}.nH_2O$) [8, 9]. Each clay can be easily synthesized by hydrothermal methods. Layered silicic acids are potentially good candidates for nanocomposite synthesis, not only because they have a platy phase morphology and an intercalation chemistry similar to smectite clays, but also because they possess high purity and structural properties that are complementary to smectite clays. An approximate structure [10, 11] has been proposed for magadiite, as illustrated in Figure 2.

1.3 TYPES OF POLYMER–CLAY NANOCOMPOSITES

From a structural point of view, polymer–clay composites can be generally classified into 'conventional composites' and 'nanocomposites'. In a conventional composite the registry of the clay nanolayers is retained when mixed with the polymer, but there is no intercalation of the polymer into the clay structure (see Figure 3a). Consequently, the clay fraction in conventional clay composites plays little or no functional role and acts mainly as a filling agent for economic considerations. An improvement in modulus is normally achieved in a conventional clay composite, but this reinforcement benefit is usually accompanied with a sacrifice in other properties, such as strength or elasticity.

Two types of polymer–clay nanocomposite are possible. Intercalated nanocomposites (Figure 3b) are formed when one or a few molecular layers of polymer are

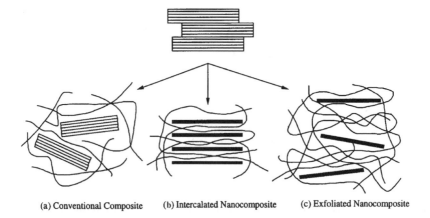

(a) Conventional Composite   (b) Intercalated Nanocomposite   (c) Exfoliated Nanocomposite

Figure 3 Schematic illustrations of the structures (a) conventional, (b) intercalated and (c) exfoliated polymer clay nanocomposites. The clay layers adopt an aggregated, intercalated and exfoliated morphology, respectively, in each type of composite. The clay interlayer spacing is fixed in an intercalated nanocomposite and independent of the polymer–clay ratio, but in an exfoliated nanocomposite the average separation between nanolayers is variable and determined by the clay loading in the polymer

inserted into the clay galleries with fixed interlayer spacings. Exfoliated nanocomposites (Figure 3c) are formed when the silicate nanolayers are individually dispersed in the polymer matrix, the average distance between the segregated layers being dependent on the clay loading. The separation between the exfoliated nanolayers may be uniform (regular) or variable (disordered). Exfoliated nanocomposites show greater phase homogeneity than intercalated nanocomposites. More importantly, each nanolayer in an exfoliated nanocomposite contributes fully to interfacial interactions with the matrix. This structural distinction is the primary reason why the exfoliated clay state is especially effective in improving the reinforcement and other performance properties of clay composite materials.

The unparalleled ability of smectite clays to boost the mechanical properties of an engineering plastic was first demonstrated in a stunning example by Toyota researchers (see Table 1) [12, 13, 14]. By replacing the hydrophilic Na^+ and Ca^{2+}

Table 1 Mechanical and thermal properties of nylon 6–clay composites [14]

Composite type	(wt %) Clay	Tensile strength (MPa)	Tensile modulus (GPa)	Impact (kJ/m^2)	HDT (°C) at 18.5 kg/cm^2
'Nanoscopic' (exfoliated)	4.2	107	2.1	2.8	145
'Micro' (tactoids)	5.0	61	1.0	2.2	89
Pristine polymer	0	69	1.1	2.3	65

exchange cations of the native clay with a more hydrophobic organic onium ion, they were able to intercalate and polymerize ε-caprolactam in the interlayer (gallery) region of the clay to form a nylon–exfoliated clay hybrid composite. At a loading of only 4.2 wt % clay the modulus doubled, the strength increased more than 50 % and the heat distortion temperature increased 80 °C compared with the pristine polymer. The key to this extraordinary performance by nylon 6–clay hybrids was the complete dispersal (exfoliation) of the clay nanolayers in the polymer matrix.

2 NANOCOMPOSITE SYNTHESIS AND CHARACTERIZATION

The nanocomposite chemistry pioneered by Toyota researchers has been extended in recent years to other thermoset and thermoplastic polymers. Although polymer–exfoliated clay nanocomposites are relatively difficult to prepare, the organocation modification approach has proven to be very successful for the design of other engineering polymer–clay nanocomposites. In the original Toyota approach, the hydrophilic inorganic exchange cations on the intragallery surfaces were replaced by cation exchange reaction with an organic cation in order to match the polarity of the prepolymer and the polymer matrix. The *in situ* polymerization of prepolymers in organoclay galleries has been especially successful for the preparation of exfoliated clay nanocomposites of polyimide, polyether, acrylonitrile rubber, epoxy, polystyrene, polysiloxane and polypropylene matrixes [15–23]. In addition to being a compatibilizing agent for the intercalation of the polymer precursor, the intragallery organic cation is potentially capable of functioning as a polymerization catalyst, as was best demonstrated by our synthesis of exfoliated clay nanocomposites of amine-cured epoxies [24, 25]. Exchange cations derived from protonated amines, in particular, were shown to acid-catalyze the intragallery curing process and cause the polymerization rate in the spatially restrictive galleries of the clay to be competitive with the extragallery polymerization rates, resulting in the formation of a monolithic polymer matrix reinforced by exfoliated clay nanolayers. This concept of making the polymerization process in intragallery space competitive with extra-gallery polymerization through intragallery catalysis has been extended recently to include the metal-catalyzed polymerization intragallery polymerization of olefins [26].

2.1 EPOXY–SMECTITE CLAY NANOCOMPOSITES

Our initial studies of thermoset epoxy systems were concerned with the ring opening polymerization of epoxides to form polyether nanocomposites [17, 27, 28]. This chemistry was followed by studies of both rubbery and glassy thermoset epoxy–clay nanocomposites using different types of amine curing agents [20, 29–34]. The mechanisms leading to nanolayer exfoliation in thermoset epoxy systems have been greatly elucidated [24, 35]. In addition, the polymer–clay interfacial properties have been shown to play a dominant role in determining the performance benefits derived from nanolayer exfoliation [34].

The synthesis of amine-cured epoxy nanocomposites through the polymerization of monomers in the galleries of protonated onium ion exchanged forms of smectite clays depends on two crucial factors: firstly, the ability of the onium ion to serve as a compatibilizing agent, which allows for cointercalation of the resin and curing agent, and secondly, the ability of the onium ion to acid–catalyze the intragallery ring opening polymerization reaction. The catalytic function of the onium ion is important because, as noted above, it allows the intragallery polymerization rate to be competitive with the extragallery polymerization rate. Consequently, the nanolayers can be completely dispersed in the polymer matrix where they can fully contribute to the reinforcement mechanism.

The dispersal of montmorillonite nanolayers in an epoxy matrix is readily achieved with acidic organophilic alkylammonium ions such as $C_{18}H_{37}NH_3^+$, henceforth abbreviated to C18A. The ion exchange reaction was accomplished using previously described methods [24]. Both a laboratory purified sodium form of montmorillonite from Wyoming (C18A-SWy) and commercially available alkylammonium exchanged montmorillonites from Nanocor, Inc., denoted as C18A-AMS and C18A-CWC clays, were used to form epoxy–clay nanocomposites. The unit cell formula and layer charge density for each inorganic clay are listed in Table 2, and the X-ray powder diffraction patterns of the organoclays are provided in Figure 4a.

The epoxide resins used for epoxy–clay hybrid composite formation were poly(bisphenol A coepichlorohydrins), more specifically, Shell EPON 828 and 826 for elastomeric and glassy epoxy matrixes, respectively:

The curing agents were poly(propylene glycol)bis(2-aminopropyl) ethers, namely Huntsman Chemical Jeffamine D-2000 and D-230 for elastomeric and glassy epoxy matrixes, respectively, with MW values of ~ 2000 and 230:

$$H_2NCHCH_2 \{ OCH_2CH \}_x NH_2$$

$$x = 33.1 \text{ or } 2.6$$

Table 2 Unit cell formulas and charge densities of smectite clays used for nanocomposite formation

Clay	Unit cell composition	CEC (meq./100 g)	Preparation method
SWy	$Na_{0.86}[Mg_{0.86}Al_{3.14}](Si_{8.00})O_{20}(OH)_4$	88	Laboratory
AMS	$Na_{0.92}^+[Mg_{0.86}Al_{3.06}](Al_{0.18}Si_{7.82})O_{20}(OH)_4$	124	Commercial
CWC	$Na_{1.08}^+[Mg_{0.54}Al_{3.38}](Al_{0.36}Si_{7.64})O_{20}(OH)_4$	147	Commercial

These curing agents afford an elastomeric epoxy polymer matrix with a subambient T_g of $-40\,°C$ ($x = 33.1$) or a glassy epoxy polymer matrix with a T_g of $\sim 82\,°C$ ($x = 2.6$).

The methodologies used for the synthesis of rubbery and glassy epoxy–clay nanocomposites differed slightly. For the rubbery systems, the desired amount of organoclay was normally added directly to the mixture of epoxide and curing agent. For the glassy systems, the organoclays were preintercalated with the epoxide resin by reaction with the onium ion clay at $50\,°C$ overnight, followed by the addition of the curing agent. The mixtures of organoclay, epoxide and curing agent were easily degassed and pourable for molding when the loading of organoclay was below 20 wt %.

The X-ray diffraction patterns of the epoxy–clay nanocomposites formed from different organoclays are shown in Figures 4b and 4c. All of these patterns are characterized by the absence of the 00l diffraction peaks, providing strong evidence that the clay nanolayers have been exfoliated in the thermoset curing process. Also, it should be noted that the two-dimensional structures of the clay nanolayers are retained in the exfoliated state, as judged by the presence of several strong in-plane peaks in the XRD patterns (not shown). The exfoliated state of clay nanolayers has been confirmed by the TEM image in Figure 5. The dark lines are the cross sections of the 10 Å thick silicate layers. A face–face layer morphology is retained, but the layers are irregularly separated by ~ 80–$150\,Å$ of polymer. This clay particle morphology is correlated with the absence of Bragg scattering in the range of low angles.

The key to achieving an exfoliated clay composite structure in an epoxy matrix is first to load the clay gallery with a hydrophobic onium ion and then to preload the gallery region with epoxide and polyoxypropyleneamine precursors. As the precursors diffuse into the gallery, the acidic onium ions catalyze intragallery polymer chain formation at a rate that is competitive with extragallery polymerization. This leads initially to gel formation and the exfoliation of the clay nanolayers. Complete network crosslinking can then be accomplished, with retention of the exfoliated clay state. The relative rates of reagent intercalation, chain formation and network crosslinking are controlled by a judicious choice of temperature for each process.

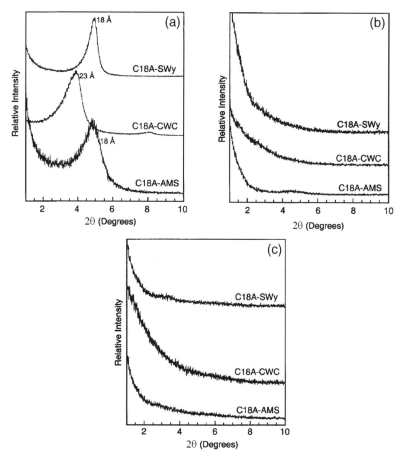

Figure 4 X-ray diffraction patterns of (a) pristine organomontmorillonites, (b) elastomeric epoxy–montmorillonite nanocomposites and (c) glassy epoxy–montmorillonite nanocomposites. In each case the gallery onium ion is octadecylammonium, abbreviated to C18A. SWy is a montmorillonite from Wyoming, as supplied by the Clay Mineral Society; montmorillonites CWC and AMS are commercial clays supplied by Nanocor, Inc. The organoclay loading for reach composite is 10 or 15 wt%

Akelah *et al.* [36] have investigated the use of 'epoxyphilic' montmorillonites for epoxy nanocomposite formation. The onium ion in these clays contained either carboxylic acid anhydride, phenolic, hydroxyl, amide or amine functionality for reaction with one or both components of an amine-cured epoxy formulation. X-ray diffraction results indicated that extensive gallery swelling occurred for both the uncured and the cured epoxy–amine–epoxyphilic clay mixtures. Interestingly, it was

Figure 5   TEM image of an ultrathin section of an elastomeric epoxy–exfoliated clay nanocomposite containing 20 wt% C18A-SWy [20]

also possible to swell even Na^+-montmorillonite with the uncured epoxy–amine mixture, but upon curing, the clay was returned to the 12 Å spacing characteristic of restacked Na^+-montmorillonite layers. On the other hand, Lee and Jang [37] showed that intercalated epoxy nanocomposites were formed from Na^+-montmorillonite through emulsion polymerization. Apparently, in the case of Akelah *et al.*, the curing rate of the epoxy was faster in the extragallery regions of the clay than in the Na^+-occupied galleries, thus allowing most of the intragallery components to migrate out of the galleries prior to becoming crosslinked.

Messersmith and Giannelis [38] independently used an epoxyphilic clay to prepare glassy epoxy nanocomposites by dispersing an ethoxylated onium ion exchange form of montmorillonite in epoxy resin and curing in the presence of nadic methylanhydride, benzyldimethylamine or boron trifluoride monoethylamine at 100–200 °C . Interlayer spacings of 100 Å or more were observed. The curing of the resin appeared to involve coupling to the hydroxy groups on the bis(2-hydroxyethyl)methyltallowammonium ions in the clay galleries. It is important to emphasize here that the epoxyphilic functional groups on the quaternary ammonium ion modifiers used both by Akelah *et al.* and by Messersmith and Giannelis play an important role in forming exfoliated clay nanocomposites. All other epoxy nanocomposites formed from quaternary alkylammonium ion clays with non-functional alkyl groups are all intercalated clay nanocomposites.

2.2 EPOXY NANOCOMPOSITE DERIVED FROM ORGANOCATION INTERCALATES OF LAYERED SILICIC ACIDS

Using similar organoclay and *in situ* polymerization techniques, we have extended the exfoliation chemistry of smectite clay in an epoxy matrix to include the layered silicate magadiite and other members of the layered silicic acid family with different nanolayer thickness for the formation of polymer–inorganic nanolayer composites [6, 35]. Magadiite has a nanolayer thickness of 11.2 Å. The reaction of the sodium ion exchange form of the clay with onium ions leads to an expansion of the gallery space, corresponding in most cases to a paraffin-like orientation of the onium ions in the gallery space. Representative X-ray diffraction patterns of organomagadiites and the corresponding epoxy–magadiite composites prepared from each organomagadiite are shown in Figure 6. The onium ion acidity again plays an important role in determining the final state of the silicate in each system. Highly acidic C18A and C18A1M intragallery ions (C18 and M, respectively, denote the presence of a octadecyl and a methyl group at the amino head group of the onium ion) afford exfoliated epoxy–magadiite nanocomposites. An intercalated nanocomposite is expected for non-acidic C18A3M-magadiite, because this non-acidic quaternary alkylammonium ion is catalytically inert and the gallery expansion is determined mainly by the initial loading of resin and curing agent in the gallery. The intercalate

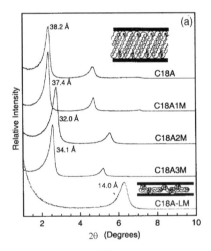 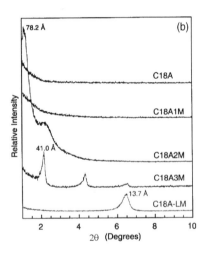

Figure 6 X-ray diffraction patterns of (a) organo magadiites and (b) cured elastomeric epoxy–magadiite nanocomposites. Primary, secondary, tertiary and quaternary long-chain alkylammonium-exchanged $CH_3(CH_2)_{17}NH_{3-n}$ $(CH_3)_n{}^+$-magadiites with $n = 0$, 1, 2 and 3 are designated as C18A-, C18A1M-, C18A2M- and C18A3M-magadiite, respectively. All of these organomagadiites have a paraffin-like gallery structure (top inset). The organomagadiite with a 14.0 Å spacing is intercalated by a lateral monolayer (bottom inset) of $CH_3(CH_2)_{17}NH_3{}^+$ ions and is denoted C18A-LM

with a lateral monolayer structure, denoted by C18A-magadiite-LM, could not be swelled by the polymer precursors. Consequently, only a conventional composite is formed. Interestingly, C18A2M-magadiite with an intermediate Brönsted acidity afforded the first example of a special type of clay nanocomposite, in which the nanolayers are exfoliated but at the same time *ordered*. That is, the interlayer spacing (78.2 Å) is far larger than the value expected for an intercalated nanocomposite, yet the nanolayer separation is sufficiently regular to give Bragg X-ray scattering along the nanolayer stacking direction. This phenomenon of ordered exfoliation can only be observed when the silicate layer charge distribution is uniform from gallery to gallery. A uniform layer charge distribution assures a uniform concentration of catalytic onium ions in the galleries and, consequently, a constant polymerization rate in all galleries.

3 NANOCOMPOSITE PROPERTIES

3.1 TENSILE PROPERTIES

The reinforcement benefits derived from the exfoliation of clay nanolayers in an epoxy polymer are indicated by the plots in Figure 7 of tensile strength and modulus (as determined from stress–strain curves) versus clay loading. The presence of exfoliated clay nanolayers substantially increases both the tensile strength and modulus relative to the pristine elastomeric polymer. Clearly, the reinforcement of the epoxy–clay nanocomposites is dependent on clay loading. Significant reinforcement is observed even at organoclay loadings ≤ 10 wt %. Also, the tensile strength

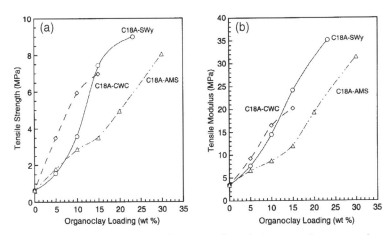

Figure 7 Comparison of the tensile properties of elastomeric epoxy–clay nano-composites prepared from a laboratory purified C18A-SWy and commercial Nanocor organoclays C18A-CWC and C18A-AMS

and modulus both increase rapidly with increasing organoclay loading. More than a ten-fold increase in strength and modulus is realized by the addition of only 15 wt % (~7.5 vol. %) of the exfoliated organoclay [20].

It is noteworthy that commercially available C18A-CWC and C18A-AMS organoclays prepared on an industrial scale afforded nanocomposites with performance properties comparable with those obtained from laboratory purified C18A-SWy. The differences observed between C18A-AMS, CWC and SWy organoclays are not unexpected because of the differences in platelet aspect ratio, layer charge density and the degree of exfoliation.

The reinforcement effect is dependent on the extent of silicate nanolayer separation, as expected. A comparison of tensile properties for the epoxy–magadiite nanocomposites prepared from C18A1M-, C18A2M-, and C18A3M-magadiite and C18A-magadiite-LM is provided in Figure 8. The tensile strengths of the microscopically homogeneous intercalated and exfoliated magadiite nanocomposites are superior to conventional composites with macroscopic homogeneity. As expected, the reinforcement benefit is substantially greater for the exfoliated nanocomposites obtained from C18A1M- and C18A2M-magadiite than for the intercalated nanocomposites derived from C18A3M-magadiite. Clearly, the tensile properties improve with increasing degree of nanolayer separation. It is noteworthy that the nanocom-

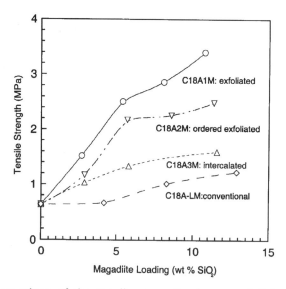

Figure 8 Comparison of the tensile strengths for an exfoliated, ordered exfoliated and intercalated epoxy–magadiite nanocomposite prepared from C18A1M-, C18A2M- and C18A3M-magadiite, respectively. The conventional composite was prepared from C18A-magadiite with a lateral nanolayer structure (cf. Figure 6). The epoxide resin and curing agent used in this system are EPON 828 and Jeffamine D-2000, respectively

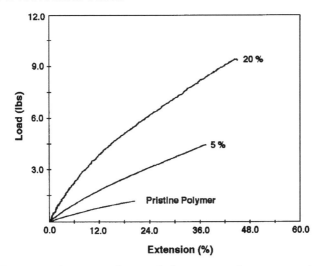

Figure 9   Stress–strain curves for a pristine epoxy elastomer and epoxy–clay nanocomposites prepared from C18A-AMS montmorillonite with organoclay loadings of 5 and 20 wt %. The load cell had a capacity of 20 lb, and the strain rate was 0.06 in/min

posites reported here exhibit an improvement in elasticity, as well as in strength and modulus. Typical stress–strain curves for the pristine epoxy elastomer and the nanocomposite prepared from C18A-AMS are shown in Figure 9. It is clear that the toughness of the exfoliated epoxy–clay nanocomposites has been significantly improved. It is very unusual to improve modulus while at the same time significantly improving the strength and toughness. The advantages of nanocomposites over conventional composites are also demonstrated in Figure 10. The strain at break is substantially increased for exfoliated and intercalated magadiite nanocomposites, whereas conventional magadiite composites exhibit normal behavior and become less elastic than the pristine polymer. The enhancement in strength and modulus is directly attributable to the reinforcement provided by the dispersed silicate nano-layers. The improved elasticity may be attributed, at least in part, to the plasticizing effect of gallery onium ions and the conformational effects on the polymer at the clay–matrix interface.

Zilg *et al.* [39] have recently characterized the tensile properties of several epoxy nanocomposites derived from hexahydrophthalic anhydride-cured bisphenol A diglycidyl ether and onium ion exchanged forms of montmorillonite, synthetic fluoromica and synthetic hectorite-type layered silicates. The onium ion modifiers included protonated alkyl amines and quaternary ammonium ion surfactants that were either non-functional (e.g. N,N,N-trimethyldodecylammonium) or epoxyphilic (e.g. N,N,-bis(2-hydroxyethyl)-N-methyldodecylammonium). In accordance with earlier results, the protonated onium ion clays resulted in a greater degree of

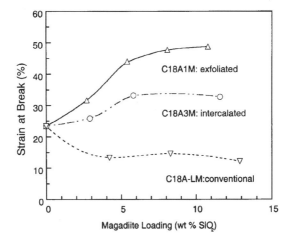

Figure 10 Comparison of the strain at break values for an exfoliated epoxy–magadiite nanocomposite prepared from C18A1M-magadiite, an intercalated nanocomposite prepared from C18A3M-magadiite and a conventional composite prepared from C18A-magadiite-LM. The epoxide resin and curing agent used in this system are EPON 828 and Jeffamine D-2000, respectively [36]

nanolayer exfoliation in the polymer matrix than the non-functional quaternary ammonium ion clays. This observation again points to the importance of catalyzing the intragallery polymerization rate in order to facilitate expansion of the gallery space between the reinforcing nanolayers. All nanocomposites derived from clays intercalated by protonated onium ions, as well as by quaternary onium ions containing functionalized hydroxyl or carboxylic acid groups, exhibited significantly lower T_g values in comparison with the nanocomposites formed from clays containing non-functional onium ions. Regardless of the decrease in T_g, all of the nanocomposites, especially those formed from protonated onium ion clays, exhibited an increase in Young's modulus, as well as an increase in the stress intensity factor, K_{Ic}, which is a measure of the energy dissipation at a crack tip during fracture, but a decrease in tensile strength. The best balance between increased stiffness and increased toughness at 5–10 wt % loadings was observed for fluoromicas intercalated by the protonated cationic forms of dodecylamine or 12-aminododecanoic acid. It was concluded that the larger anisotropic intercalated particles contributed to increased toughness, whereas exfoliated nanolayers were the key to increased stiffness.

3.2 COMPRESSIVE PROPERTIES

The reinforcement properties of smectite clay nanolayers in a glassy epoxy matrix were evaluated in part by measuring the compressive yield strengths and moduli as a

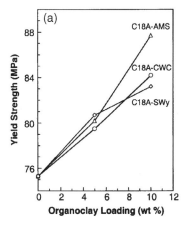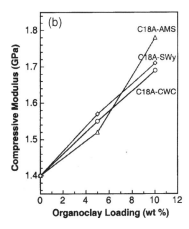

Figure 11 (a) Compressive yield strength and (b) moduli for the pristine epoxy polymer and the exfoliated epoxy–clay nanocomposites prepared from C18A-AMS, C18A-CWC and C18A-SWy. The epoxide resin and curing agent used in this system are EPON 828 and Jeffamine D-230, respectively [33, 40]

function of clay loading. The results for the exfoliated nanocomposites are shown in Figure 11. The exfoliated nanocomposites show a substantially greater compressive yield strength and modulus relative to the pristine polymer. This is the first evidence for clay nanolayer reinforcement of a glassy epoxy matrix under compressive strain [40]. Clearly, the degree of reinforcement increases with increasing clay loading, as expected. On the other hand, the intercalated nanocomposite prepared from a quaternary onium ion exchanged organoclay and the conventional composite were completely ineffective in providing reinforcement to the matrix under compressive strain.

Although the exfoliated state of the clay is preferred for reinforcement, differences in the degree of reinforcement are observed for the different exfoliated clay nanocomposites. The differences observed between C18A-AMS, CWC and SWy organoclays, which appear to be statistically significant especially at 10 wt% loading, most likely arise because of differences in interfacial interactions, platelet aspect ratio, layer charge density and other factors that influence interfacial interactions between the nanolayers and the glassy epoxy matrix.

3.3 DYNAMIC MECHANICAL ANALYSIS MEASUREMENTS

Dynamic mechanical analysis (DMA) was used to investigate the thermomechanical properties for our glassy epoxy–exfoliated clay nanocomposites [40]. Figure 12 shows the temperature dependence of the storage modulus, E', for the pristine polymer and the epoxy–clay nanocomposites containing 5 and 10 wt% C18A-AMS. Below T_g, the nanocomposites exhibit higher storage modulus than that of the

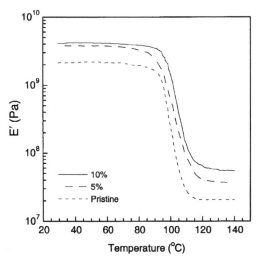

Figure 12 Storage moduli curves obtained by three-point bending dynamic mechanical analysis for a pristine glassy epoxy polymer and the exfoliated clay nanocomposites of the same polymer containing 5 and 10 wt % C18A-AMS montmorillonite. The epoxide resin and curing agent used in this system are EPON 826 and Jeffamine D-230, respectively [33, 40]

pristine polymer. In fact, at 50 °C , still well within the glassy region, E' of the 5 % nanocomposite is approximately 72 % higher compared with the pristine epoxy polymer. In the rubbery state above T_g, the 5 and 10 % nanocomposites exhibit storage moduli that are 76 and 164 % higher, respectively, than that of the pristine polymer.

Using DMA methods, Messersmith and Giannelis [19] found that the glassy epoxy–exfoliated clay nanocomposites formed from epoxyphilic exchange forms of montmorillonite exhibited a broadened T_g at slightly higher temperature than the unmodified polymer. This result is in contrast to the observed lowering of glass transition temperature by Zilg et al. [39]. The dynamic storage modulus at 4 vol % loading of silicate was 58 % higher in the glassy region and 450 % higher in the rubbery region. In related studies, Kelly et al. [41] also showed that epoxy nanocomposites formed from epoxyphilic clays exhibited improvements in glass transition temperature and dynamic behavior, together with a lowering of residual stress.

The chemistry of protonated onium ion clays for the preparation of intercalated and exfoliated onium clay nanocomposites of epoxy resins has been put to good use by Kornmann [42] in an effort to determine the effect of clay nanolayers on the segmental motions of the polymer matrix. By controlling the cure temperature of the composites formed through crosslinking of a EPON-828 resin with Jeffamine D-230 diamine and 3,3′-dimethylmethylene dicyclohexylamine (3DCM), he was able to form intercalated and exfoliated composites using the octadecylammonium exchange form of montmorillonite. Dynamic mechanical thermal analysis indicated the evolution of tan δ corresponding to T_g near $187\,^{\circ}$C and a secondary transition near $-60\,^{\circ}$ for the pure polymer cured at $160\,^{\circ}$C with 3DCM. The same features were observed for a conventional composite formed from an inorganic exchange form of the clay. The intercalated and exfoliated organoclay nanocomposites showed low-temperature shifts for the two transitions, most likely due to the known thermal degradation of the onium ion at the high curing temperature. However, the composites cured with Jeffamine D-230 at $110\,^{\circ}$C, where the onium ion is thermally stable, showed little or no shift in the T_g ($\sim 95\,^{\circ}$C) for the conventional intercalated clay nanocomposites in comparison with the pure polymer. Thus, it appears that the observed reduction in T_g for the 3DCM-cured system is almost certainly due to onium ion decomposition.

3.4 DIMENSIONAL STABILITY

The effect of exfoliated smectite clay platelets on the thermal expansion coefficient over the temperature range 40–$120\,^{\circ}$C is shown in Figure 13. As expected on the basis of the reinforcement effects described earlier, the expansion coefficients for the nanocomposite samples are lower than that of the pristine polymer, both above and

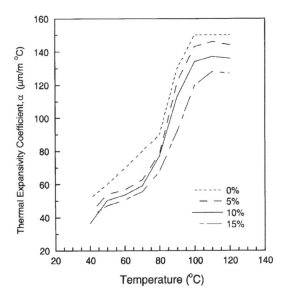

Figure 13 Thermal expansivity coefficients obtained for the pristine polymer and the 5, 10 and 15 wt % C18A-AMS epoxy–clay nanocomposites. The epoxide resin and curing agent used in this system are EPON 826 and Jeffamine D-230, respectively [33, 40]

below the glass transition temperature. In the region below T_g, the expansion coefficient at $70\,^\circ$C for the 5 % nanocomposite is as much as 27 % lower than for the pure polymer. Above T_g, the thermal expansivity progressively decreases with increasing clay content. The expansion coefficient of the nanocomposite with 15 wt % C18A-AMS is approximately 20 % lower than that of the pristine polymer.

3.5 CHEMICAL STABILITY AND SOLVENT RESISTANCE

Exfoliated epoxy–clay nanocomposites exhibit not only superior mechanical properties but also exceptional chemical stability and solvent resistance. The results are listed in Table 3. Since the pristine epoxy polymer already exhibits good inertness to basic and aqueous solution uptake, we did not expect an impressive improvement towards aqueous and basic solutions for the nanocomposites. Nevertheless, for both inorganic and organic acid solutions, the exfoliated nanocomposite containing 9.1 wt % C18-magadiite did show a significant reduction in solution uptake. The uptake of methanol and toluene was also reduced substantially for the exfoliated nanocomposites. On the other hand, we observed that barrier to solvent uptake by the conventional and intercalated composites was not as significant as for the exfoliated nanocomposites.

Nanolayer reinforcement also reduces the solvent swelling of a glassy epoxy matrix [40]. For instance, the uptake of methanol and propanol (Figures 14a and b) is faster for the pristine polymer than for the nanocomposite materials. In the case of methanol, the equilibrium absorption after 30 days is almost equal to that of the pristine polymer. However, the pristine sample after being submerged 30 days in methanol was rubbery, whereas the composite material appeared unaffected by the solvent. After 50 days in propanol, the pristine polymer absorbs 2.5 times more than the nanocomposite, and at this time the pristine sample began to crack and break up, whereas the shape and texture of the nanocomposite sample appeared unchanged.

The absorption of toluene and chloroform are shown in Figures 14c and d. In chloroform, the pristine specimen disintegrated into many pieces after only a few hours, whereas the nanocomposite survived 2 days exposure before any fractures

Table 3 Chemical stability and solvent resistance of elastomeric Epoxy–exfoliated magadiite nanocomposites prepared from C18A-magadiite[a]

Materials	10 % NaOH[b]	Distilled H$_2$O[b]	30 % H$_2$SO$_4$[b]	5% acetic acid[b]	Methanol[c]	Toluene[d]
Pristine polymer	1.6	2.5	16.7	13.4	76.5	189
C18A-magadiite	1.5	1.6	7.2	9.6	58.5	136

[a] The organomagadiite loading was 9.1 wt %. Values reported are immersion weight gains (wt %) after specified uptake periods.
[b] Weight gain after 15 days.
[c] Weight gain after 48 h.
[d] Weight gain after 24 h.

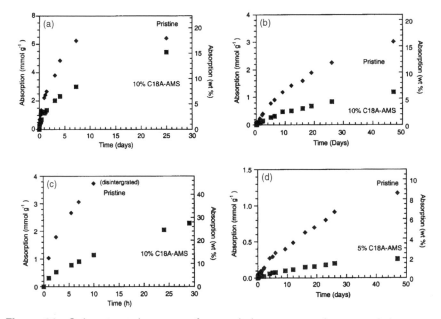

Figure 14   Solvent uptake curves for a pristine epoxy polymer and the corresponding C18A-AMS nanocomposite: (a) methanol; (b) propanol; (c) chloroform; (d) toluene. The epoxide resin and curing agent used in this system are EPON 826 and Jeffamine D-230, respectively [33, 40]

were observed in the sample. Toluene absorption by the pristine sample was almost 5 times greater than the amount absorbed by the nanocomposite with only 5 wt % by weight clay loading. The pristine sample swelled and became rubbery, whereas the size and shape of the nanocomposite sample was unchanged.

3.6   THERMAL STABILITY

We have also obtained useful insights into the nature of polymer–clay nanocomposites from TGA studies of thermal stability [6]. Figure 15 compares the TGA curves for an exfoliated nanocomposite prepared from C18A1M-magadiite and an intercalated nanocomposite prepared from C18A3M-magadiite. Included in the figure are the TGA curves for the pristine organomagadiites and epoxy polymer. As has been observed previously for nylon 6–exfoliated clay nanocomposites, the thermal stability of the polymer matrix is not sacrificed by nanocomposite formation. It is particularly noteworthy, however, that the lower temperature weight loss for the intercalated C18A3M-magadiite nanocomposite is indicative of the decomposition of the quaternary alkylammonium cations on the magadiite basal planes, because an analogous weight loss is observed for pristine C18A3M-magadiite. In comparison,

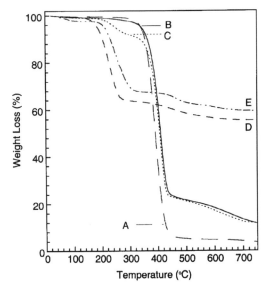

Figure 15   Thermogravimetric analysis curves for (A) a pristine epoxy polymer and for epoxy nanocomposites formed from (B) C18A1M-magadiite and (C) C18A3M-magadiite. The loading of organomagadiite was 20 wt for each epoxy-magadiite nanocomposite. Curves (D) and (E) are for the pristine C18A1M- and C18A3M-magadiites, respectively. The epoxide resin and curing agent used in this system are EPON 828 and Jeffamine D-2000, respectively [36]

the TGA curve for the exfoliated C18A1M-magadiite nanocomposite does not show a similar low temperature weight loss for the decomposition of surface onium ions, verifying that the secondary onium ions are indeed incorporated into the polymer network.

3.7 OPTICAL TRANSPARENCY

We have learned that the improvement in tensile properties provided by exfoliated magadiite nanolayers is not quite as good as that afforded by exfoliated smectite clays, particularly with regard to tensile modulus at higher loadings. However, the most significant property of our epoxy–layered silicic acid nanocomposites is their high optical transparency [35]. As shown in Figure 16, the layered silicic acid nanocomposites are almost as transparent as pristine epoxy polymer. It is very difficult to achieve this performance in the system of smectite clay at the same loading. This result suggests that the refractive index of the layered silicic acid mineral family more nearly matches that of the organic matrix which can be partially attributed to the high purity of synthetic clays.

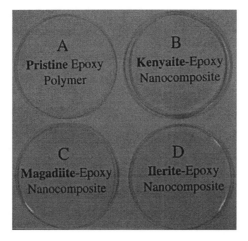

Figure 16 Optical transparency properties of (A) a pristine elastomeric epoxy polymer and of epoxy–layered silicic acid nanocomposites prepared from (B) C18A-kenyaite, (C) C18A-magadiite and (D) C18A-ilerite. Each specimen has a 2 in diameter and 1/16 in thickness with a silicate loading of 5 wt% SiO_2. The epoxide resin and curing agent used in this system are EPON 828 and Jeffamine D-2000, respectively

4 CONCLUSIONS

Layered silicate clays intercalated by protonated ammonium ions or epoxyphilic (functionalized) quaternary ammonium ions are especially effective in forming epoxy–exfoliated clay nanocomposites. The exfoliated forms of the silicate nanolayers in both rubbery and glassy epoxy matrixes provide effective reinforcement when a tension or a compressive load is applied. The presence of clay nanolayers also improves the dynamic storage modulus and reduces the thermal expansivity coefficient of the epoxy polymer matrix. Epoxy–clay nanocomposites are substantially more resistant to chemical attack and solvent swelling than the pristine polymer. The knowledge generated through studies of epoxy nanocomposites can be expected to apply to the rational design of other thermoset nanocomposites. The concept of achieving nanolayer exfoliation by tailoring the intra- and extragallery polymerization rates through the intercalation of a gallery catalyst can also be expected to apply to design of thermoplastic polymer nanocomposites. Nanolayer reinforcement will continue to make it possible to use conventional engineering polymers in entirely new situations, including conditions sufficiently hostile to cause failure in conventional composite forms of the polymers. In particular, the enhanced mechanical properties, dimensional stability, thermal stability, chemical stability, resistance to solvent swelling and excellent optical properties together with the anticipated barrier film properties and reduced flammability of these materials should make them especially attractive for packaging materials and protective films.

5 ACKNOWLEDGEMENTS

The partial financial support of the Michigan State University Center for Fundamental Materials Research and the Composite Materials and Structure Center, Nanocor, Inc. (Arlington, IL) and Claytec, Inc. (East Lansing, MI) is gratefully acknowledged.

6 REFERENCES

1. Messersmith, P. B. and Stupp, S. I., *J. Mater. Res.*, **7**, 2599 (1992).
2. Okada, A. and Usuki, A., *Mater. Sci. Eng.*, **C3**, 109 (1995).
3. Giannelis, E. P., *Adv. Mater.*, **8**, 29 (1996).
4. Novak, B. M., *Adv. Mater.*, **5**, 422 (1993).
5. Pinnavaia, T. J., *Science*, **220**, 365 (1983).
6. Wang, Z. and Pinnavaia, T. J., *Chem. Mater.*, **10**, 1820 (1998).
7. Lagaly, G., Beneke, K. and Weiss, A., *Am. Miner.*, **60**, 642 (1975).
8. Beneke, K. and Lagaly, G., *Am. Miner.*, **62**, 763 (1977).
9. Beneke, K. and Lagaly, G., *Am. Miner.*, **68**, 818 (1983).
10. Garces, J. M., Rocke, S. C., Crowder, C. E. and Hasha, D. L., *Clays Clay Miner.*, **36**, 409 (1988).
11. Schwieger, W., Heidemann, D. and Bergk, K. H., *Rev. Chim. Minér.*, **12**, 639 (1985).
12. Usuki, A., Kawasumi, M., Kojima, Y., Okada, A., Kurauchi, T. and Kamingaito, O., *J. Mater. Res.*, **8**, 1174 (1993).
13. Usuki, A., Kojima, Y., Kawasumi, M., Okada, A., Fukushima, Y., Kurauchi, T. and Kamigaito, O., *J. Mater. Res.*, **8**, 1179 (1993).
14. Kojima, Y., Usuki, A., Kawasumi, M., Okada, A., Fukushima, Y., Kurauchi, T. and Kamigaito, O., *J. Mater. Res.*, **8**, 1185 (1993).
15. Yano, K., Usuki, A., Okada, A., Kurauchi, T. and Kamigaito, O., *J. Polym. Sci.*, *Part A*, *Polym. Chem.*, **31**, 2493 (1993).
16. Kojima, Y., Fukumori, K., Usuki, A., Okada, A. and Kurauchi, T., *J. Mater. Sci. Lett.*, **12**, 889 (1993).
17. Wang, M. S. and Pinnavaia, T. J., *Chem. Mater.*, **6**, 468 (1994).
18. Lan, T., Kaviratna, P. D. and Pinnavaia, T. J., *Chem. Mater.*, **6**, 573 (1994).
19. Messersmith, P. B. and Giannelis, E. P., *Chem. Mater.*, **6**, 1719 (1994).
20. Lan, T. and Pinnavaia, T. J., *Chem. Mater.*, **6**, 2216 (1994).
21. Burnside, S. D. and Giannelis, E. P., *Chem. Mater.*, **7**, 1597 (1995).
22. Vaia, R. A., Jandt, K. D., Kramer, E. J. and Giannelis, E. P., *Chem. Mater.*, **8**, 2628 (1996).
23. Usuki, A., Kato, M., Okada, A. and Kurauchi, T., *J. Appl. Polym. Sci.*, **63**, 137 (1997).
24. Lan, T., Kaviratna, P. D. and Pinnavaia, T. J., *Chem. Mater.*, **7**, 2144 (1995).
25. Pinnavaia, T. J., Lan, T., Wang, Z., Shi, H. and Kaviratna, P. D., *ACS Symp. Ser.*, **622**, 250 (1996).

26. Heinemann, J., Reichert, P., Thomann, R. and Mulhaupt, R., *Macromol. Rapid Commun.*, **20**, 423 (1999); Bergman, J., Chen, H., Giannelis, E. P., Thomas, M. G. and Coates, G. W., *Chem. Commun.*, 2179 (1999).
27. Pinnavaia, T. J., Lan, T., Kaviratna, P. D. and Wang. M. S., *Mater. Res. Soc. Symp. Proc.*, **346**, 81 (1994).
28. Lan, T., Kaviratna, P. D. and Pinnavaia, T. J., *J. Phys. Chem. Solids*, **57**, 1005 (1996).
29. Lan, T., Kaviratna, P. D. and Pinnavaia, T. J., *Polym. Mater. Sci. Eng.*, **71**, 528 (1994).
30. Lan, T., Wang, Z., Shi, H. and Pinnavaia, T. J., *Polym. Mater. Sci. Eng.*, **73**, 296 (1995).
31. Lan, T. and Pinnavaia, T. J., *Mater. Res. Soc. Symp. Proc.*, **435**, 79 (1996).
32. Pinnavaia, T. J. and Lan, T., *Proc. Am. Soc. Compos. Tech. Conf.*, **11**, 558 (1996).
33. Massam, J., Wang, Z., Pinnavaia, T. J., Lan, T. and Beall, G., *Polym. Mater. Sci. Eng.*, **78**, 274 (1998).
34. Shi, H., Lan, T. and Pinnavaia, T. J., *Chem. Mater.*, **8**, 1584 (1996).
35. Wang, Z., Lan, T. and Pinnavaia, T. J., *Chem. Mater.*, **8**, 2200 (1996).
36. Akelah, A., Kelly, P., Qutubuddin, S. and Moet, A., *Clay Miner.*, **29**, 169 (1994).
37. Lee, D. C. and Jang, L. W., *J. Appl. Polym. Sci.*, **68**, 1997 (1998).
38. Messersmith, P.B. and Giannelis, E.P., *Chem. Mater.*, **6**, 1717 (1994).
39. Zilg, C., Müllhaupt, R. and Finter, J. *Macromol. Chem. Phys.*, **200**, 661 (1999).
40. Massam, J. and Pinnavaia, T. J., *Mater. Res. Soc. Symp. Proc.*, **520**, 223 (1998).
41. Kelly, A., Akelah, A., Qutubuddin, S. and Moet, A., *J. Mater. Sci.*, **29**, 2274 (1994).
42. Kornmann, X. Licentiate thesis, Luleå, Sweden (1999).

8

Polypropylene–Clay Nanocomposites

A. OYA

Faculty of Engineering, Gunma University, Kiryu, Gunma 376-8515, Japan

1 INTRODUCTION

The Toyota group has developed a clay/nylon nanocomposite with excellent mechanical properties [1, 2]. This success has aroused much attention for the use of clay as a reinforcement material for polymers. The nanocomposite has been used to fabricate parts of automobiles, but this application was stopped because of the high cost. The high cost was caused by the time-intensive preparation process and the high price of nylon. This is why we aimed to develop nanocomposites using polypropylene, one of the conventional polymers, as a matrix.

In order to develop polymer/clay nanocomposites as engineering plastics with high mechanical properties, at least two requirements must be satisfied. One is to separate the stacked layer structure of clay, into monolayers if possible, in a polymer matrix. The aspect ratio of the filler particle increases and, as a result, the reinforcement effects are enhanced [3, 4]. Another is to control the interfacial affinity between a clay particle surface and the matrix polymer [5, 6]. We also thought that the type of clay would influence the mechanical properties of the resulting nanocomposites.

Until now, two techniques have been used to prepare useful clay/polymer nanocomposites. In the technique developed by the Toyota group, the polymerizing monomer is intercalated in the interlayer space of clay and is polymerized there [1,2]. Another technique developed by Giannelis *et al.* [7, 8] uses the intercalation of polymers between clay layers directly through heating the mixture. However, only polymerizing cations or polar monomers are applicable for the former technique, and only polymers with large polarities are suitable for the latter. The development

Polymer–clay nanocomposites Edited by T. J. Pinnavaia and G. W. Beall
© 2000 John Wiley & Sons Ltd

of new techniques was required for the preparation of a clay/polypropylene nanocomposite because of the low polarity of polypropylene.

We have now achieved polypropylene/clay nanocomposites with high mechanical properties using clays belonging to the smectite group of minerals. The ion exchange property of these minerals was the most important consideration in reaching our goal. The preparation methods and processing work leading to the development of the nanocomposites is described in this chapter.

2 METHOD USING MECHANICAL SHEAR

The ion exchange property of smectite clays may suggest a relatively weak attraction force between the clay layer and the cation intercalated between them. Therefore, the stacked layer structure of the clay might be expected to separate when subjected to mechanical shear, particularly after expanding the interlayer space through intercalating bulky cations. This was the approach explored in the mechanical shear method of nanocomposite formation.

2.1 PREPARATION PROCEDURE

Figure 1 shows the experimental procedure [9]. The clay used is montmorillonite (Kunimine Ind. Co. Ltd, trade name Kunipia-F, abbreviation MM) and its analytical data are shown in Table 1. Dimethyloctadecyl ammonium chloride was added to a 3 wt % MM aqueous suspension at twice the cation exchange capacity (CEC) of the MM used, and kept for 19 h at 80 °C with stirring to complete the ion exchange reaction. The ion-exchanged MM (abbreviation Org-MM) was washed repeatedly with methanol until chlorine ions were not detected. The powder after freeze-drying was blended with polypropylene (Sumitomo Chemical Ind. Co., trade name Noblen abbreviation PP) in a Laboplastomill (Toyo Seiki) according to the conditions shown in Figure 1. The filler content was controlled at 4 wt % as an inorganic component (pristine MM base). The composite was finally subjected to hot pressing at 10 MPa for 1 min and 15 MPa for 3–4 min to prepare films of ca 0.8 mm thickness. As references, PP films with and without 1 wt % pristine MM were prepared according to the procedure shown in Figure 1.

2.2 STRUCTURE OF COMPOSITE

Figure 2 shows X-ray diffraction profiles of the films. The PP film containing 1 wt % MM showed a weak and broad peak at 5.8 ° (2θ, CuK_α), corresponding to a 1.52 nm interlayer distance, d_{001}, of MM. This peak results from MM containing interlayer cations (Na^+) coordinated with water molecules. The d_{001} value of MM was expanded to 3.84 nm after ion-exchange with the dimethyloctadecyl ammonium cation (not shown in Figure 2.). Thus, we expected the expanded layer structure to be

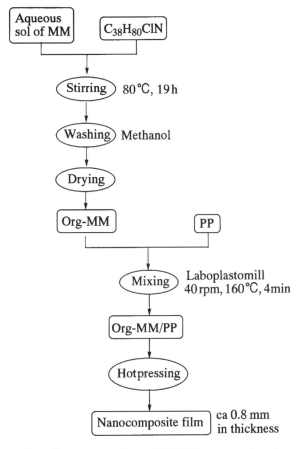

Figure 1 Procedure for preparation of MM/PP composites by a mechanical blending method

Table 1 Chemical analysis (wt %) of Kunipia-F clay

SiO_2	61.3	MgO	3.43
TiO_2	0.15	CaO	0.46
Al_2O_3	21.9	Na_2O	4.06
Fe_2O_3	1.90	K_2O	0.12
FeO	0.31	P_2O_5	0.02
MnO	0.01	Ig. loss	6.24
CEC	115 meq./100 g		

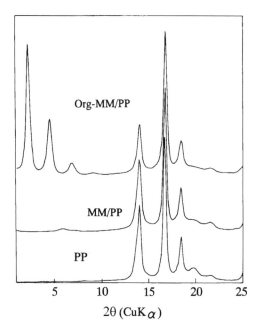

Figure 2 X-ray diffraction profiles of PP, MM/PP and Org-MM/PP

separated by mechanical shear after blending with PP with a Laboplastomill. However, the PP composite prepared with a Laboplastomill clearly showed three strong and sharp (001) peaks, corresponding to the d_{001} value of the Org-MM. It was concluded that the layer structure of MM, even if the interlayer distance was expanded greatly, cannot be separated by mechanical shear. In other words, the technique using mechanical shear will not be effective for the preparation of a clay/polymer nanocomposite.

3 METHOD USING SOLVENT

It is known that clay intercalated with organic cations is hydrophobic and disperses thoroughly in some organic solvents to give a transparent sol. We thought that a clay/PP nanocomposite would form after PP was dissolved together with the organoclay in the suspension, with subsequent removal of the solvent. We supposed, in addition, that the chemical affinity between the clay surface and the matrix PP would be low because of the low polarity of PP in contrast to the high polarity of the clay surface. This might lead to poor mechanical properties, even if a nanocomposite were prepared. Consequently, we also examined how a methyltrimethoxysilane

(MTMS) treatment of the clay surface influenced the mechanical properties of the resulting composite.

3.1 PREPARATION PROCEDURE

The preparation procedure is shown in Figure 3. The clays used were a hydrophobic hectorite (Corp Chemical Co., abbreviation SAN) and a hydrophilic hectorite (Corp Chemical, abbreviation SWN). The analytical data for these clays are shown in Table 2. The ideal chemical formula of SWN is $Na_{0.33}(Mg_{2.67}Li_{0.33})Si_4O_{10}(OH)_2$. SAN was prepared through ion exchange of Na^+ in SWN by $[(CH_3)_3NR]^+$. Here R indicates $-C_{16}H_{33}$ and $-C_{18}H_{37}$, but the ratio of the alkyl chains is not specified.

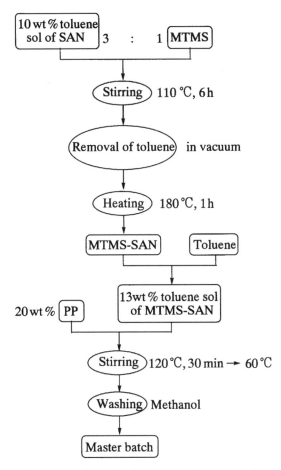

Figure 3 Preparation procedure for the composite master batch, consisting of MTMS-treated SAN and PP, by using a solvent

Table 2 Chemical analysis (wt %) of SWN, SAN and MAE clays

	SWN	SAN	MAE	MAN
SiO_2	54.3	38.6	41.2	35.8
MgO	27.9	18.7	18.8	2.0
Al_2O_3	—	—	—	12.8
Fe_2O_3	—	—	—	1.1
Na_2O	3.4	—	1.3	—
Li_2O	1.3	1.0	—	—
F	—	—	4.9	—
Organic	—	38.3	33.8	46.0

After drying at 60 °C, 10 wt % SAN was thoroughly dissolved in toluene to give a transparent sol. At higher temperatures the SAN changed from white to pale brown. The MTMS was added to the sol at one-third the sol by weight, followed by stirring at 110 °C for 6 h for MTMS treatment. In this process, $CH_3(OCH_3)_3Si$ is expected to react with hydroxyl (–OH) groups on the SAN particle edges to form a $CH_3(OCH_3)_2Si$–O–SAN bond, lending a hydrophobic property to the SAN surface. The MTMS-treated SAN (abbreviation MTMS-SAN), after removal of toluene under reduced pressure, was heated at 180 °C for 1 h to remove unreacted polar –OH groups on the SAN surface.

Figure 4 Composite specimens prepared for measurements

The MTMS-SAN was dissolved in toluene again to give a 13 wt % sol, mixed with 20 wt % conventional PP (1 : 1 mixture of J700GP and H700 supplied by Calp Corporation) and stirred at 120 °C for 30 min to homogenize the mixture. The solution was dropped into methanol after cooling to 60 °C with stirring, and finally washed with methanol repeatedly to remove the unreacted MTMS. The MTMS-SAN/PP composite thus prepared was used as a 'master batch'.

The master batch after drying was kneaded with the PP and shaped into molds (shown in Figure 4) under 350 kPa at 210 °C using an injection molding machine (Nissei-Jushi Kogyo FS-150). The MTMS-SAN content in the composite was adjusted to 3 wt %. Two samples were prepared as references. They were the PP reinforced with pristine SAN and SWN, respectively. The former was prepared by almost the same procedures used for the MTMS-SAN/PP composite, and the latter was blended mechanically without the solvent.

3.2 STRUCTURE AND PROPERTIES OF THE COMPOSITE

3.2.1 Methyltrimethoxysilane Treatment

Figure 5 shows IR spectra of the SAN before and after MTMS treatment. Some characteristic absorptions are assigned. The absorptions assigned to v CH ($2848 \, cm^{-1}$, $2920 \, cm^{-1}$) were observed in SAN but not in the SWN. This difference shows that SAN includes quaternary ammonium cations. It is also clear from the appearance of absorptions at $780 \, cm^{-1}$ (v Si–C), $1270 \, cm^{-1}$ (v Si–C) and $856 \, cm^{-1}$ (δ Si–O) in the MTMS-SAN that the MTMS treatment was successfully carried out. The absorption for δ Si–O (possibly assigned to δ Si–OH) in MTMS-SAN disappeared after heating to 180 °C, as expected.

3.2.2 Separation of the Stacked Layer Structure in a Solvent

Figure 6 shows the X-ray diffraction profiles of SAN before and after dissolving in toluene. The former SAN showed a strong (001) diffraction peak, corresponding to 2.26 nm in d_{001}. After dissolving in toluene (SAN content 7.8 wt %), the strong (001) peak of the SAN disappeared and a broad peak around 18 ° (2θ, CuK_{α}) appeared instead. In this work, a thin polyethylene terephthalate film (Mylar film) was used to keep the sol in the sample holder. The SAN/toluene xerogel (SAN content 96 wt %) derived from the sol through evaporating toluene showed a strong (001) diffraction peak, at 2.12 nm, together with some unknown peaks. In order to obtain further information on this structure, the SAN/toluene sol was subjected to X-ray diffraction at a lower angle. As shown in Figure 7, three peaks were observed at angles lower than ~ 5 °. On the basis of the d values at 5.19, 2.54 and 1.64 nm [3], they were identified as the (001), (002) and (003) peaks from the expanded layered structure, respectively.

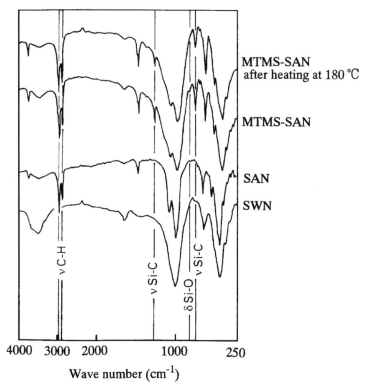

Figure 5   IR spectra of SWN and SAN before and after MTMS treatment

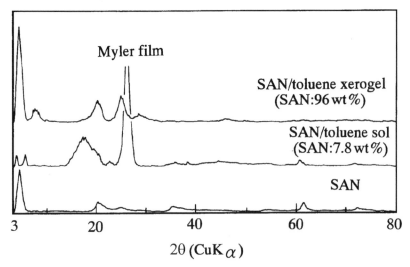

Figure 6   X-ray profiles of SAN, SAN/toluene sol and SAN/toluene gel

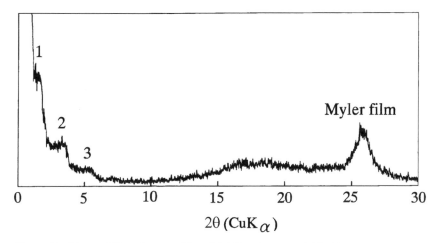

Figure 7 X-ray diffraction profile of SAN/toluene sol (SAN content 10 wt%)

3.2.3 Structure of the Composite

Table 3 includes the total light transmission value which reflects intimately the dispersion state of filler particles in the matrix PP. This property was measured using a color computer (Suga Shikenki Co., SH-3) according to Japan Industrial Standard K7105. The MTMS-SAN gave a large value similar to that of the PP without filler. The lowest value was obtained for SWN/PP, which means that the mechanical blending is less effective than that using a solvent.

Figure 8 shows EPMA photographs of the PP with and without filler. The white spots in the composites indicate silicon constituting the clay mineral. Large SWN particles were seen in the SWN/PP sample. As suggested from the small total light transmission value, the SAN particles in the other two specimens dispersed more homogeneously throughout the PP matrix.

The problem is how to disperse the SAN particles in the PP by a simple technique. The SAN dissolved readily in toluene, resulting in a transparent sol, but it still retained its stacking layered structure with a large interlayer spacing of ~ 5 nm. This structure returned to the original stacking structure after toluene was removed.

Table 3 Properties of PP with and without fillers

	PP	SWN/PP	SAN/PP	MTS-SAN/PP
Total light transmission (%)	80.1	49.4	66.2	76.0
Tensile strength (MPa)	35.9	34.4	36.1	34.3
Tensile modulus (MPa)	1570	1630	1720	1620
Bending strength (MPa)	40.3	42.8	45.3	43.1
Bending modulus (MPa)	1490	1640	1800	1650

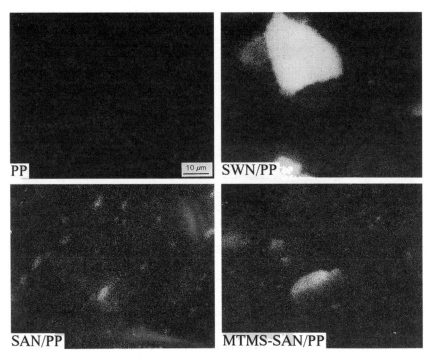

Figure 8   EPMA photographs of PP and its composites

We expected that the PP molecules would penetrate into the interlayer space expanded by toluene and remain there after removal of toluene, but this was not the result. It is not clear whether the PP molecule could not enter into the interlayer space, or if, upon entering, the PP molecule was removed from the interlayer space together with the toluene.

3.2.4   Mechanical Properties of the Composite

The bending and tensile strengths of the specimens were measured using a Bendgraph (Toyo Seiki Co.) and a Strograph (Toyo Seiki Co.) according to ASTM-D790 and ASTM-D638, respectively.

Figure 9 shows load/elongation curves of the PP plates with and without the fillers. No difference was observed among the curves. Some mechanical properties are listed in Table 3. The fillers did not influence the tensile properties favorably, a result different from the case of bending properties. The SAN without MTMS treatment gave the most favorable reinforcing effect.

The SAN was confirmed to have been trimethylsilylated through the appearance of absorptions assigned to v Si–C vibrations. Also, the resulting MTMS-SAN

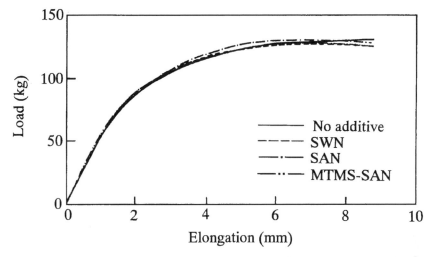

Figure 9   Load/elongation curves of plates of PP and its composites

dispersed more finely in the PP than the SAN without the treatment, as suggested from its larger total light transmission value. The mechanical properties of the SAN composite, nevertheless, were superior to those of the MTMS-SAN composite, as shown in Table 3, completely different from our expectations. We have no data evaluating the affinity of the polymer for the clay surface, but we expect that the predominant effects will appear when the MTMS-SAN particle is separated into the more thin flake-like nanoparticles because of the increasing interfacial contact area with the PP matrix. Such a state should also result in an increase in the aspect ratio of the filler particle, leading to enhancement of mechanical properties of the composite [3].

The dispersion state of the SAN by the present method was superior to that by the conventional mechanical blending technique. Nevertheless, there were almost no differences among their mechanical properties. The far finer separation of the SAN particles, if possible into single nanolayers, is required to improve the mechanical properties of the resulting composite.

4   NEWLY DEVELOPED METHODS

As stated above, a stacked layer structure of clay is so difficult to separate in a non-polar PP matrix that we designed the following method to reach our goal [10, 11]. The process consists of (i) intercalating a small amount of polymerizing polar monomer between the clay layers to lengthen the interlayer distance and then to polymerize therein to avoid the polymer escaping from the interlayers, (ii) intercalating polar maleic acid-modified polypropylene into the interlayer space to make a composite used as a 'master batch' and finally (iii) blending the master batch with a conventional PP to prepare a nanocomposite. Maleic acid-modified polypropylene (m-PP) was expected to lengthen the interlayer distance further and also to improve the affinity between the clay and PP, possibly leading to an improvement in mechanical properties. Later it was revealed that a nanocomposite can be prepared simply by mixing clay with m-PP in solvent under heating. The structures and properties of the composites prepared are described herein. In particular, the effects of a small amount of polymerizing monomer and the kind of clay on the mechanical properties of the resulting composites are noted.

4.1 PREPARATION PROCEDURES

Two clays in addition to SAN were used. A hydrophobic montmorillonite (Kunimine Ind. Co., abbreviation MAN), shown in Table 2, was derived from natural montmorillonite Kunipia-F (cf. Table 1). This clay contained 46 wt % of the same quaternary ammonium cations used to prepare the SAN through ion-exchange reactions. Also, a hydrophobic mica (abbreviation MAE) was prepared by Corp Chemical Ind. Co. The interlayer K^+ ions in the synthesized mica with an ideal chemical formula $K_xMg_{3-x/2}Si_4O_{10}F_2$ ($x = 0.5$–0.8) was ion exchanged with the same quaternary ammonium cations used in the preparation of SAN. The analytical data are shown in Table 2.

Four clay/PP composites were prepared by the two preparation methods shown in Figure 10, in addition to the conventional PP without the filler and the talc/PP composite used for the reference. The composites prepared by the two methods are distinguished by appendixes (A) or (B):

(a) SAN–PP(A): method (A) is characterized by the use of a polymerizing polar monomer, diacetone acrylamide [$CH_2 = CHCONHC(CH_3)_2CH_2COCH_3$, abbreviation DAAM] and maleic acid-modified PP with 10 wt % maleic acid (YUMEX 1010 supplied by Sanyo Chemical Industries, abbreviation m-PP) [10]. A toluene solution of DAAM was mixed with a toluene solution of SAN in which a polymerization catalyst [2,2′-azobis(isobutyronitrile $(CH_3)_2C(CN)N=NC(CN)(CH_3)_2$, abbreviation AIBN] was intercalated in advance, kept for 1 h at 30 °C for intercalation of DAAM and subsequently kept at 75 °C for 1 h to polymerize DAAM between the clay layers. During the treatment, the solution was stirred under an N_2 atmosphere. The toluene solution after the polymerization treatment was mixed with a toluene solution of m-PP and heated to 100 °C with stirring for intercalation of m-PP. The solution after cooling to 60 °C was dropped into methanol, and the deposited pasty material was washed repeatedly with methanol to remove free DAAM and m-PP, and finally dried to obtain a 'master batch'. The master batch was blended with conventional polypropylene (a 1 : 1 mixture of J700GP and H700 supplied by

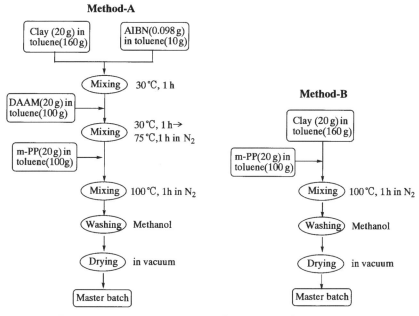

Figure 10 Two preparation procedures for a clay/PP nanocomposite master batch

Calp Corporation, abbreviation PP) with a twin screw extruder (Ikegami, PCM30), followed by molding into specimens as shown in Figure 4 with a Nissei Jyushi Kogyo FE120S.

(b) SAN/PP(B): toluene solutions of SAN and m-PP were mixed at 100 °C for 1 h under an N_2 atmosphere. The subsequent processes were the same as used for the SAN/PP(A) composite.

(c) MAN/PP(A): the composite was prepared according to the same processes for SAN/PP(A) except for the use of MAN instead of SAN [6].

(d) MAE/PP(B): MAE was used as a filler. The procedures were the same as those used for SAN/PP(B).

4.2 COMPOSITE STRUCTURE AND PROPERTIES

4.2.1 Structures of Pristine Clays

The X-ray diffraction profiles of the clays are shown in Figure 11, and the X-ray parameters calculated from the (001) peaks are summarized in Table 4. SAN gave the broadest (001) peak among the three clays, and its (002) peak was very equivocal. Before starting the present work, we supposed the sharpest diffraction

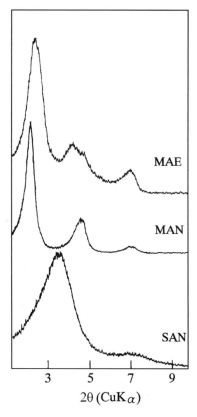

Figure 11 X-ray diffraction profiles of the pristine clays

Table 4 Interlayer distances, d_{001}, and crystalline thicknesses, L_c, of pristine clay minerals and those in the nanocomposites

	d_{001} (nm)	L_c (nm)
SAN	2.49	10.0
SAN/PP(A)	3.98	11.4
MAN	3.80	35.3
MAN/PP(A)	3.80	18.5
SAN	2.49	10.0
SAN/PP(B)	2.26	9.1
MAE	3.43	19.6
MAE/PP(B)	2.69	18.6

peak would be given by MAE, but MAN showed the sharpest peak, as shown in Figure 11.

There were large differences among the d_{001} values of the three clays in spite of being intercalated with identical quaternary ammonium cations. A clay with a larger d_{001}, in general, has a larger crystallite thickness, L_c. The average number of layers constituting a clay crystallite, L_c/d_{001}, is 4.0, 9.3 and 5.7 for SAN, MAN and MAE, respectively.

Figure 12 shows the TEM images of the pristine clays. Well-developed crystalline layer structures were observed in both MAN and MAE, whereas the SAN showed a loosely stacked layer structure. Comparison of the three micrographs suggests that the SAN layers are somewhat smaller in width and more flexible, in other words less stiff, than the other two layers.

4.2.2 Structures of Nanocomposites

Figure 13 shows X-ray diffraction profiles of the prepared nanocomposites. The diffraction peaks of clays in the composites are broader than those of the corresponding pristine clays. This difference means that the stacked layer structures of pristine clay were separated into thinner platelets through the composite preparation processes. This figure also indicates that the (001) X-ray diffraction profile became broader in the order MAN/PP(A) < MAE/PP(B) < SAN/ PP(A) < SAN/PP(B), the last one showing no peak. This order indicates that a lower crystalline clay structure is separated more easily.

Table 4 shows X-ray parameters of the clays in the composites. Since the content of clay was controlled to be just 3 wt %, the diffraction peaks were quite weak, leading to somewhat unreliable X-ray parameters. Nevertheless, one tendency can be deduced from Table 4, that is, the clays in the composites by method A have larger or nearly equal d_{001} values compared with those of the pristine ones. Opposite results were observed in the case of method B. The formation of polymerized DAAM between the layers suppressed the shrinkage of the interlayer distance during the nanocomposite preparation process.

Figure 14 shows the TEM images of the nanocomposites. As suggested from the X-ray diffraction profiles, well-separated layers are observed in both SAN/PP(A) and SAN/PP(B). Many of them are stacked in two or three layers, and monolayers are also observed. The layers are nearly equal in length to the pristine clay, so that the SAN layers were not cut, though separated, through the preparation processes. The SAN layers in the composite were relatively shorter and seemed to be less stiff than those of MAN and MAE as well as in the cases of the pristine clays.

The TEM micrographs showed that the thickly stacked layer structures were separated into thinner ones in MAN/PP(A) and MAE/PP(B) through the nanocomposite preparation processes, though not so perfectly as in SAN/PP(A) and SAN/PP(B). These separated layer structures are strictly flat, without bending or wrinkling, and seem to be stiffer than the SAN layer.

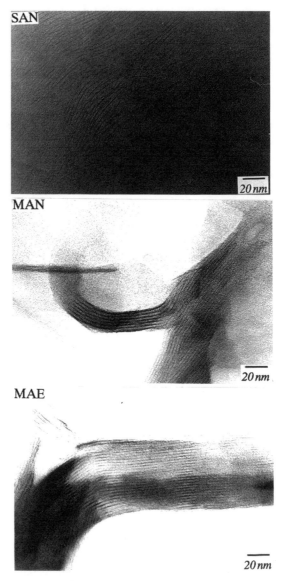

Figure 12 TEM micrographs of the pristine clays

4.2.3 Behavior of Diacetone Acrylamide

Figure 15 shows IR spectra of DAAM after the treatments for SAN/PP(A) preparation. The DAAM monomer intercalated in SAN has two characteristic absorptions at $3088 \, \text{cm}^{-1}$ (ν C–H) and $1624 \, \text{cm}^{-1}$ (ν C=C). They were not observed

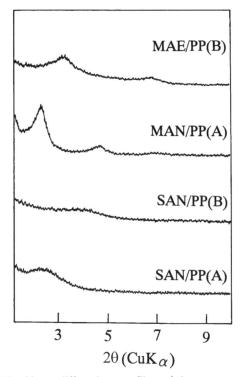

Figure 13 X-ray diffraction profiles of the nanocomposites

after the polymerization treatment, showing the success of polymerization. It is certain that polymerized DAAM exists between the clay layers, because (i) the absorptions of v C=O (1710 and 1680 cm^{-1}) and δ N–H (1540 cm^{-1}) were observed and (ii) the d_{001} of DAAM-intercalated SAN after polymerization was larger than that for SAN by 1.14 nm. Both characteristic adsorptions became weak after blending with m-PP and washing with methanol. In fact, about 6.5 wt % of the polymerized DAAM was obtained in the extract after evaporation of methanol, but no change was observed in X-ray diffraction profiles. We think that m-PP was intercalated instead of the polymerized DAAM.

4.2.4 Properties of Nanocomposites

The properties of the nanocomposite, together with those of the conventional PP and a talc/PP composite, are summarized in Table 5. The filler content was controlled to be 3 wt % in all composites. It should be emphasized that clays exhibited more desirable reinforcement effects than talc which is a conventional reinforcement material. Maiti and Sharma [12] reported that the tensile strength and tensile

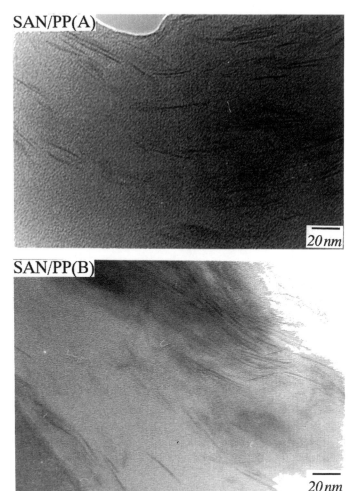

Figure 14 TEM micrographs of the nanocomposites

modulus of a conventional talc/PP composite decreased and increased with an increase in the talc content, respectively, although values were not reported concretely. The data in Table 5 are discussed from the two points below:

(a) *Effects of polymerized DAAM.* As explained by the results in Table 4, the polymerized DAAM resulted in a large interlayer distance for the clay after nanocomposite preparation. This behavior, we expected initially, leads to a preferable separation of the stacked layers in the resulting nanocomposite. Based on the data shown in Figure 13, however, method B showed a more diffused

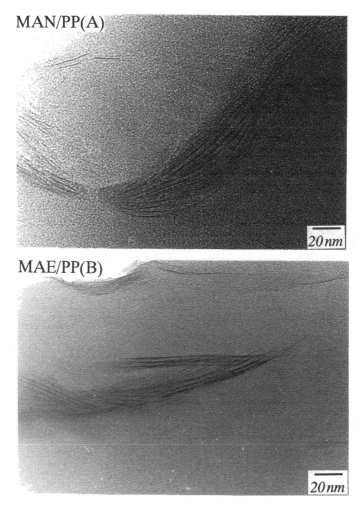

Figure 14   (*Continued*)

X-ray diffraction profile. Thus, we concluded that the presence of polymerized DAAM may or may not lead to a more preferable separation of clay stacked layers. Another point to clarify is the effect of polymerized DAAM on the properties of the nanocomposite, which will be revealed through a comparison between SAN/PP(A) and SAN/PP(B). Both composites gave almost equal values in all properties except for notched IZOD impact strength. It can be concluded that the presence of polymerized DAAM has no substantial effect on the properties of the nanocomposite. This technique could be used for preparation of a clay/polymer nanocomposite.

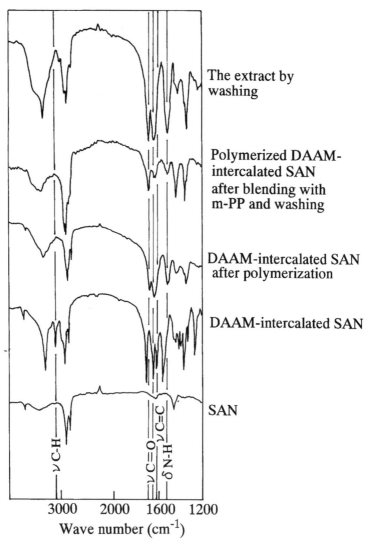

The extract by washing

Polymerized DAAM-intercalated SAN after blending with m-PP and washing

DAAM-intercalated SAN after polymerization

DAAM-intercalated SAN

SAN

Figure 15   IR spectra of SAN before and after intercalation with DAAM

(b) *Clay effects.* Higher reinforcement effects were achieved for both MAN/PP(A) and MAE/PP(B) than for the SAN-reinforced composites in spite of the poorer separation of the stacked layer structures. These results are completely opposite to our initial expectation. There must be other stronger factors influencing the properties of nanocomposites. The effects of size and stiffness of a clay layer were suggested.

Table 5 Properties of PP with and without fillers

	PP	Talc/PP	SAN/PP(A)	MAN/PP(A)	SAN/PP(B)	MAE/PP(B)
Content of inorganic matter (wt %)	0	3	3	3	3	3
DAAM	—	No	Yes	Yes	No	No
Density ($g\,cm^{-3}$)	0.91	0.92	0.93	0.93	0.93	0.93
Tensile strength (MPa)	31	35	37	39	35	39
Bending strength (MPa)	38	45	48	53	49	55
Bending modulus (Mpa)	1500	1900	2000	2100	2020	2500
IZOD impact strength (with notch) ($kJ\,m^2$)	2.0	2.1	2.3	3.4	2.9	3.9
HDT ($^\circ C$)[a]	120	—	125	130	125	132

[a] Heat distortion temperature.

It is well known that a clay filler with a larger aspect ratio exhibits more preferable reinforcement effects on the resulting composite [4], but we could not measure the crystallite size, L_c, from X-ray diffraction profiles of the composites. Another way is to use the TEM micrographs. Both the MAN and MAE have larger layer lengths but simultaneously have thicker layer stacking. Therefore, it was also difficult to evaluate the aspect ratios of clays in the composites. Another possibility is the stiffness of clay layers. No data on the stiffness of clay layers have been available until now. However, it is reasonable to assume from the results stated above that the MAN and MAE layers will be more stiff than the SAN layer. The properties of the nanocomposite are controlled more strongly by the stiffness of the clay layer. This idea will be supported by a large bending modulus of MAE/PP(B), because this property, in general, is improved remarkably by using a stiff reinforcement material. As stated above, data on the stiffness of clays are not currently available. When the data in Table 5 are considered on the basis of the above discussion, MAE must be stiffer than MAN.

As a result, a nanocomposite with excellent mechanical properties is prepared through preferable separation of stiff clay layers throughout the matrix.

5 ACKNOWLEDGEMENTS

This work was carried out with Mr Y. Kurokawa, H. Yasuda and M. Kashiwagi, whom the present author thanks. Also, the author wishes to thank Kunimine Ind. Co. Ltd and Corp Chemical Co. for supplying clay samples.

6 REFERENCES

1. Fukushima, Y., Okada, A., Kawasumi, M., Kurauchi, T. and Kamigaito, O. *Clay Miner.*, **23**, 27 (1988).
2. Usuki, A., Kojima, Y., Kawasumi, M., Okada, A., Fukushima, Y., Kurauchi, T. and Kamigaito, O. *J. Mater. Res.*, **8**, 1179 (1993).
3. Padawer, G. E. and Beecher, N. *Polym. Eng. Sci.*, **10**, 139 (1970).
4. Lusis, J., Woodhams, R. T. and Xanthos, M. *Polym. Eng. Sci.*, **13**, 139 (1973).
5. Cai, S., Ji, G., Fang, J. and Xue, G. *Angew. Macromol. Chem.*, **179**, 77 (1990).
6. Garton, A., Kim, S. W. and Wiles, D. W. *J. Polym. Sci., Polym. Lett. Ed.*, **20**, 273 (1982).
7. Giannelis, E. P. *Adv. Mater.*, **8**, 29 (1996).
8. Messersmith, P. B. and Giannelis, E. P. *Chem. Mater.*, **6**, 1719 (1994).
9. Furuichi, N., Kurokawa, Y., Fujita, K. and Oya, A. *J. Mater. Sci.*, **31**, 4307 (1996).
10. Kurokawa, Y., Yasuda, H. and Oya, A. *J. Mater. Sci. Lett.*, **15**, 1481 (1996).
11. Kurokawa, Y., Yasuda, H., Kashiwagi, M. and Oya, A. *J. Mater. Sci., Lett.*, **16**, 1670 (1997).
12. Maiti, S. N. and Sharma, K. K. *J. Mater. Sci.*, **27**, 4605 (1992).

9

Polyethylene Terephthalate–Clay Nanocomposites

T.-Y. TSAI

Inorganic and Solid State Chemistry Department, Applied Chemistry Division, Union Chemical Laboratories (UCL) of Industrial Technology Research Institute (ITRI), Hsinchu, Taiwan

1 INTRODUCTION

Nanocomposites are composite materials that contain particles in the size range 1–100 nm. These materials bring into play the submicron structural properties of molecules. These particles generally have a high aspect ratio and a layered structure that maximizes bonding between the polymer and particle. Adding a small quantity of these additives (0.5–5 %) can increase many of the properties of polymer materials, including higher strength, greater rigidity, higher heat resistance, higher UV resistance, lower water absorption rate and lower gas permeation rate, while at the same time maintaining transparency and allowing for recycling. This chapter deals with the synthesis and characterization of polyethylene terephthalate (PET)–clay nanocomposites.

Commercial nanocomposite products first appeared in 1990. Commercially produced products include the nylon series nanocomposites of Ube and Unitika which are used as engineering plastics and low-permeability packaging films. Other product applications include use as crystal nucleating agents, compatibilizers, heat-resistant additives, compounding aids for polymer alloys, filaments for tire cord, reclosable fasteners and industrial brushes.

Nanocomposite technology has developed very quickly all around the world. In terms of patents, before 1995 only around 10 patents were issued every year. Since 1995 the number has increased to 80–120 patents. The world's leading petrochemical, plastic, rubber and resin companies have energetically engaged in research and

Polymer–clay nanocomposites Edited by T. J. Pinnavaia and G. W. Beall
© 2000 John Wiley & Sons Ltd

development. For example, Toyota Central R&D Labs of Japan has developed nanoacrylic paint and nanorubber [1]. Allied Signal of the US has developed a compounding process for nanodispersed organoclay to produce nylon 6/clay nanocomposites [2]. Showa Denko has developed engineering nylon 66/clay nanocomposites [3]. Mitsubishi Chemical, inc. has developed a nanodispersed manufacturing process for engineering plastics based on a metallocene catalyst carrier [4]. GE Plastics Company has focused on developing a compounding process for nanodispersed clay in a PP matrix [5]. Eastman Chemical Company [6], The Dow Chemical Company and the National Institute of Standards and Technology (NIST) have studied polyester/clay nanocomposites for engineering plastic applications. Clay minerals companies, such as Taiwan's Pai-Kong Ceramic Materials Company, Amcol [7–13] and Southern Clay Products [14, 15] of the US, have also devoted research on treated clays to pursue nanocomposite projects and other applications.

Union Chemical Laboratories (UCL) of the Industrial Technology Research Institute (ITRI) has also been working on polymer/clay nanocomposites to develop a novel process for nylon 6/clay nanocomposites, and polyaniline/clay, PET/clay, polystyrene/clay, ABS/clay, polyvinylester/clay and epoxy/clay nanocomposites.

2 DEVELOPMENT REVIEW AND PATENT ANALYSIS OF PET/CLAY NANOCOMPOSITES

Research on polyester/clay nanocomposites is at an early stage and to date no products have entered the market. However, a number of companies are busy with development and have been issued numerous patents over the last few years (Table 1).

There are two methods for preparation of PET/clay nanocomposites. One is where the smectites are treated and added to the PET monomer, followed by polymerization. The other is to melt-mix PET with treated clays by compounding processes. Because the polymerization of PET is a two-phase reaction, inserting the monomer into the interlayer of clays to carry out the reaction of polymerization is very difficult. PET monomers, ethylene glycol (EG) and terephthalic acid (TPA), are polar chemical compounds. In the polymerization process, the polarity decreases as the molecular weight increases, which results in phase separation of the clay and polymer; therefore, increasing the compatibility between the clays and polymer is important. The scope of patents includes clay treatment methods, nanocomposite formation and manufacturing process and processing application. One of the central issues in all of these efforts is the modification technology of the inorganic layered materials.

There are three main types of clay treatment method. The first is the ion exchange method. Using a swelling agent such as an onium ion, ammonium salts or organometallic compounds, by means of ion exchange the clay is expanded. Then

Table 1 PET/clay nanocomposite patent survey

Patent No.	Issued date	Assignee	Intercalation method	Process	Polymer matrix	Properties
JP3-62846 [16]	03/1991	Toyota Central R&D	Onium ion	Melt/synthesis	PBT	HDT: 72–100 °C
JP7-166036 [17]	06/1995	Mitsubishi Chemical	Ammonium ion	Melt	Aromatic polyester	
JP7-252725 [20]	10/1995	Unitika	Fluoromica	Dry blend and spinning	Polyester	Fiber
JP7-316927 [21]	12/1995	Unitika	Fluoromica	Polyester/fluoromica fiber	Polyester	
JP8-73710 [22]	03/1996	Unitika	Fluoromica	Synthesis	Polyester	HDT: 98–110 °C (PET)
JP8-120071 [23]	05/1996	Unitika	Fluoromica	Synthesis	PET	HDT: 116 °C
JP8-187774 [24]	07/1996	Unitika	Fluoromica	Synthesis	PET	Film
JP8-199048 [18]	08/1996	Mitsubishi Chemical	Organic onium ion	Melt	PET	
JP9-48856 [25]	02/1997	Mitsubishi	Fluoromica	Melt	PBT	HDT = 87 °C
JP9-53004 [26]	02/1997	Nippon Ester Co. Ltd	Fluoromica	Synthesis	Polyester	T_m: 250–254 °C T_{cc}: 130–133 °C b/a: 2–4
JP9-77961 [39]	03/1997	Nippon Ester Co. Ltd	Muscovite/sericite	Dry blend/melt spinning	Polyester	Good IR radiation
JP9-143359 [40]	06/1997	Kaneka Co.	Fluoromica	Melt	PET/PC blends	Processing improvement
JP9-169893 [41]	06/1997	Unitika	Layered silicates	Synthesis	Aliphatic polyester	$T_m = 103$ °C $T_{cc} = 59$ °C
JP9-176461 [42]	07/1997	Unitika	Fluoromica	Synthesis	PET	Transparent; better mechanical and processing properties

(Continued)

Table 1 PET/clay nanocomposite patent survey (*continued*)

Patent No.	Issued date	Assignee	Intercalation method	Process	Polymer matrix	Properties
JP9-208811 [43]	08/1997	Kaneka Co.	Fluoromica/additives	Melt	Polyester	HDT: 91–120°C
JP9-208813 [44]	08/1997	Kaneka Co.	Fluoromica	Melt	Polyester copolymer	
US5514734 [19]	05/1996	Allied Signal	Organometallic species			
US5530052 [5]	06/1996	General Electric. Co.	Layered minerals/organic			
US5552469 [7]	09/1996	AMCOL	Water-soluble polymer/water: pyrrolidone	Melt	PA; PVA, PET; PBT; PVI	
US5567758 [45]	10/1996	Toyo Boseki Kabushiki Kaisha and Co-Op Chemical	Fluoromica/sulfonate terminal group		Polyester	
US5578672 [8]	11/1996	AMCOL	Intercalant polymers: aromatic ring; carboxyl; hydroxyl; amine; amide; ether; ester/water/extrusion	Melt	PA; PVA, PET; PBT; PVI	
US5669824 [9]	12/1997	AMCOL	Water-insoluble polymer	Melt	PA, PET, PBT, PVI	XRD: no peak when $d > 12.4$
US5760121 [10]	06/1998	AMCOL	Intercalant/water : polyvinyl pyrrolidone; polyvinyl alcohol; polyvinylimine	Melt	PA, PET, PBT, PVI	XRD: no peak when $d > 12.4$
US5804613 [11]	09/1998	AMCOL	Intercalant monomer (a carbonyl; a carboxylic acid or salt)/water		PA;PVA; PC; PVI; PET; PBT	
US5830528 [12]	11/1998	AMCOL	Intercalant monomer (hydroxyl; a polyhydroxyl)/water			

US5849830 [13]	12/1998	AMCOL	Intercalant [(1) an N-alkenyl amide and an allylic monomer; (2) an oligmer formed by copolymerizing an N-alkenyl amide monomer and an allylic monomer; (3) a polymer formed by copolymerizing an N-alkenyl amide monomer and an allylic monomer] /solvent	Melt	PA, PET, PBT, PVI
US5876812 [27]	03/1999	Tetra Laval		Container	PET
WO97/43343 [46]	11/1997	Kaneka	Silane	Melt	PET
WO98/29499 [6]	07/1998	Eastman Chemical		Synthesis	PET

polymerization with the monomer is carried out, or else a compounding process to mix the clay with the molten polymer. The second method uses synthetic clay or a sintered clay-like fluoromica. The third method involves the molten polymer directly diffusing into the intercalated clay layers to achieve the objective of exfoliated clays in the polymer matrixes. Patents relating to the various clay treatment methods are summarized below.

2.1 ION EXCHANGE METHODS

Japanese patent 3-62846 uses an organic onium ion to expand and swell the layered clay, which is then mixed with molten PET (or PBT) or added to an aromatic polyester monomer and then polymerized to make a polyester/clay nanocomposite. The clay used includes montmorillonite and saponite. Water absorption decreases from 1.35 to 0.34 %, and the heat distortion temperature increases from 78 to 92 °C for a PET/clay nanocomposite containing 2.5 wt % treated montmorillonite [16].

Japanese patent 7-166036 uses ammonium ions as a swelling agent to treat the layered silicates. Layered silicates include smectites (such as montmorillonite and hectorite) and synthetic fluoromica (ME100) with a cation exchange capacity of over 30 meq./100 g clay. The modified layered silicates are mixed with an aromatic polymer (such as polyalkylene terephthalate) by a molten blending method to form the nanocomposite, which has comparatively higher rigidity and toughness and a smoother surface. The proportion of layered silicates in the aromatic polyester is about 0.5–10 wt % [17].

Japanese patent 8-199048 involves exchanging organic ammonium acids into fluoromica and then compounding with PET. The result is that the crystallization rate increases. The thermal properties are shown in Table 2 [18].

US patent 5514734 (Allied Signal Inc.) uses quaternary ammonium salts in conjunction with silane coupling agents to expand the clay and then form nanocomposites with the polymer. The types of applicable polymers include polyamide polyester, halogenated polyolefin, polycarbonate, polyurethane, polyether, polyolefin, polyvinyl, polyamide, etc. [19].

Table 2   Thermal properties of PET [18]

	T_{ch} (°C)	ΔH_{cc} (J/g)	T_{cc} (°C)	ΔH_{cc} (J/g)	ΔH_m (J/g)	Ashes (wt %)	Clay type
Example 1	119	4.9	208	69	48	0.91	Hectorite
Example 2	120	6.2	207	67	50	0.95	ME-100
Example 3	121	11	212	65	52	0.92	Hectorite
Comparison 1	136	32	199	55	49	—	—
Comparison 2	131	36	201	61	53	—	—
Comparison 3	122	22	209	60	49	1.1	ME-100
Comparison 4	116	24	214	55	42	0.72	Hectorite

US patent 5530052 (General Electric Co.) exchanges a cation in the clay with a heteroaromatic cation and then adds the expanded clay to the monomer where a reaction resulting in a nanocomposite takes place. The claims of the patent cover macrocyclic oligomers, linear polymers and branched polymers [5].

2.2 SYNTHETIC CLAY METHODS

In Japanese patents 7-252725 [(2)] and 7-316927 [21], PET and fluoromica are melt mixed and then extruded into filaments; the properties are shown in Table 3.

In Japanese patents 8-73710 [22] and 8-120071 [23], fluoromica is directly added to the reacting tank where the polymerization takes place. The claims of the patents include the composition and the manufacturing process of polyester/clay nanocomposites. The composition of the polyester/clay nanocomposite contains 0.01–10 wt % expandable mica. The preparation of polyester/mica involves adding synthetic fluoromica to ethylene glycol and paraterephthalic acid, carrying out a reaction for 2 h at 255 °C and then, after formation of oligomers, reacting for 4 h at 275 °C to form PET/mica nanocomposites. The same manufacturing process can be used for PBT. The polyester/mica nanocomposites obtained using this method have better mechanical strength, toughness, heat distortion temperature and size stability, as shown in Tables 4 and 5.

Japanese patent 8-187774 [24] is a processing patent for forming film PET/fluoromica nanocomposites. The properties of PET/fluoromica are shown in Table 6.

Japanese patents 9-48856 [25] and 9-53004 [26] show that the crystallization rate of polyester increases with the amount of expandable fluoromica dispersed in the polymer matrix. After the fluoromica is treated with an onium ion and then mixed with molten polyester, the nanocomposites formed have lower T_{cc} and T_m.

2.3 DIFFUSION METHODS

These clay treatment methods have been filed mainly in a series of patents claimed by AMCOL International Corporation. US patent 5552469 show that, after disper-

Table 3 Properties of PET nanocomposite fibers [20]

	1	2	3	4	5	6
Fluoromica content (wt %)	0.1	0.5	1.2	0	0.1	0.1
Drawing speed (m/min)	3000	3000	3000	3000	1500	4000
Draw ratio	2.0	1.9	1.8	2.0	2.6	1.5
Strength (g/d)	7.4	7.2	6.7	7.3	8.5	5.7
Elongation (%)	13.0	13.3	14.0	13.5	12.5	22.0
Shrinkage ratio (%)	3.9	3.7	3.5	4.3	5.4	3.0

Table 4 Properties of PET nanocomposite (Japanese Pat. 8-73710) [22]

	PET	PET	Ex. 1	Ex. 2	Ex. 3	Ex. 4	Ex. 5
Clay		Glass fibre	M-1	M-1	M-1	M-2	M-3
Clay content (wt %)	0	43	2	5	1	2	2
Flexural strength (kg/cm^2)	730	2300	870	910	850	880	870
Flexural modulus (kg/cm^2)	23 000	100 000	31 000	36 000	30 000	31 000	32 000
Izod impact strength (kg cm/cm)	2.8	7.1	5.6	5.3	5.4	5.1	5.4
HDT ($^\circ$C)							
18.6 kg/cm^2	71	231	104	110	98	100	102
4.5 kg/cm^2	142	246	177	192	175	175	176
Shrinkage (%)	1.2	0.6	0.8	0.7	0.9	0.8	0.8
Thermal expansion coefficient ($\times 10^{-5}/\,^\circ$C)	9.1	3.1	7.6	6.3	8.3	7.6	7.4
Warpage (mm)	0.6	1.3	0.4	0.4	0.5	0.4	0.4
Gloss (%)	91.6	82.4	91.2	91.3	91.3	91.5	91.4
Recrystallization temperature ($^\circ$C)	140	134	125	120	128	124	123

Table 5 Properties of PET nanocomposite (Japanese Pat. 8-73710) [22]

	PBT	PBT	Ex. 1	Ex. 2	Ex. 3	Ex. 4	Ex. 5
Clay	No	Talc	M-1	M-1	M-1	M-5	M-6
Clay content (wt %)	0	2	2	5	1	2	2
Flexural strength (kg/cm^2)	850	840	980	1050	970	940	980
Flexural modulus (kg/cm^2)	24 000	25 000	33 000	41 000	33 000	33 000	32 000
Izod impact strength (kg cm/cm)	5.2	3.1	6.4	6.5	6.4	6.2	6.4
HDT ($^\circ$C):							
18.6 kg/cm^2	54	55	124	141	121	121	122
4.5 kg/cm^2	153	155	188	193	185	190	185
Shrinkage (%)	1.7	1.8	1.3	1.1	1.3	1.3	1.3
Thermal expansion coefficient ($\times 10^{-5}/\,^\circ$C)	10.6	10.0	7.8	7.3	7.7	7.7	7.6
Warpage (mm)	0.8	0.9	0.6	0.4	0.5	0.6	0.6
Gloss (%)	92.3	90.1	92.1	91.8	92.0	91.8	92.4
Recrystallization temperature ($^\circ$C)	—	—	—	—	—	—	—

Table 6 Properties of PET nanocomposite film [24]

	PET	PET	Ex. 1	Ex. 2	Ex. 3	Ex. 4	Ex. 5
Clay	No	Talc	M-1	M-1	M-1	M-5	M-6
Clay content (wt %)	0	2	2	5	1	2	2
Tensile strength MD/TD (kg/mm^2)	22/23	21/22	24/25	22/24	23/24	25/27	24/25
Elongation (%)	115/100	100/80	110/90	100/100	120/110	110/90	110/100
CO_2 permeation rate (cm^3/m^2, 24 h)	60	30	25	20	30	25	25
Thermal shrinkage (%)	1.5/0.5	1.0/1.0	0.7/0.5	0.5/0.5	1.0/0.5	0.7/0.4	0.7/0.5
Hz	2.0	8.0	2.4	2.8	1.6	2.4	2.4
Puncture strength (kg/mm)	60	55	80	85	75	80	75
Friction coefficient	> 1	0.5	0.3	0.2	0.4	0.3	0.3

sing layered clay in a water-soluble polymer, nanocomposites are formed by mixing the molten thermoplastics or thermoset plastics with these base materials. In the process of expanding the clay, water or another solvent as a carrier is required. Water-soluble polymers include polyvinyl pyrrolidone, polyvinyl alcohol and polyvinylimine [7].

US patent 5578672 relates to mixtures of phyllosilicate, water and polymer with functional groups (such as aromatic ring, carboxyl, hydroxyl, amine, amide, ether and ester, etc.) that are extruded through a die. This type of nanocomposite has the functional group intercalated into the clay layers. The d spacing between the clay layers can increase by 20 Å to 100 Å. With clay as the standard, the water content is at least 20 % and the high polymer content is at least 16–80 %. Although in this patent the melt-mixing method is used to make nanocomposites, nanocomposite mechanical properties or data on other properties are provided [8].

US patent 5698624 monomers with a benzene ring, hydroxyl group, carboxyl group, or low molecular weight polymers are intercalated into the clay layers using non-aqueous solvents. After the clay is expanded, it is mixed with the polymer. The range of suitable polymers includes polyamide polyesters. XRD results illustrated that the clay basal diffraction peak disappeared [9].

US patent 5760121 is the extension of 5578672. The clay is expanded with oligomers or polymers in the solvent and then mixed with molten thermoplastics or thermoset plastics to form nanocomposites. The claims of the patent include composition and manufacturing methods. The polymer expanders include polyvinyl, pyrrolidone, polyvinyl alcohol and polyvinylimine [10].

US patent 5804613 mixes carbonyl and carboxylic acid functional monomers with water and clay, and this is intercalated into the clay. The expanded clay is then mixed with molten thermoplastics or thermoset polymers to form nanocomposites [11].

US patent 5830528 mixes hydroxyl or polyhydroxyl functional monomers with water and clay, and this is intercalated into the clay. The expanded clay is then mixed with molten thermoplastics or thermoset polymers [12].

Table 7   Oxygen permeation of PET/clay nanocomposites [27]

Clay content (%)	Clay aspect ratio	O_2 permeation rate $(cm^3/pack\ 24\ h\ 0.2\ atm\ O_2)$
0	N/A	0.090
1	500	0.015
1	1000	0.008
3	500	0.005
5	1000	< 0.002

Table 8   Oxygen permeation of bottles fabricated from various materials [27]

Bottle type	O_2 permeation index
Monolayer PET	100
Multilayer (3–7 layers) PET with coinjected EVOH (max. 5 %)	30–60
Multilayer (3–7 layers) PET with coinjected nanocomposite PET (max. 5 %)	30–60
Monolayer nanocomposite PET	< 30

US patent 5849830 mixes special functional monomers with water and clay, swelling the clay. The special functional monomers include:

(a) N-alkenyl amide monomer and an allylic monomer;
(b) an oligomer formed by copolymerizing an N-alkenyl amide and an allylic monomer;
(c) a polymer formed by copolymerizing an N-alkenyl amide monomer and an allylic monomer.

The clay is expanded and mixed with thermoplastics or thermoset polymers to form nanocomposites [13].

Recently, Tetra Laval Holdings and Finance Corporation issued a patent for the container processing of nanocomposite. The patent covers the container processing manufacturing process. The gas permeability rate of the clay will be lowered (as shown in Tables 7 and 8) [27].

The patent results show that PET/clay nanocomposites have outstanding mechanical properties, heat resistance and gas blocking properties. However, many bottlenecks remain (for example, color problems and clay dispersion control) which have delayed the products from reaching the market and are still to be solved.

3 CHARACTERIZATION AND MODIFICATION OF THE CLAYS

Most nanocomposite materials are initially prepared by transforming the hydrophilic clay to a hydrophobic clay. Related investigations emphasize the compatibility between clay and polymer, but overlook the diffusion of the monomer into the interlayer to proceed with polymerization. This causes most polymer/clay nanocomposites to be mainly intercalated dispersions of clay instead of exfoliated dispersions. Therefore, we have used a catalyst in a unique polymerization process to make the stratiform inorganic mineral disperse uniformly in the polymer materials and form nanocomposites. Doing so significantly enhances the mechanical properties, thermal deformation temperature, gas barrier and moisture barrier of polymer/clay nanocomposites [28].

3.1 HIGH-PURITY CLAYS

Clays from different mines in Japan (CL4), America (CL22), Taiwan (CL26), Australia (CL39) and China (CL42) and synthetic mica CL31 differ in properties. Some characteristic data are shown in Table 9 [29]. After the purifying process, three highly pure clays (CL26, CL39 and CL42), will be introduced into the market by PAI-KONG Ceramic Materials Co., Ltd. There is a large range of cation exchange capacity (CEC), from 77 to 140 meq./100 g clay, among the different mines. The cation exchange capacity of clays relates to the capacity for intercalating the modifier into the clay. It is an important factor in controlling the amount of intercalated modifier in each treated clay since the polymerization will cause a side reaction if there is too much modifier in the treated clay. The purity of the clays also requires that some metal oxides be absent since they will cause degradation during the polymerization process.

3.2 MODIFICATION OF THE PURE CLAYS

The thermal stability of the modifier is another factor determining the properties of polymer/clay nanocomposites. Different polymer/clay nanocomposites are made

Table 9 Properties of various high purity clays [29]

Sample	CEC[a]	Swelling ability[b]	pH[c]	Bulk density	d spacing (Å)	Brightness	Particle size (μm)	Surface area (m²/g)	Thermal stability (°C)
CL04	114	100	7.8	2.42	12.44	61.5	1.58	3.84	500
CL22	140	100	8.6	2.38	12.61	63.4	2.72	3.17	500
CL26	107	100	9.6	2.58	12.27	67.3	X	19.68	425
CL31	84.5	100/12	8.9	2.83	12.62	95.5	6.54	4.89	700
CL39	98	100	9.8	2.37	12.62	57.9	3.25	23.7	500
CL42	77	100	9.7	2.41	12.58	85.5	3.31	1.65	500

[a] Unit: meq. of cation exchange capacity/100 g clay.
[b] Unit: ml height/2 g clay in 100 ml H_2O.
[c] Measured from 2 g clay in 100 ml H_2O.

from different treated clays. The following is an example for polyester/clay nanocomposites [30].

A series of modified clays are prepared from cocoamphopropionate (SB) and terephthalic acid bis(2-hydroxyethylester) (BHET), as described in previous work [31]. PAI KONG Ceramic Materials Co., Ltd and Nanocor Co. provided highly pure clays for this study. Three types of clays were used: PK-802 (CL42), PK-805 (CL39) and CWC (CL22). These treated clays dispersed uniformly in the polymer materials and formed nanocomposites.

XRD analysis of these samples can be seen in Figure 1. There are impurities in clays PK-805 and CWC, such as albite and quartz. When PK-802, PK-805 and CWC are treated with antimony acetate [Sb(OAc)$_3$], the interlayer spaces are expanded by 18.18, 16.17 and 15.22 Å, respectively, as shown in Figures 2a to c. Different d spacings are observed because of the different levels and type of substitution of the natural clays. The cation exchange capacities (CEC) of PK-802, PK-805 and CWC are 77, 98 and 140 meq./100 g clay, respectively. PK-805 is

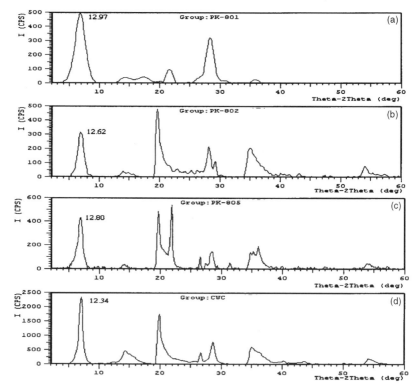

Figure 1 Powder XRD profiles for pure clays: (a) PK-801; (b) PK-802; (c) PK-805; (d) CWC [31]

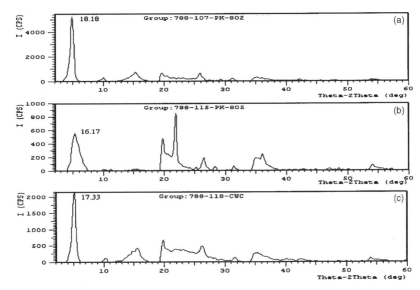

Figure 2 Powder XRD profiles for Sb(OAc)$_3$-treated clays: (a) 788-107-PK-802; (b) 788-115-PK-805; (c) 788-118-CWC [31]

nontronite [M$_x$(Fe^{3+}, Al$_2$)(Al$_x$Si$_{4-x}$O$_{10}$)(OH)$_2$.zH$_2$O], where the negative charge is distributed on the surface of the tetrahedral sheet. On the other hand, PK-802 and CWC are montmorillonite [Na$_x$(Al$_{2-x}$Mg$_x$)(Si$_4$O$_{10}$)(OH)2.zH$_2$O], so that the negative charge is distributed on the octahedral sheet. However, CWC has larger interaction with the clay interlayer than PK-802 because of the larger CEC.

Figure 3 presents the XRD patterns of the clays treated with cocoamphopropionate (SB) and Sb(OAc)$_3$. The d spacings of samples 462-037-PK-802 and 788-086-CWC are 13.19 and 12.72 Å. Some Sb(OAc)$_3$ is leached from the clay layers when SB is intercalated between the clay layers. However, for sample 788-010-PK-805 the d spacing increased to more than 23.7 Å compared with sample 788-115-PK-805 treated only with Sb(OAc)$_3$. This means that no Sb(OAc)$_3$ leached out of the clay and that SB was intercalated. Sample 788-010-PK-805 has stronger bonding strength between the clay layers and the modifying agent compared with samples 462-037-PK-802 and 788-086-CWC, because PK-805 has a tetrahedral layer charge and easily forms crosslinking with the modifying agent in the interlayer. However, PK-802 and CWC are octahedral substituted minerals and form poor crosslinking with the modifying agents. Small-angle XRD was applied to characterize PK-805 treated with Sb(OAc)$_3$ and SB, as shown in Figure 4. The d spacing was 44.1 Å.

Figure 5 compares the XRD patterns for the modified clays dispersed in the PET/clay nanocomposites. Figure 5(a) shows that an exfoliated PET/PK-805 nanocomposite is made from modified clay 788-010-PK-805. Figure 5(b) shows that an intercalated PET/CWC nanocomposite is made from SB modified clay

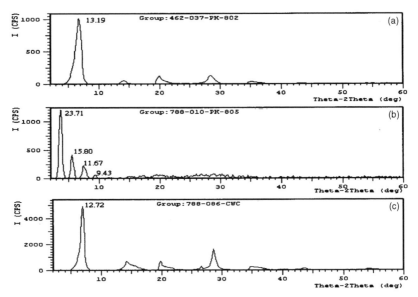

Figure 3   Powder XRD profiles for Sb(OAc)₃- and SB-treated clays: (a) 462-037-PK-802; (b) 778-010-PK-805; (c) 788-086-CWC [31]

without $Sb(OAc)_3$ treatment under the same unique polymerization process. According to Figure 5(c), modified clay 788-086-CWC, when applied to the same unique polymerization process, affords an exfoliated inorganic mineral nanocomposite.

In order to measure the clay dispersion, transmission electron micrographs of the 677-159-PK-805-5 (PET/clay) nanocomposite made from modified clay 788-010-PK-805 were obtained as shown in Figure 6. The scale bar is 475 nm. From this TEM image, we conclude that the modified clay is well dispersed and exfoliated in the PET matrix.

The function of the $Sb(OAc)_3$-treated clays is to create active sites between the clay layers for the polymerization process. Also, the purpose of the SB-modified clay

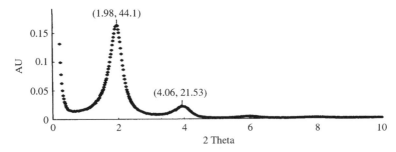

Figure 4   Small-angle XRD profile for 788-010-PK-805 [31]

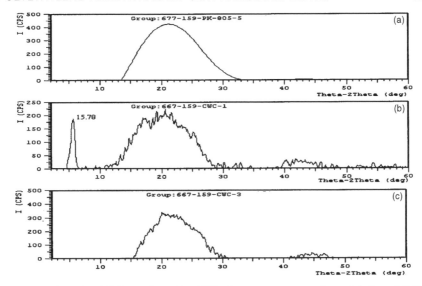

Figure 5 Powder XRD profiles for PET/clay nanocomposites: (a) 677-159-PK-805-5; (b) 667-159-CWC-1; (c) 667-159-CWC-3 [31]

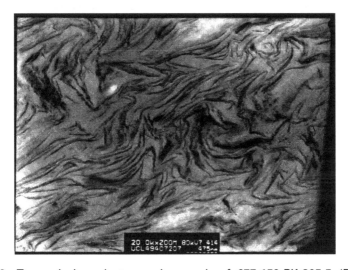

Figure 6 Transmission electron micrograph of 677-159-PK-805-5 (PET/clay) nanocomposite. Scale bar 475 nm [31]

is to increase the compatibility and to provide an opportunity for bond formation between the clay and polymer. Doing so significantly enhances the mechanical properties, thermal deformation temperature, gas barrier and moisture barrier of the PET/clay nanocomposites.

4 CONCLUSION

ITRI has played an important role in the development of PET–clay nanocomposites in Taiwan. Since 1996, the Taiwanese government has supported UCL in its polymer/clay nanocomposites program. During the period 1996–2000 there have been several important technologies established at ITRI, including process technology for clay purification, clay modification, and clay nanocomposite formation for PET and other polymer systems such as PS, ABS and epoxy.

5 ACKNOWLEDGEMENTS

The author is grateful to the Ministry of Economic Affairs of ROC for granting this nanocomposite program for four years. The author is also grateful to the program director, Dr Mao-Song Lee, for his advise, and to colleagues Dr Chien-Shiun Liao, Mr Shyh-Yang Lee, Dr Sung-Jeng Jong, Ms Chih-Lan Hwang, Dr Ren-Jye Wu, Mr Wen-Faa Kuo, Mr Gow-Long Chen, Ms Li-K. Lin, and Yi-Tsai Lai, for all their support and assistance. The author would like to thank her family (husband Chung-Chiang Lin and son Andy Lin) for all their understanding and encouragement during these four years.

6 REFERENCES

1. Okada, A. and Usuki, A. *Mater. Sci. Eng.*, **C3**, 109–115 (1995).
2. Maxfield, M., Christiani, B. R., Murthy, S. N. and Tuller, H. US Pat. 5385776 (1995).
3. Japanese Pat. 10-87878 (1998).
4. Japanese Pat. 10-60036 (1998).
5. Takekoshi, T., Khouri, F. F., Campbell, J. R., Jordan, T. C. and Dai, K. H. US Pat. 5530052 (1996).
6. James, C., Sam, R., Bobby, J., Gary, W., Joh, W. and Robert, B. WO Pat. 9829499 (1998).
7. Beall, G. W., Tsipursky, S., Sorokin, A. and Goldman, A. US Pat. 5552469 (1996).
8. Beall, G. W., Tsipursky, S., Sorokin, A. and Goldman, A. US Pat. 5578672 (1996).
9. Beall, G. W., Tsipursky, S., Sorokin, A. and Goldman, A. US Pat. 5698624 (1997).
10. Beall, G. W., Tsipursky, S., Sorokin, A. and Goldman, A. US Pat. 5760121 (1998).
11. Beall, G. W., Tsipursky, S., Sorokin, A. and Goldman, A. US Pat. 5804613 (1998).
12. Beall, G. W., Tsipursky, S., Sorokin, A. and Goldman, A. US Pat. 5830528 (1998).
13. Tsipursky, S., Beall, G. W. and Vinokour, E. I. US Pat. 5849830 (1998).
14. Dennis, H. R. WO 96/35764 (1996).

15. Gadberry, J. F., Hoey, M. and Powell, C. E. WO 97/09285 (1997).
16. Japanese Pat. 3-62846 (1991).
17. Japanese Pat. 7-166036 (1995).
18. Japanese Pat. 8-199048 (1996).
19. Maxfield, M., Christiani, B. R. and Sastri, V. R. US Pat. 5514734 (1996).
20. Japanese Pat. 7-252725 (1995).
21. Japanese Pat. 7-316927 (1995).
22. Japanese Pat. 8-73710 (1996).
23. Japanese Pat. 8-120071 (1996).
24. Japanese Pat. 8-187774 (1996).
25. Japanese Pat. 9-48856 (1997).
26. Japanese Pat. 9-53004 (1997).
27. Frisk, P. and Laurent, J. US Pat. 5876812 (1999).
28. Tsai, T.-Y., Hwang, C.-L., Wu, R.-J., Liao, C.-S. and Lee, M.-S. Taiwanese Pat., US Pat., Japanese Pat., filed (1999).
29. Tsai, T.-Y., Hwang, C.-L. and Cheng, K.-L. *UCL Annual Technol. Forum*, 93–95 (1999).
30. Tsai, T.-Y., Lee, S.-Y., Wu, R.-J. and Lee, M.-S. in Proc. 1999 Annual Meeting of Polymer Society, Kaohsiung, in press.
31. Tsai, T.-Y. and Hwang, C.-L. in Society of Plastics Engineers-Annual Technical Conference 2000 Proceedings, No. 248, in press.
32. Yeh, J.-M., Liou, S.-J., Lai, C.-Y., Wu, P.-C. and Tsai. T.-Y. *Enhancement of Corrosion Protection Effect in Electroactive Polyaniline via the In-Situ Formation of Polyaniline–Clay Nanocomposites*, submitted for publication.
33. Liao, C.-S. and Lin, L.-K. Taiwanese Pat., US Pat., Japanese Pat., pending.
34. Liao, C.-S., Kuo, W.-F. and Lin, L.-K. Taiwanese Pat., US Pat., Japanese Pat., pending.
35. Kuo, W.-F., Lee, M.-S. and Liao, C.-S. Taiwanese Pat., in press and US Pat., Japanese Pat., pending.
36. Lee, M.-S., Kuo, W.-F., Jong, S.-J. and Lee, S.-Y. in 23rd ROC Polymer Symposium, Conference Proceedings, 2000, in press.
37. Tsai, T.-Y., Jong, S.-J. and Lai, Y.-T. *Preparation and Characterization of Thermoset/Clay Nanocomposite*, in preparation.
38. Tsai, T.-Y. and Hwang, C.-L. Unpublished results.
39. Japanese Pat. 9-77961 (1997).
40. Japanese Pat. 9-143359 (1997).
41. Japanese Pat. 9-169893 (1997).
42. Japanese Pat. 9-176461 (1997).
43. Japanese Pat. 9-208811 (1997).
44. Japanese Pat. 9-208813 (1997).
45. Kinami, N., Okamoto, M., Shinoda, Y., Sekura, T. and Yamaguchi, A. US Pat. 5567758 (1996).
46. Suzuki, N. and Oohara, Y. WO Pat. 97/43343 (1997).

PART III

Special Properties and Applications

10

Polymer–Layered Silicate Nanocomposites with Conventional Flame Retardants

J. W. GILMAN AND T. KASHIWAGI

Fire Science Division, Building and Fire Research Laboratory, National Institute of Standards and Technology, Gaithersburg, MD, USA

1 INTRODUCTION

In this chapter we review the current literature associated with the inherent flammability properties of polymer–layered silicate nanocomposites, with a special focus on the use of nanocomposites combined with conventional flame retardants to affect both reduced flammability *and* improved physical properties.

1.1 NANOCOMPOSITES

Polymer–layered silicate nanocomposites were first reported in 1961, when Blumstein demonstrated polymerization of vinyl monomers intercalated into montmorillonite (MMT) clay [1]. Several other groups have developed this approach [2, 3] and other methods to prepare polymer–layered silicate nanocomposites. In general, these methods achieve nanometer-scale incorporation of the layered silicate [e.g. MMT clay (see Figure 1), or synthetic layered silicates] in the polymer matrix by addition of a modified silicate either to a polymerization reaction (*in situ* method) [4–9], to a solvent-swollen polymer (solution blending) [10], or to a polymer melt (melt blending) [11–13]. Additionally, a method has been developed to prepare nanocomposites by polymerizing layered silicate precursors in the presence of a polymer [14].

Polymer–clay nanocomposites Edited by T. J. Pinnavaia and G. W. Beall
© 2000 John Wiley & Sons Ltd

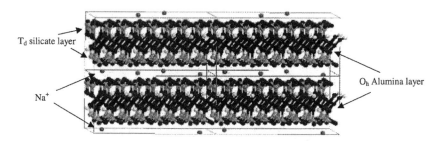

Figure 1 Molecular representation of sodium montmorillonite, showing two aluminosilicate layers with the Na$^+$ cations in the interlayer gap or gallery

Two terms, *intercalated* and *delaminated*, are used to describe the two general classes of nanomorphologies that can be prepared (see Figure 2). *Intercalated* nanocomposites contain self-assembled, well-ordered multilayered structures where the extended polymer chains are inserted into the gallery space (2–3 nm) between parallel individual silicate layers. The *delaminated* (or *exfoliated*) nanocomposites result when the individual silicate layers are no longer close enough to interact with gallery cations of the adjacent layers [15]. In this case the interlayer spacing can be of the order of the radius of gyration of the polymer; therefore, the silicate layers may be considered well dispersed in the polymer. Both of these hybrid structures can coexist in the polymer matrix; this mixed nanomorphology is very common for composites based on smectite silicates and clay minerals [16].

Polymer–layered silicate nanocomposites have unique properties [11]. For example, the mechanical properties of a PA-6–layered silicate nanocomposite, with a silicate mass fraction of only 5 %, show substantial improvement over those for the pure PA-6. The nanocomposite exhibits increases of 40 % in tensile strength, 68 % in tensile modulus, 60 % in flexural strength, and 126 % in flexural modulus. The heat distortion temperature (HDT) is also increased, from 65 to 152 °C, and the impact strengths are lowered by just 10 % [17]. Recently, Wang and Pinnavaia have reported a 400 % improvement in tensile modulus and tensile strength, and a substantial

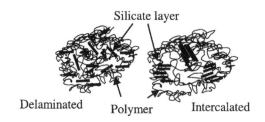

Figure 2 Schematic representation of the delaminated (or exfoliated) and intercalated morphologies

increase in the strain at break of aliphatic amine-cured epoxy–layered silicate nanocomposites [18]. Decreased gas permeability and increased solvent resistance also accompany the improved physical properties [11]. Finally, as will be discussed below, polymer–layered silicate nanocomposites exhibit reduced flammability [19–24] and often show increased thermal stability [25, 26].

1.2 THERMAL STABILITY AND FLAMMABILITY PROPERTIES

The improved thermal stability of a polymer–layered silicate nanocomposite was first reported by Blumstein [27]. He showed that PMMA polymerized in the gallery space of MMT clay PMMA (refluxing decane, 215 °C, N_2, 48 h). Blumstein argued that the stability of the PMMA to thermal degradation under conditions that would otherwise completely degrade pure nanocomposite is due to its different structure, and the restricted thermal motion of PMMA in the gallery. Nyden has observed this effect in molecular dynamics simulations of the thermal degradation of nanoconfined polypropylene [28].

The first mention of the potential flame retardant (FR) properties of these types of material appears in a 1976 Unitika [29] patent application on PA-6–MMT nanocomposites [30]. However, not until more recent studies did the serious evaluation of the flammability properties of these materials begin.

Giannelis and coworkers have reported self-extinguishing flammability for a polycaprolactone–layered silicate nanocomposite [26] and aliphatic polyimide–layered silicate nanocomposites [12]. The polyimide nanocomposites also showed improved thermal stability. In view of the improved barrier properties observed for other polymer–layered silicate nanocomposites, the increased thermal stability was attributed to hindered diffusion of volatile decomposition products within the nanocomposite.

Recent work done in our laboratories, using cone calorimetry [31], has quantitatively characterized the reduced flammability of a number of other polymer–layered silicate nanocomposites. We first observed reduced flammability, in the form of a lower peak heat release rate (HRR), for *delaminated* polyamide-6–MMT nanocomposites [19]. Subsequent investigations established that this approach was a general phenomenon, which reduced the flammability of both thermoplastics, such as polypropylene (see Figure 3) and polystyrene, and thermoset resins, such as vinyl ester– and epoxy–layered silicate nanocomposites (see Tables 2 and 3) [20, 32]. Not only was reduced flammability attained at very low loading levels (2–5 %), without increasing carbon monoxide or smoke yields, but this approach simultaneously improves the physical properties of the polymer. This combination of effects makes this approach to flame retarding polymers a unique one, since most flame retardant approaches come at the expense of physical properties.

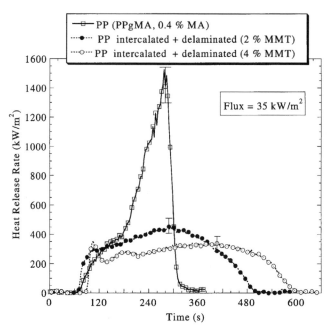

Figure 3 Comparison of the heat release rate (HRR) plots for pure PPgMA, and two PPgMA–layered silicate (MMT) nanocomposites, at 35 kW/m² heat flux, showing a 70–80 % reduction in peak HRR for the nanocomposites with a mass fraction of only 2 or 4 % layered silicate, respectively

2 NANOCOMPOSITE FLAME RETARDANT MECHANISM

By studying the combustion residues using transmission electron microscopy (TEM) and X-ray diffraction (XRD), we have found evidence for a common mechanism of flammability reduction [32]. *The nanocomposite structure appears to collapse during combustion. The multilayered carbonaceous–silicate structure appears to enhance the performance of the char through structural reinforcement.* The multi-layered carbonaceous–silicate structure is similar to that prepared from the pyrolysis of polymers intercalated into MMT by Sonobe and coworkers [33]. This silicate-rich char may act as an excellent insulator and mass transport barrier, slowing the escape of the volatile products generated as the polymer decomposes [34]. Indeed, the cone calorimeter data shown in Tables 1 to 3 reveal that the only parameter, aside from HRR, that changes is the mass loss rate (MLR).

Table 1 Cone calorimetry data for PA-6, PS, PP, and the respective MMT nanocomposites

Sample (structure)	Residue yield (%) ±0.5	Peak HRR (kW/m²) (Δ%)	Mean HRR (kW/m²) (Δ%)	Mean MLR (g/s m²) (Δ%)	Mean H_c (MJ/kg)	Mean SEA m²/kg	Mean CO yield (kg/kg)
PA-6	1	1010	603	29	27	197	0.01
PA-6–MMT, 2% delaminated	3	686 (32%)	390 (35%)	17 (41%)	27	271	0.01
PA-6–MMT, 5% delaminated	6	378 (63%)	304 (50%)	12 (58%)	27	296	0.02
PS	0	1120	703	29	29	1460	0.09
PS-MMT, 3% immiscible	3	1080	715	28	29	1840	0.09
PS–MMT, 3% intercalated/delaminated	4	567 (48%)	444 (38%)	17 (41%)	27	1730	0.08
PPgMA	5	1525	536	13	39	704	0.02
PPgMA–MMT, 2% intercalated/delaminated	6	450 (70%)	322 (40%)	8 (38%)	44	1028	0.02
PPgMA–MMT, 4% intercalated/delaminated	12	381 (75%)	275 (49%)	6 (46%)	44	968	0.02

Heat flux 35 kW/m²; H_c = specific heat of combustion; SEA = specific extinction area. Peak heat release rate, mass loss rate, and SEA data, measured at 35 kW/m², are reproducible ±10% (one sigma). The carbon monoxide and heat of combustion data are reproducible ±15% (one sigma).

Table 2 Cone calorimeter data for MBA and BAN vinyl esters

Sample	Residue yield (%) ± 0.5	Peak HRR (kW/m^2) (Δ%)	Mean HRR (kW/m^2) (Δ%)	Mean MLR (g/s m^2)	Mean H_c (MJ/kg)	Mean SEA (m^2/kg)	Mean CO yield (kg/kg)
Mod-bis-A vinyl ester	0	879	598	26	23	1360	0.06
Mod-bis-A vinyl ester[a]	8	656 (25%)	365 (39%)	18 (30%)	20	1300	0.06
Bis-A/novolac vinyl ester	2	977	628	29	21	1380	0.06
Bis-A/novolac vinyl ester[a]	9	596 (39%)	352 (44%)	18 (39%)	20	1400	0.06

Heat flux 35 kW/m^2; H_c = specific heat of combustion; SEA = specific extinction area. Peak heat release rate (peak HRR) and mass loss rate (MLR) data, measured at 35 kW/m^2, are reproducible $\pm 10\%$ (one sigma). The carbon monoxide and heat of combustion data are reproducible $\pm 15\%$ (one sigma).
[a] 6% intercalated MMT nanocomposite.

3 NANOCOMPOSITES AND CONVENTIONAL FLAME RETARDANTS

These promising results have inspired other researchers to apply this unique approach to improve the physical properties *and* reduce the flammability of formulated polymer products. Indeed, several recent patents and papers reveal the use of nanodispersed layered silicates in combination with other flame retardants; they claim a simultaneous reduction in flammability and improvement in mechanical properties. The general approach in these studies starts with a conventional flame-

Table 3 Cone calorimeter data for DGEBA/MDA and DGEBA/BDMA epoxies

Sample	Residue yield (%) ± 0.5	Peak HRR (kW/m^2) (Δ%)	Mean HRR (kW/m^2) (Δ%)	Mean MLR (g/s m^2)	Mean H_c (MJ/kg)	Mean SEA (m^2/kg)	Mean CO yield (kg/kg)
Epoxy resin DGEBA/MDA	11	1296	767	36	26	1340	0.07
Epoxy resin DGEBA/MDA[a]	19	773 (40%)	540 (29%)	24 (33%)	26	1480	0.06
Epoxy resin DGEBA/BDMA	3	1336	775	34	28	1260	0.06
Epoxy resin DGEBA/BDMA[a]	10	769 (42%)	509 (35%)	21 (38%)	30	1330	0.06

Heat flux 35 kW/m^2; H_c = specific heat of combustion; SEA = specific extinction area. Peak heat release rate (peak HRR) and mass loss rate (MLR) data, measured at 35 kW/m^2, are reproducible $\pm 10\%$ (one sigma). The carbon monoxide and heat of combustion data are reproducible $\pm 15\%$ (one sigma).
[a] 6% intercalated MMT nanocomposite.

retarded polymer formulation, which already shows good flammability properties (i.e. low HRR or a V-0 rating in the UL 94 flammability test) [35]. Then, incorporation of a nanodispersed layered silicate allows a significant portion of the conventional flame retardant to be removed from the formulation. This maintains, or improves, flammability performance and enhances the physical properties. These improvements result from reduction in the amount of conventional flame retardant used, which often degrades properties, and from the normal reinforcing effect of the nanodispersed layered silicate.

3.1 NANOCOMPOSITES AND INTUMESCENT FLAME RETARDANT

Bourbigot *et al.* have studied the substitution of a PA-6–MMT (2% MMT) nanocomposite for pentaerythritol in a typical ammonium polyphosphate (APP) based intumescent flame retardant formulation [36]. This work demonstrated measurable improvement in the mechanical *and* flammability properties of an ethylene vinyl acetate formulation (EVA-24, 24% vinyl acetate). The heat release rate data are shown in Figure 4. The mechanical properties data are shown in Figure 5. Furthermore, the EVA-24/APP/PA-6 sample, which contains no MMT, has a UL 94 V-0 rating; use of PA-6 with MMT (2%) not only reduced the heat release rate but also provided the option of removing some of the APP. As much as one-third of the APP can be removed while a V-0 rating is maintained, and the elongation is increased from 800 to 850% (data not shown).

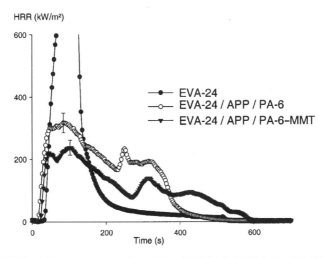

Figure 4 HRR values versus time for EVA24/APP/PA-6 (60:30:10) and EVA24/APP/PA 6–MMT (60:30:10) in comparison with pure EVA24 (heat flux $50 \, kW/m^2$)

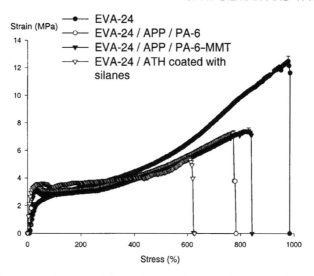

Figure 5   Mechanical properties of the EVA24/APP/PA-6 and EVA24/APP/PA-6–MMT formulations in comparison with pure EVA24 and silane-treated EVA24–aluminum trihydrate (ATH)

3.2 NANOCOMPOSITES WITH BROMINATED POLYCARBONATE/Sb_2O_3 AND POLYTETRAFLUOROETHYLENE

A recent General Electric Company patent uses a similar approach for polybutylene terephthalate (PBT, Valox 315) [23]. An organic modified MMT in combination with polytetrafluoroethylene (PTFE), dispersed in a styrene–acrylonitrile (SAN) copolymer (50 % PTFE), was used to replace 40 % of the brominated polycarbonate/Sb_2O_3 flame retardant in PBT. The main claim of this patent is the synergistic interaction of the PTFE and the organo-MMT; as examples 3 and 5 (Table 4) show, without both additives present the PBT formulations fail the UL 94 vertical test. However, when both are present in the formulation, as in example 4, a V-0 rating is achieved. The total additive fraction in the PBT formulation is lower by 12 % in example 4; this additional PBT in the formulation, along with the usual nanocomposite effects, should yield significant improvements in the mechanical properties of the system. However, no mechanical property data are given in the patent.

3.3 POLYETHYLENE–LAYERED SILICATE NANOCOMPOSITES AND CONVENTIONAL FLAME RETARDANT

A Japanese patent recently issued to Sekisui Chemical Inc. presents data on the use of organic modified layered silicates (OMLS) in polyethylene (PE), in combination

Table 4 PBT fire retardant blends data

	Example No.				
	1	2	3	4	5
Blend composition (%):					
Valox 315	69.12	76.74	81.32	81.32	81.4
BC-58	20	15	12	12	12
Sb_2O_3 (conc.)	10.5	7.88	6.3	6.3	6.3
T-SAN	0.08	0.08	0.08	0.08	—
Organoclay	None	None	None	Clayton HY	Clayton HY
(% of PBT)	—	—	—	2	2
UL 94 Test	V-0	V-0	F	V-0	F
Total flame out time (s)	8.8	10.6	—	10.2	—

Valox 315—polybutylene terephthalate (weight-average molecular weight $M_w = 105\,000$); BC-58—brominated bis-phenol A polycarbonate oligomers (58 % bromine); T-SAN—polytetrafluoroethylene dispersion in styrene acrylonitrile copolymer (>50 % PTFE); Clayton HY—MMT (dimethyldi(hydrogenated tallow)ammonium ion); PBT—poly(1,4-butylene terephthalate).

with a variety of conventional flame retardants and metal oxides [24]. Flammability of the PE formulations is determined using the cone calorimeter with a heat flux of $50\,kW/m^2$. The heat release data for PE with three different OMLS (SBAN-400, SBAN D, SBAN X) at 10 phr [37] and 20 phr loading levels are shown in Table 5. The peak HRRs for the PE/OMLS are between 50 and 60 % lower than the peak HRRs for a series of control samples presented in Table 5. These controls include pure PE and PE with 10 phr pristine (untreated) MMT. Comparison of the peak HRRs for the PE/OMLS samples (Table 5) to the peak HRR for the PE/pristine MMT sample (Table 6, example 4) shows that, without a suitable organic treatment, MMT mixed with PE produces little effect on flammability. This is similar to the results we reported for PS, where we found that MMT must be nanodispersed to

Table 5 Elongation and cone calorimeter data for PE with three different OMLSS

	1	2	3	4	5
Sample Composition (phr):					
Polyethylene	100	100	100	100	100
SBAN N-400	10	—	—	20	—
SBAN D	—	10	—	—	—
SBAN X	—	—	10	—	20
Peak heat release rate (kW/m²)	687	616	593	602	537
% Elongation	900	920	950	850	880

Sample size $100 \times 100 \times 3$ mm; flux $50\,kW/m^2$; phr = parts 100 parts resin.

Table 6 Elongation and cone calorimeter data for comparison samples

	1	2	3	4	5
Sample composition (phr):					
Polyethylene	100	100	100	100	100
Bengal (pristine MMT) A	—	—	—	10	—
Bengal D15 (organic modified MMT)	—	—	—	—	10
DBDPO	—	7.5	—	—	—
Sb$_2$O$_3$	—	2.5	—	—	—
APP	—	—	10	—	—
Peak heat release rate (kW/m^2)	1327	1309	1272	1067	1202
% Elongation	980	830	590	820	770

Sample size $100 \times 100 \times 3$ mm; flux 50 kW/m^2; phr = parts per 100 parts resin; DBDPO = decabromodiphenyl oxide; APP = ammonium polyphosphate.

realize the reduction in flammability (see Table 1). Presumably, the PE/OMLS is nanodispersed and the PE/pristine MMT is an immiscible, or conventional, mixture.

It is interesting to note that at the 10 phr loading level, where the OMLS reduces the peak HRR substantially, the DBDPO (decabromodiphenyl oxide)/Sb$_2$O$_3$ FR has little effect on peak HRR. The weak effect for the 10 phr APP sample is not as surprising, since no oxygen functionality is present in the formulation. In contrast, the data in Table 7 show an additional 30 % reduction in the peak HRR when only 5 phr APP is used in conjunction with the OMLS, SBAN N-400 (example 6), and a 20 % lower peak HRR when 5 phr of a phenyl phosphate is combined with SBAN N-400 in PE (example 7).

Table 7 Elongation and cone calorimeter data

	Comparison example				
	6	7	6	7	8
Sample composition (parts):					
Polyethylene	100	100	100	100	100
SBAN N-400	10	10	—	—	—
APP	5	—	—	—	15
Phenyl phosphate	—	5	—	—	—
DBDPO	—	—	—	11.25	—
Sb$_2$O$_3$	—	—	—	3.75	—
Peak heat release rate (kW/m^2)	493	543	1327	1189	989
% Elongation	900	930	980	720	490

Sample size $100 \times 100 \times 3$ mm; flux 50 kW/m^2; DBDPO = decabromodiphenyl oxide; APP = ammonium polyphosphate.

For more direct comparison with 15 phr conventional FR formulations, the peak HRR data for PE with 15 phr DBDPO/Sb_2O_3 and for PE with 15 phr APP are also included in Table 7. Again, only small (10–20 %) reductions are found in the peak HRRs. However, more importantly, at these loading levels the mechanical properties suffer, especially for the APP sample where the % elongation has been reduced by half (980–490 %), while the PE/SBAN N-400 samples (with or without other FR) exhibit no loss of the % elongation.

The patent also shows data for PE formulations that combine the OMLS, SBAN X, and a series of metal oxides. Some additional reductions in peak HRRs are seen for the Cu and Zn oxide samples. Again, this is accomplished without a significant effect on the % elongation. The mechanism is not discussed. However, in studies by Thomas *et al.* of MMT as cracking catalysts, various metal cation intercalates were examined; Cu salts were found to have significant catalytic activity in condensation and cracking reactions [38].

3.4 NANOCOMPOSITES AND MELAMINE

Another recent Japanese patent issued to Inoue and Hosokawa of Showa Denko reports the use of silicate–triazine intercalation compounds in fire resistant polymeric composites [22]. By combining the known FR properties of the triazine, melamine, and those of polymer–layered silicate nanocomposites, the inventors produced V-0 ratings in the UL 94 flammability test (see Table 8), while increasing both the bending modulus and the heat distortion temperature. The inventors state that the quaternary alkylammonium treatment, typically used to facilitate intercalation and delamination of layered silicates in polymers, increases the flammability of the nanocomposite. Their solution is the use of the ammonium salt of melamine, which has known FR properties [39]. PA-6, PBT, polyoxymethylene (POM), and polyphenylene sulfide (PPS) were prepared as silicate nanocomposites using the synthetic silicate, fluorinated synthetic mica (FSM). FSM is chemically similar to MMT but the aspect ratio of the individual silicate layers is 5–10 times larger than that of MMT. Various melamine–ammonium salts were used to treat the FSM, and 8–15 % total mass fraction of melamine and other melamine salts were also added to the polymer. Inoue and Hosokawa showed the importance of nanodispersing the FSM. Using TEM, the spacing between the FSM layers was characterized; they found that, without a uniform dispersion (>50 % delamination, Table 8, No. 5) of the silicate layers in the polymer, only an HB rating (self-extinguishing in a horizontal burn) was obtained in the UL 94 test. They also established that both a melamine treatment on the FSM and melamine or a melamine salt (melamine cyanurate) compounded into the resin were necessary to achieve V-0 performance (see Table 8, Nos 4, 7, 8, 9, and 12).

Table 8   UL 94 data for FSM nanocomposites

Experiment No.	Nanocomposite	Silicate (%)	Additive	% added	Delamination	UL 94
1	PA-6/OM-FSM[a]	5	Melamine	3.3	80	HB
2	PA-6/MEL-FSM	5	—	—	80	V-2
3	PA-6/MEL-FSM	5	Melamine	3.3	80	V-2
4	PA-6/MEL-FSM	5	Melamine	10	80	V-0
5	PA-6/MEL-FSM	5	Melamine cyanurate	3.3	50	V-2
6	PA-6,6/OM-FSM	5	Melamine cyanurate	3.3	>50	HB
7	PA-6,6/MEL-FSM	5	Melamine	3.3	>50	V-0
8	PA-6,6/MEL-FSM	5	Melamine cyanurate	3.3	>50	V-0
9	PA-6,6/MEL-FSM	3	Melamine	5	>50	V-0
10	PBT/OM-FSM	5	Melamine cyanurate	6	80	HB
11	PBT/MEL-FSM	5	—	—	80	V-2
12	PBT/MEL-FSM	5	Melamine cyanurate	5	80	V-0

[a]OM-FSM = DODDMA-FSM = dioctadecyldimethylammonium fluorinated synthetic mica.

4 SUMMARY

All MMT-based nanocomposite systems reported so far show reduced flammability. The delaminated versions of MMT-based nanocomposites also offer measurable improvements in a variety of physical properties. Many issues are unresolved as to the mechanism of these property enhancements. Once resolved, nanocomposites may fulfill the requirements for a high-performance additive-type flame retardant system, i.e. one that reduces flammability while improving the other performance properties of the final formulated product. As the patents and publications discussed above have demonstrated, the nanocomposites approach has become another tool for the flame retardant chemist, which allows successful improvement of both the flammability and the mechanical properties of many different polymer systems [22–24, 36].

5 ACKNOWLEDGEMENTS

The authors would like to thank Dr Richard Lyon and the Federal Aviation Administration for partial funding of this work through Interagency Agreement DTFA03-99-X-90009. We would also like to thank Dr Alexander Morgan, Mr Richard Harris, Mr James Brown, and Ms Lori Brassell for sample preparation and data analysis, Mr Michael Smith for cone calorimeter analysis, Dr James Cline for use of XRD facilities, Dr Marc Nyden for the MMT structure, and Dr Catheryn Jackson for TEM. We would also like to express our gratitude to Sekisui Corporation for supplying a copy of their patent, and to Southern Clay Products for organic-modified layered silicates and translation of the Showa Denko patent. A portion of this work was carried out by the National Institute of Standards and Technology (NIST), an agency of the US government, and by statute is not subject to copyright in the United States.

6 REFERENCES AND NOTES

1. Blumstein, A. *Bull. Chim. Soc.*, 899 (1961).
2. Hawthorne, D. G., Hodgkin, J. H., Loft, B. C. and Solomon, D. H. *J. Macro. Mol. Sci., Chem.* **A8**(3), 649–657 (1974).
3. Dekking, H. G. G. *J. Appl. Polym. Sci.*, **11**(1), 23–36 (1967).
4. Fujiwara, S. and Sakamoto, T. Kokai Pat. Application SHO 51(1976)-109998 (1976).
5. Usuki, A., Kojima, Y., Kawasumi, M., Okada, A., Fukushima, Y., Kurauchi, T., Kamigaito, O. *J. Mater. Res.*, **8**, 1179 (1993).
6. Lan, T. and Pinnavaia, T. J. *Chem. Mater.*, **6**, 2216 (1994).
7. Usuki, A., Kato, M., Okada, A., and Kurauchi, T. *J. App. Polym. Sci.*, **63**, 137 (1997).
8. Meier, L. P., Shelden, R. A., Caseri, W. R. and Suter, U. W. *Macromolecules*, **27**, 1637–1642 (1994).
9. Reichert, P., Kressler, J., Thomann, R., Mulhaupt, R. and Stoppelmann, G. *Acta Polym.*, **49**, 116–123, (1998).
10. Jeon, H. G., Jung, H. T., Lee, S. D. and Hudson, S. *Polym. Bull.*, **41**, 107 (1998).
11. Maxfield, M., Christiani, B. R., Murthy, S. N. and Tuller, H. US Pat. 5,385,776 (1995).
12. Giannelis, E. *Advd. Mater.*, **8**, 29 (1996).
13. Fisher, H., Gielgens. L., Koster, T. *Nanocomposites from Polymers and Layered Minerals*, TNO-TPD report, 1998.
14. Carrado, K. A. and Langui, X. *Microporous and Mesoporous Mater.*, **27**, 87 (1999).
15. Lan, T. and Pinnavaia, T. J. *Chem. Mater.*, **6**, 2216 (1994).
16. For definitions and background on layered silicate and clay minerals, see: Kroschurtz J. S. (Ed.), *Kirk-Othmer Encyclopedia of Chemical Technology*, 4th edition, Vol. 6, John Wiley & Sons, New York, 1993.
17. Kojima, Y., Usuki, A., Kawasumi, M., Okada, A., Fukushima, Y., Kurauchi, T. and Kamigaito, O. *J. Mater. Res.*, **8**, 1185 (1993).
18. Wang, Z. and Pinnavaia, T. J. *Chem. Mater.*, **10**, 1820 (1998).
19. Gilman, J. W., Kashiwagi, T. and Lichtenhan, J. D. *SAMPE J.*, **33**, 40 (1997).
20. Gilman, J., Kashiwagi, T. and Lichtenhan, J. in Proceedings of 6th European Meeting on *Fire Retardancy of Polymeric Materials, FRPM '97*, September 1997, p. 19.
21. Gilman, J., Kashiwagi, T., Lomakin, S., Giannelis, E., Manias, E., Lichtenhan, J. and Jones, P. *Fire Retardancy of Polymers: the Use of Intumescence*, The Royal Society of Chemistry, Cambridge, 1998, pp. 203–221.
22. Inoue, H. and Hosokawa, T., Japanese Pat. Application (Showa Denko K. K., Japan) Jpn Kokai Tokkyo Koho 10 81,510 (98 81,510) (1998).
23. Takekoshi, T., Fouad, F., Mercx, F. P. M. and De Moor, J. J. M. US Pat. 5,773,502, issued to General Electric Company (1998).
24. Okada, K. (Sekisui) Japanese Pat. 11-228748 (1999).
25. Burnside, S. D. and Giannelis, E. P. *Chem. Mater.*, **7**, 1597 (1995).
26. Lee, J., Takekoshi, T. and Giannelis, E. *Mat. Res. Soc. Symp. Proc.*, **457**, 513 (1997).
27. Blumstein, A. *J. Polym. Sci.*, **A3**, 2665 (1965).
28. Nyden, M. R. and Gilman, J. W. *Comp. Theor. Polym. Sci.*, **7**, 191–196 (1997).

29. Certain commercial equipment, instruments, materials, services or companies are identified in this paper in order to specify adequately the experimental procedure. This in no way implies endorsement or recommendation by NIST.
30. Fujiwara, S. and Sakamoto, T. Kokai Pat. Application SHO 51(1976)-109998 (1976).
31. Babrauskas, V. and Peacock, R. D. *Fire Safety J.*, **18**, 255 (1992).
32. (a) Gilman, J. W., Kahsiwagi, T., Brown, J. E. T., Lomakin, S., Giannelis, E. P. and Manias, E. in Proceedings of 43rd International SAMPE Symposium and Exhibition, May 1998, p. 1053; (b) Gilman, J. W., Kashiwagi, T., Nyden, M., Brown, J. E. T., Jackson, C. L., Lomakin, S., Giannelis, E. P. and Manias, E. *Chemistry and Technology of Polymer Additives*, Blackwell Scientific, Oxford, 1999, pp. 249–265.
33. Sonobe, N., Kyotani, T. and Tomita, A. *Carbon*, **29**(1), 61–67 (1991).
34. Gilman, J. W. *Appl. Clay Sci.*, **15**, 31 (1999).
35. Tests for Flammability of Plastic Materials—UL 94, 1997, Underwriters Laboratories Inc. The UL-94 test is performed on a plastic sample (125 mm by 13 mm, with various thicknesses up to 13 mm) suspended vertically above a cotton patch. The plastic is subjected to two 10 s flame exposures with a calibrated flame in a unit that is free of air currents. After the first 10 s exposure, the flame is removed, and the time for the sample to self-extinguish is recorded. Cotton ignition is noted if polymer dripping ensues, dripping is permissible if no cotton ignites. Then the second ignition is performed on the same sample, and the self-extinguishing time and dripping characteristics are recorded. If the plastic self-extinguishes in less than 10 s after each ignition, with no dripping, it is classified as V-0. If it self-extinguishes in less than 30 s after each ignition, with no dripping it is classified as V-1, and if the cotton ignites then it is classified as V-2. If the sample does not self-extinguish before burning completely it is classified as failed (F).
36. Bourbigot, S., Le Bras, M., Dabrowski, F., Gilman, J. and Kashiwagi, T. *Fire Mater.*, submitted.
37. phr = parts per 100 parts resin, e.g. 10 phr is a mass fraction of 9.1 %.
38. Tricker, M. J., Tennakoon D. T. B., Graham, and Thomas, J. M. *Nature (Lond.)*, **253**, 110 (1975).
39. Weil, E. D. and Zhu, W. in Nelson, G. (Ed.), *Fire and Polymers*, ACS Symposium Series 599, American Chemical Society, Washington, DC, 1995, pp. 199–215.

11

Nanocomposite Technology for Enhancing the Gas Barrier of Polyethylene Terephthalate

J. C. MATAYABAS JR AND S. R. TURNER

Polymers Technology Group, Eastman Chemical Company, Kingsport, TN, USA

1 INTRODUCTION

Polyethylene terephthalate (PET) is a large volume plastic that has been in use for fibers and films for many years. Today, the fastest growing area for PET is in food and beverage packaging. This growth is based on important properties of PET such as gas barrier, processability in high throughput stretch blow molding, clarity, color, and resistance to flavor scalping. Thus, PET has become the preferred packaging material for a wide variety of products including carbonated soft drinks, water, sport drinks, fruit juices, and peanut butter. However, the barrier of PET to oxygen and carbon dioxide is not sufficient to package many large-volume products that are sensitive to oxygen and loss of carbon dioxide. Examples include beer, wine, and tomato-based products. Even small soft drink containers are problematic owing to the high surface–volume ratio, which leads to unacceptable carbon dioxide loss and consequently loss of shelf life. The discoveries that the exfoliated platelets of aluminosilicate clays could enhance barrier properties of polymeric materials have led to considerable research on applying this technology to improve the barrier properties of PET. This chapter reviews efforts to exfoliate clay in PET and the barrier results that have been obtained.

The need for higher barrier in PET has been recognized for many years and numerous PET backbone modifications have been accomplished to enhance barrier

Polymer–clay nanocomposites Edited by T. J. Pinnavaia and G. W. Beall
© 2000 John Wiley & Sons Ltd

[1, 2]. However, all these approaches have suffered from the high cost and investment required to build new monomer and polymer plants and to obtain food contact regulatory clearances for new polymers. Currently, the use of multilayer processing techniques, where high-barrier polymers such as poly(m-xylylene adipamide) and EVOH are sandwiched between PET layers, are being employed to meet the opportunities for enhanced barrier products [3].

The ability of exfoliated clays to improve the gas barrier characteristics of resins, as exemplified with poly(caprolactone) [4], prompted a considerable amount of work to apply this technology to PET for barrier enhancement. The conceptual advantage of this approach is based on the use of conventional inexpensive and available PET as the polymer backbone, with barrier improvement arising from dispersion of low to modest levels of relatively inexpensive clay materials. This, coupled with the potential to process such a material in conventional stretch blow molding processing equipment, makes this approach very attractive.

2 BARRIER ENHANCEMENT FROM PLATELET MATERIALS

2.1 THEORY

Barrier is defined in this chapter as the ratio of the permeability of PET divided by the permeability of the polymer being compared. Platelet particles enhance barrier (reduce gas permeability) of polymers according to a tortuous path model, developed by Neilson [5], in which the platelets obstruct the passage of gases and other permeants through the matrix polymer. The barrier improvement is predicted by this model to be a function of the volume fraction of platelets, ϕ, and a function of the aspect ratio of the platelets, α, with higher aspect ratios providing greater barrier improvement according to the following equation for permeability:

$$P_{\text{nanocomposite}} = \frac{(1 - \phi)P_{\text{matrix}}}{1 + \alpha\phi/2} \tag{1}$$

where $P_{\text{nanocomposite}}$ represents the permeability of the resulting nanocomposite, and P_{matrix} represents the permeability of the matrix polymer. An illustration of the tortuous path model for barrier enhancement is presented in Figure 1.

2.2 EXFOLIATED NANOCOMPOSITES

Polymeric nanocomposites are materials comprising nanometer sized smectic clay platelets that are dispersed in a polymer matrix. Exfoliated polymeric nanocomposites are those where the layered clay platelets (Figure 2) are delaminated and dispersed as single platelets in the matrix polymer. In the fully exfoliated state, individual clay platelets will have the highest aspect ratio possible, and thus the highest barrier improvement is expected. Furthermore, exfoliating the clay into

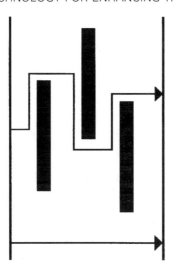

Figure 1 Illustration of Neilson's tortuous path model for barrier enhancement of nanocomposites. P_{matrix} represents the permeability of the matrix polymer in the absence of the platelets, and $P_{nanocomposite}$ represents the tortuous path created by the platelets

individual platelets yields the smallest particle size, thereby providing the maximum clarity of films. In addition, exfoliated polymer nanocomposites are typically characterized by higher strength, stiffness, dimensional stability, and heat resistance than the unfilled polymer. Therefore, nanocomposite technology has the potential to expand PET into a wide variety of applications including those where higher barrier is needed.

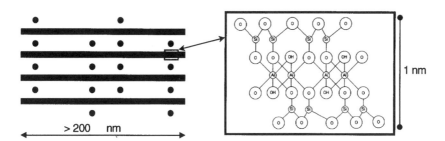

Figure 2 Illustration of the tactoid structure of montmorillonite clay and chemical representation of the clay platelet. The height of the clay platelet, which is about 1 nm, is dictated by the crystal structure of the platelet, and the width is typically greater than 200 nm; therefore, the clay platelets typically have an aspect ratio greater than 200. The solid circles represent the exchangeable cations that are present between the clay platelets

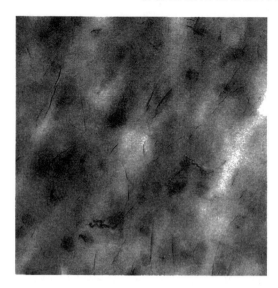

Figure 3 Transmission electron micrograph (TEM) image (×100 000) of an exfoliated polyethylene terephthalate nanocomposite. The edges of the clay particles are revealed as black bands uniformly dispersed throughout the polyethylene terephthalate matrix

The transmission electron micrograph (TEM) image (×100 000) in Figure 3 is of an exfoliated PET nanocomposite. The edges of the clay platelets appear as black bands uniformly dispersed throughout the PET. Most of the clay is in the form of individual platelets, and thus this image of an actual PET nanocomposite closely resembles the illustration presented in Figure 1.

Exfoliated PET–clay nanocomposites have barrier properties that follow the tortuous path model. Very small amounts of exfoliated clay in PET, typically less than about 5 wt% ash, give significant barrier improvement to gases such as oxygen, carbon dioxide, and water vapor. The exfoliated clay was found to improve the barrier of a number of different polyesters including polyethylene naphthalate (PEN), PET, and amorphous glycol modified PETs (PETG). Figure 4 presents the oxygen permeabilities of 10 mil thick amorphous compression molded films as a function of the clay loading, as measured by wt% ash, for a series of polyester nanocomposites based on the three polyester compositions above. In each case, the data can be fit to equation (1) and an effective aspect ratio can be calculated from each curve fit. Typical values for the effective aspect ratios are 150–200 for amorphous unoriented nanocomposites. The effective aspect ratio has proven to be a good indicator of the extent of clay exfoliation, with higher effective aspect ratios indicating better exfoliation of the clay.

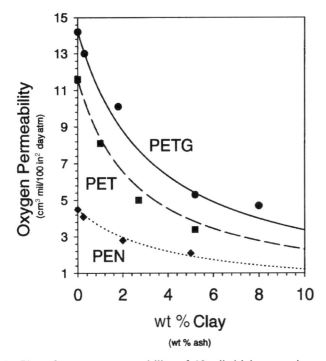

Figure 4 Plot of oxygen permeability of 10 mil thick amorphous compression molded films versus clay loading, as measured by wt% ash, for a series of polyester nanocomposites: glycol-modified polyethylene terephthalate (PETG), polyethylene terephthalate (PET), and polyethylene naphthalate (PEN). The data are fit to equation (1) for each polyester nanocomposite, resulting in a calculated effective aspect ratio of 150–200 for these amorphous unoriented polyester nanocomposites

The remainder of this chapter will discuss processes and mechanistic interpretations involved in incorporating clay into PET and obtaining an exfoliated system. The application of polyamide–clay nanocomposites in multilayer PET configurations will also be discussed.

3 PREPARATION AND PROPERTIES OF PET NANOCOMPOSITES

Achieving the degree of exfoliation presented in Figures 3 and 4 requires the selection and optimization of many variables. Some of the major variable categories include the choice of matrix, the process of incorporating clay, the choice of clay, the clay treatment, and the optional use of dispersing aids. Achieving the proper balance of the many variables to give exfoliation of the clay into individual platelets requires a significant amount of research. There are several potential methods by which a clay might be incorporated into a polymer such as PET. These methods are typically separated into the two broad categories of melt compounding and incorporation during the PET polymerization or *in situ* incorporation.

3.1 *IN SITU INCORPORATION OF CLAY INTO PET*

PET for containers is typically made by a three-step process:

(a) transesterification of ethylene glycol and dimethyl terephthalate or esterification of ethylene glycol and terephthalic acid to prepare bishydroxyethyl terephthalate and its oligomers;
(b) melt phase polycondensation at 260–300 °C to inherent viscosities of 0.50–0.70 dL/g;
(c) solid state polymerization at 180–230 °C to inherent viscosities of 0.65–1.0 dL/g.

There are many possible variations for *in situ* incorporation of clay, depending upon when during the polymerization the clay is added. Two of the most convenient points of addition are 'up front' with the monomers or during the transition to poly-condensation (after transesterification or esterification).

There were several literature reports on preparing and studying PET nanocomposites prior to the research that we report in this chapter. One report describes several variations for the incorporation of sodium montmorillonite into PET and the subsequent formation of stretch blow molded containers that reportedly have improved barrier to oxygen [6]. A second report concerns the preparation of a PET nanocomposite by *in situ* polymerization of a dispersion of organoclay in water; however, characterization of the resulting composite was not reported [7]. This report claims that water serves as a dispersing aid for intercalation of the monomers into the galleries of the organoclay and discloses that a wide variety of small molecules can serve as dispersing aids in place of, or in combination with, water. Certain proprietary treated clays in PET, where the clays were incorporated by an *in situ* process, have been described to be good nucleating agents for the polymer [8].

Two general routes exist for *in situ* clay incorporation. The first is based on novel exfoliation of clays into ethylene glycol, one of the basic monomers for PET. This technology was developed by Nanocor, Inc. [9, 10]. The keys to this approach include finding a clay treatment that gives exfoliation of the clay into ethylene glycol and finding polymerization conditions that permit polymerization while maintaining the exfoliation of the clay. One approach to modifying the clay involves intercalation of sodium montmorillonite with a polar polymer, such as polyvinylpyrrolidone (PVP) or polyvinyl alcohol (PVOH) [9]. Sodium montmorillonite intercalated with PVP or PVOH will exfoliate into a wide variety of media, including ethylene glycol, to give viscous gels [10]. We have observed that the introduction of these EG gels

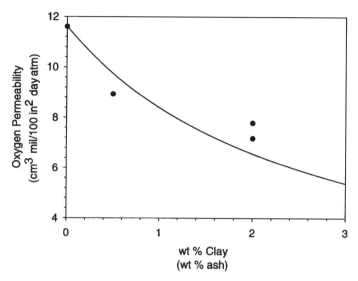

Figure 5   Oxygen permeability versus wt% clay, as measured by wt% ash, for 10 mil thick compression molded films of polyester

directly into the charge of a PET melt polymerization with DMT is an effective method to prepare PET nanocomposites with improved oxygen barrier, as shown in the plot of oxygen permeability versus clay loading in Figure 5 [11]. In these polymerizations a significant increase in the low-shear melt viscosity was observed with increasing clay content. This limits the molecular weight of the PET that can be prepared directly in the melt phase polymerization, but the PET nanocomposites were discovered to solid-state polymerize to PET with molecular weights that were suitable for conventional polymer processing steps such as stretch blow molding [11].

The second approach is by *in situ* incorporation of an organoclay. Modification of clay by exchanging the sodium atoms with alkylammonium compounds to form an organoclay has been practiced for many years. The ammonium becomes tethered to the surface of the clay while the long alkyl chains swell the clay, as indicated by the increase in interlayer spacing observed by X-ray analysis, and improve its dispersibility into organic materials. Ideally, by proper selection of the alkylammonium tether, the organoclay can be tuned for a specific polymer, such as PET. For example, Figure 6 presents a plot of the oxygen permeabilities of 10 mil thick amorphous compression molded films of PET nanocomposites prepared by *in situ* incorporation of varying amounts, as measured by wt% ash, of four different organoclays [12]. The oxygen permeabilities decrease and the calculated effective aspect ratios increase in the order of trimethyloctadecylammonium-treated sodium montmorillonite (I); Claytone APA (II), a commercial organoclay with an unknown alkylammonium

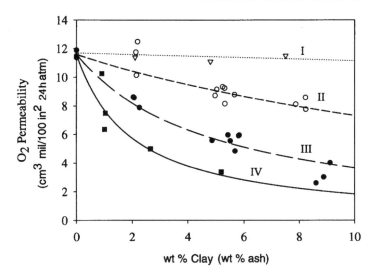

Figure 6 Plot of oxygen permeability versus wt% clay, measured by wt% ash, of 10 mil thick amorphous compression molded films of polyethylene terephthalate nanocomposites prepared by *in situ* incorporation of organoclays: trimethyloctadecylammonium-treated sodium montmorillonite (I, open triangles), Claytone APA (II, open circles), bishydroxyethylmethyltallowammonium-treated sodium montmorillonite (III, closed circles), and bisethoxylated (15)methyloctadecylammonium-treated sodium montmorillonite (IV, closed squares)

tether from Southern Clay Products; bishydroxyethylmethyltallowammonium-treated sodium montmorillonite (III); and bisethoxylated (15)methyloctadecylammonium-treated sodium montmorillonite (IV). Similarly, visual inspections, measurements of haze, and evaluations by transmission electron microscopy all show that, as permeability decreases, the exfoliation of the clay in the nanocomposite is improved. Thus, the selection of alkylammonium compound has a major effect on the exfoliation of the organoclay, which is reflected in the resulting barrier properties and the visual appearance of the polymer nanocomposites.

Modeling can be helpful in guiding selection of the tether and the clay treatment selection for experimentation. A mean field theory approach (Flory–Huggins) has been proposed by Giannelis to model polymer nanocomposites [13, 14]. This model suggests that the alkylammonium tether alters the enthalpy of interaction between the clay and polymer and compensates for the reduction in entropy owing to confining the polymer in the gallery of the clay. Further improvements in this model have been achieved using molecular dynamics simulations, including the constraint of the clay surface and giving a better representation of the tethered alkylammonium surfactant [15]. A more recent model by Balazs has resulted in a number of interesting hypotheses for experimental verification, including the suggestion that low molecular weight polymers may more readily intercalate the organoclay [16].

3.2 MELT COMPOUNDING APPROACHES TO PET NANOCOMPOSITES

It is desirable to have the option of using melt compounding as a method of incorporating clays into polymers for several reasons. Firstly, many polymers, such as polyethylene terephthalate, are produced on a large scale and it is desirable to be able to use these materials as they are currently produced. Secondly, melt compounding is a convenient and flexible process capable of producing a variety of formulations on a variety of product volume scales. And thirdly, the high-shear environment of the melt extruder may permit the incorporation of significantly higher concentrations of clay compared with the concentrations that can be achieved by a commercial *in situ* polymerization process. The presence of exfoliated clay greatly increases the low-shear melt viscosity of nanocomposites compared with the unmodified polymers, and this may cause limitations for commercial-scale *in situ* processes which are typically low-shear environments. Interestingly, it has been reported that for nylon 6,6 nanocomposites the high low-shear melt viscosity becomes an advantage when injection molding the nanocomposites, because they typically can be injection molded using faster cycle times, by using higher injection speeds and pressures, without flashing [17].

There are very few examples in the literature of the preparation of PET nanocomposites by melt compounding. Claytone APA, a commercial organoclay available from Southern Clay Products, was melt compounded into PET and an improved modulus was observed [18]. Another report describes the incorporation of sodium montmorillonite into PET by a melt compounding process, but few details are provided [6].

In order to study PET nanocomposites prepared by melt compounding, we melt-compounded varying amounts of Claytone APA with PET-9921, which is a 1,4-cyclohexanedimethanol-modified PET with an inherent viscosity of about 0.72 dL/g, as measured at a concentration of 0.5 g/dl in a 40/60 wt% mixture of phenol and 1,1,2,2-tetrachloroethane at 25 °C [11]. Physical mixtures of the two components were thoroughly dried in a vacuum oven at 120 °C and then extruded at a melt temperature of 280 °C on a Leistritz Micro 18 twin-screw extruder using general-purpose screws and vacuum ventilation. The extrudate was quenched in water and chopped into pellets as it exited the die. As shown in Figure 7, the inherent viscosities of the polyester nanocomposites decreases between 0.04 and 0.26 dL/g from the original inherent viscosity of PET-9921, and this degradation increases in severity as the loading of platelet particles increases from an ash value of 0.36–6.7 wt%.

In an attempt to compensate for the degradation, a high molecular weight PET resin with an inherent viscosity of 0.98 dL/g was used [11]. Unfortunately, the

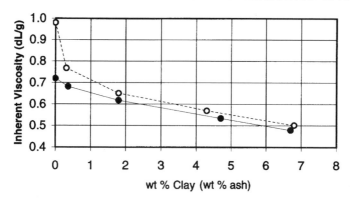

Figure 7 Plot of inherent viscosity versus wt% clay, as measured by wt% ash, of polyethylene terephthalate nanocomposites prepared by the melt compounding of PET-9921 (inherent viscosity 0.72 dL/g, closed circle) and PET-13339 (inherent viscosity 0.98 dL/g, open circle) with Clayton APA. The inherent viscosity values for 0 wt% clay (0 wt% ash) are the inherent viscosities of the initial polyethylene terephthalate

amount of degradation that was observed was even more severe, as shown in Figure 7. The inherent viscosities of the polyester nanocomposites decreased between 0.21 and 0.48 dL/g from the original inherent viscosity of 0.98 dL/g. Interestingly, at a concentration of Claytone APA as little as about 0.7 wt% ash (0.67 and 0.68 in the examples), the resulting inherent viscosities are about the same (0.48 and 0.50 dL/g in the examples) for both polyester nanocomposites, regardless of the inherent viscosity of the original polyester. Thus, these experiments demonstrate that:

(a) degradation upon the melt compounding of organoclay with PET is severe and
(b) the degradation cannot be overcome simply by increasing the inherent viscosity of the PET.

3.3 EXPANDED ORGANOCLAYS

We found that many organic, oligomeric, and polymeric materials can be added to organoclays to further expand the basal spacing between the clay layers and further improve exfoliation during melt compounding with polymers [19]. This provides yet another method of fine tuning a clay for exfoliation into a target matrix, such as PET. It was found that the beneficial use of the expanding agents can be achieved in a wide variety of processes involving addition of expanding agents:

(a) before cation exchange,
(b) during cation exchange,
(c) after cation exchange, and
(d) during melt compounding.

In addition, it was found that the expanding agent can be incorporated by processes involving:

(a) dispersions of the organoclay and expanding agent in water, organic solvents, and their mixtures, and

(b) melt compounding of the organoclay and expanding agent, either with simultaneous or subsequent melt compounding with polymer.

Table 1 lists a wide variety of materials that were found to be useful expanding agents for organoclay III and gives the intitial and final basal spacings of clay compositions that were prepared by mixing 40 wt% expanding agent with 60 wt% organoclay III in methylene chloride. As shown in Table 1, basal spacings of up to 4.5 nm have been achieved, and these expanded organoclays have been found to be useful for the preparation of polymer nanocomposites by melt compounding. One useful process involves the expansion of organoclay III with AQ-55, which is a water-dispersible polyester based on sodiosulfoisophthalic acid produced by Eastman Chemical Company. In this process the clay is dispersed in an aqueous solution of the water soluble polymer. This is followed by evaporation of the dispersion in the

Table 1   Effect of expanding agents on the basal spacing of organoclay III[a]

Expanding agent	Basal spacing (nm)
Poly(dimethylsiloxane), carbinol terminated	4.5
Polyethylene glycol distearate	4.2
Zonyl A	3.8
Polysar 101 polystyrene	3.7
Vitamin E	3.6
Ethoquad 18-25	3.5
Polyglycidylacrylate PD7610	3.4
AQ55	3.2
PETG 6763	3.1
Epon 828	3.1
Polycaprolactone	3.0
Polymethacrylate SCX800B	3.0
Poly(vinylpyrrolidone)	2.9
Makrolon 2608 polycarbonate	2.9
Poly(ethylene oxide), MW 3350	2.4
None	2.0

[a] The expanded organoclays were prepared by mixing organoclay III with 40 wt% expanding agent in methylene chloride. Ethoquad 18-25 is commercially available from AKZO Chemical Company. PD7610 is commercially available from Anderson Chemical Company. Epon 828 is available from Shell Chemical Company. SCX800 is made by S. C. Johnson Wax, Co. Polydimethylsiloxane, carbinol terminated, is available from Petrach Systems, Inc. AQ55 and PETG 6763 are commercial available polyesters made by Eastman Chemical Company.

presence of PET and completed by melt compounding of the material [20]. PET nanocomposites prepared by this process were found to have significantly improved color and clarity compared with PET nanocomposites prepared by other processes.

3.4 MECHANISM FOR IN SITU EXFOLIATION IN PET

Although they exhibit significantly improved barrier properties, none of the PET nanocomposites prepared by the above melt compounding processes have yielded polyester nanocomposites having gas barrier properties as high as those exhibited by PET nanocomposites prepared by *in situ* processes. These results indicate that the exfoliation level achieved in the *in situ* process exceeds that achieved by melt compounding. Given the desirability of achieving high exfoliation by melt compounding, we found it important to study the differences and mechanisms involved in these processes. In particular, we wanted to understand the mechanism for exfoliation by *in situ* processes. In the case of the polymer-intercalated clay–ethylene glycol gel approach to PET nanocomposites, the mechanism for exfoliation is easily understood in that exfoliation is achieved as a dispersion in ethylene glycol. However, the exfoliation of organoclay IV during polymerization of ethylene glycol with dimethyl terephthalate is less easily understood. Therefore, comparison with known exfoliated clay/polymer nanocomposites prepared by direct polymerization is helpful in gaining an understanding of the PET nanocomposite system.

Nylon 6, the first known exfoliated polymer nanocomposite, was prepared by *in situ* incorporation of a 12-aminododecanoic acid treated organoclay during caprolactam polymerization [21]. It is proposed that exfoliation was achieved by:

(a) swelling of the organoclay by caprolactam monomer (but not exfoliation of the organoclay into the caprolactam),
(b) polymerization initiated by the addition of caprolactam to the end of the dodecanoic acid chain that is tethered to the clay by the ammonium group,
(c) swelling of the organoclay by more caprolactam monomer,
(d) polymerization of the caprolactam, accompanied with further intercalation of the monomer, and
(e) finally, near complete consumption of the caprolactam, exfoliation [22]. Similar mechanisms have been proposed for exfoliation of 12-aminododecanoic acid treated clay in polycaprolactone from caprolactone polymerization [4] and an exfoliated epoxy resin prepared by epoxy polymerization in the presence of a functionalized organoclay [23]. In the epoxy system it was demonstrated, by X-ray analysis (WAXS), that the basal spacing of the clay increases during polymerization and eventually disappears completely, indicating complete exfoliation of the clay into individual platelets [24].

There are several implications if this proposed mechanism is applicable in general to many polymer systems. Firstly, it can be concluded that swelling of the clay by monomer is required, which means that it is necessary to select clay treatments to achieve swelling by the monomer and if necessary use low molecular weight dispersing aids further to improve intercalation of the monomer into the gallery of the organoclay. Secondly, polymerization in the gallery of the clay is required, which means that one should design the clay so that the tether will participate in polymerization as either an initiator, a monomer, or a catalyst. Thirdly, it is obvious that compounding routes to polymer nanocomposites will not be adaptable to this mechanism. Therefore, this may be the basis for the observations that incorporation of organoclays by melt compounding does not achieve the degree of property improvements achieved by incorporation of organoclays by the *in situ* process.

These successful *in situ* exfoliation processes have influenced a large amount of nanocomposite research, with the development of dispersing aids for monomers [6] and tethers functionalized with a wide variety of polymerizable groups [25], initiators [22], and catalysts [23]. Unfortunately, these efforts to achieve exfoliation add complexity and cost which are barriers to the eventual commercialization of polymer nanocomposites.

Experimental observations regarding *in situ* incorporation of organoclay IV into polyesters do not fit this general mechanism for the *in situ* exfoliation of 12-aminododecanic acid-treated organoclay in nylon 6. For example organoclay IV is not intercalated by ethylene glycol, dimethyl terephthalate, or PET. Nevertheless organoclay IV is essentially completely exfoliated after melt phase polymerization, as indicated by oxygen permeability measurement and analyses by transmission electron microscopy and X-ray. However, some samples, especially those comprising high loadings of organoclay IV, show a basal spacing that is the same as the original basal spacing of organoclay IV (3.4 nm). Moreover, equivalent nanocomposites are obtained whether organoclay IV is added before or after transesterification of the ethylene glycol and dimethyl terephthalate.

In order better to understand the mechanism for exfoliation of organoclay IV, polymerizations were stopped for analyses after transesterification at 220 °C and at varying times during polycondensation at 280 °C, ranging from 0 to 45 min. The target composition was a glycol modified PET containing 10 wt % organoclay IV, which gives a target ash value of about 5 wt %. This value was selected because of the desire to have a high enough concentration of organoclay IV to give reliable X-ray information, while having a low enough concentration to avoid potential concentration effects. Wide-angle X-ray scattering (WAXS) analyses of the samples and of a physical mixture of 10 wt % of organoclay IV with PET showed that:

(a) the X-ray intensity of the original 3.4 nm basal spacing of organoclay IV after transesterification is about the same as that of the physical mixture,

(b) the X-ray intensity at about 3.4 nm decreases to about 10 % of the original value very early during polycondensation and remains constant throughout the polycondensation time, and

(c) none of the samples exhibit any other clay basal spacings.

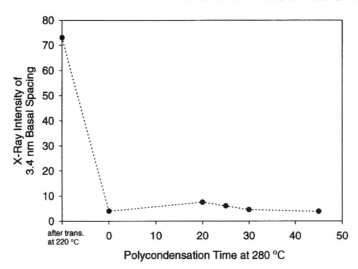

Figure 8 Plot of X-ray intensity of the 3.4 nm basal spacing of organoclay IV in polyethylene terephthalate nanocomposites prepared by *in situ* incorporation of organoclay IV and stopped after transesterification of 220 °C (− 10), and at various times during polycondensation at 280 °C. The X-ray intensity of a 10 wt% physical mixture of organoclay IV with polyethylene terephthalate is about 72

Figure 8 presents a plot of X-ray intensity versus the polycondensation time at 280 °C. Analysis of the samples by transmission electron microscopy (TEM) showed that:

(a) after transesterification no individual platelets are observed,
(b) all of the polycondensation samples have high concentrations of individual platelets along with some tactoids and agglomerates, and
(c) qualitatively, the concentration of individual platelets increases with increasing polycondensation time.

Then, as a final experiment on these samples, each was solid-state polymerized to increase the molecular weight sufficiently to permit the preparation of 10 mil thick compression molded films. The resulting oxygen permeabilities of the films were essentially equivalent for each of the samples.

Based on the above results, the following description is proposed for the process of exfoliation of organoclay IV during polymerization of PET. Firstly, upon reaching a critical molecular weight, oligomeric PET swells about 90 % of the organoclay IV into a disordered, essentially exfoliated, state. Although not known, the cause of the attraction may be a favorable balance of repeat units and hydroxyl end groups, favorable kinetics for intercalation of the low molecular weight species into organoclay IV, or, most likely, their combination. Then, during the rest of polycondensation, the exfoliated platelets continue to separate and disperse uniformly throughout the matrix, while the ordered tactoids remain unchanged. The critical hypothesis is that there exists a low molecular weight PET for which exfoliation of organoclay IV is favorable.

Thus, the process for the exfoliation of organoclay IV in PET appears to operate by a different mechanism to the exfoliation of clay in nylon 6 since the swelling of the clay by the monomers is not required and since the polymerization does not have to occur in the gallery of the clay. The exfoliation process of organoclay IV in PET may also apply to other condensation polymers including polyesters, polyamides, polyimides, and polyurethanes.

In order to test the critical hypothesis of this mechanism, 10 wt% of organoclay IV was melt compounded with glycol modified PETs with inherent viscosities ranging from 0.08 to 0.50 dL/g. As shown in Figure 9, the X-ray intensity of the 3.4 nm basal spacing of organoclay IV decreases as the inherent viscosity of the polymer decreases. These results are consistent with the critical hypothesis that there exists a low molecular weight PET for which exfoliation of organoclay IV is favorable.

These results show that an oligomer concentrate process to prepare PET nanocomposites is possible. This process is a specific case of the expanded

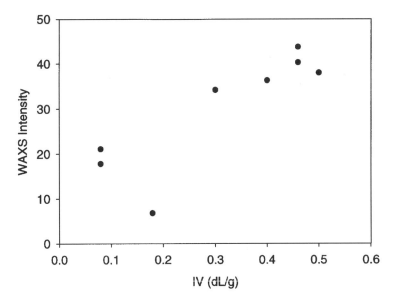

Figure 9 Plot of X-ray intensity of the 3.4 nm basal spacing of organoclay IV in polyethylene terephthalate nanocomposites prepared by the melt compounding of polyethylene terephthalate with 10 wt% organoclay IV. The X-ray intensity of a 10 wt% physical mixture of organoclay IV with polyethylene terephthalate is about 72

organoclay approach in which the expanding agent is an oligomeric version of the matrix polymer [19]. For PET this involves the use of oligoethylene terephthalate to prepare PET nanocomposites by the following steps:

(a) melt compounding of oligomeric PET with organoclay and
(b) increase in the molecular weight of the nanocomposite by one of several methods including solid state polymerization, chain extension, compounding into high molecular weight polymer, and their combinations and variations.

PET nanocomposites prepared by the oligomer concentrate process have achieved significantly improved barrier properties compared with those of nanocomposites prepared by the melt blending of organoclay with high molecular weight PET and have, in some cases, achieved essentially equivalent barrier results to those achieved by PET nanocomposites prepared by the *in situ* process.

4 MULTILAYER NANOCOMPOSITE CONTAINERS

In addition to monolayer PET containers, polymer nanocomposite multilayer food and beverage containers are of considerable research and commercial interest in the packaging industry. There are several potential approaches for the use of nanocomposite technology in multilayer containers. One approach involves the use of PET nanocomposites as the barrier layer in multilayer containers. A second approach involves the use of polymer nanocomposites comprising a high-barrier matrix polymer, such as high-barrier nylon, as the barrier layer. A third approach involves the application of a nanocomposite coating to preformed PET film and bottles.

4.1 PET NANOCOMPOSITES

Placing a PET nanocomposite into an internal layer of a multilayer container has some advantages. One advantage is that multilayer containers have neat polyethylene terephthalate as the food contact surface. Another potential advantage is that using a high-barrier PET nanocomposite as the barrier layer should significantly improve recycle compared with typical multilayer containers comprising other high-barrier polymers. Another advantage, one that is specific for polymer nanocomposites, is that multilayer containers comprising polymer nanocomposites have lower haze than monolayer containers comprising polymer nanocomposites [26]. It is believed that sandwiching the nanocomposite between two layers of unfilled polymer masks light-scattering surface roughness caused by clay that is present at or near the surface of the polymer upon stretching. One disadvantage is that placing the PET nanocomposite in a multilayer structure dilutes the barrier improvement achieved by the clay.

4.2 HIGH-BARRIER NYLON NANOCOMPOSITES

Nylon nanocomposites are ideal for multilayer containers because they have high gas barrier, they process similarly to PET, and they provide adhesion to PET [27]. In addition, the nylon nanocomposites are designed to be a good match for existing multilayer technologies, including preform injection molding machines and bottle stretch blow molding machines.

Figure 10 presents a plot of the whole-bottle oxygen transmission rates for 0.5 L, 29 g trilayer bottles comprising varying vol % of a center layer of a developmental nylon nanocomposite containing 1–3 wt % clay, as measured by wt % ash [28]. Also represented on the plot are the oxygen transmission rates of similar bottles comprising a high-barrier nylon. As is evident from the plot in Figure 10, the nylon nanocomposite can be used at a similar thickness to high-barrier nylon to achieve a significant reduction in oxygen transmission. Alternatively, the nylon nanocomposite can be used as a significantly thinner layer to achieve the same oxygen transmission while minimizing the amount of non-PET material, which may be important for cost and recycle considerations. The nylon nanocomposites can also be at intermediate thicknesses to optimize the barrier properties and the material usage for the target application. Work is in progress to improve barrier performance, and such materials may see commercial applications in the near future.

The nylon nanocomposites provide oxygen barrier that is significantly improved compared with high-barrier nylon. The actual performance in bottles depends on many factors including the bottle weight and design, the vol % and placement of the barrier layer, selection of the nanocomposite, and the clay loading. Using these variables, containers can be designed to meet the barrier needs of a wide variety of targeted applications. The nylon nanocomposites provide a wide variety of bottle

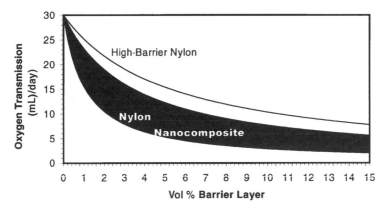

Figure 10 Plots of whole-bottle oxygen transmission rate versus vol % barrier layer consisting of a high-barrier nylon and nylon nanocomposites containing 1–3 wt % silicate (ash). The data were measured on 0.5 L, 29 g, three-layer bottles comprising the barrier material in the center layer

designs using existing processing windows. Photographs of some of the containers of various sizes, shapes, and colors that have been produced during the course of the evaluations are presented in Figure 11, see facing page [27]. As can be seen from the photographs, the appearance is good for colored containers, such as amber and green bottles. Still, haze remains a central issue for the use of polymer nanocomposites in stretch blow molded containers, and the level of haze currently exhibited by trilayer bottles comprising nylon nanocomposites is typically about three to 15 %. This level of haze is good for amber containers, but may not be sufficient for clear, colorless containers where the increase in haze is more apparent. Considerable research work is needed to understand and reduce haze in order for these materials to enjoy commercial success.

5 NANOCOMPOSITE COATINGS

Another multilayer approach for using nanocomposite technology for improving the gas barrier properties of PET is to apply a very thin layer of a nanocomposite coating after forming the article. A highly loaded thermoset melamine formaldehyde nanocomposite coating that imparts very high barrier to PET film has been developed [28]. Another approach is the use of clay and other platelet materials as additives that impart high barrier to the epoxy resin coatings applied to PET bottles prior to filling [29]. It is reported that these highly loaded nanocomposite coatings have very high gas barrier, and thus it is understandable that a considerable amount of research is being devoted to these nanocomposite coatings.

6 SUMMARY

The platelet structure of aluminosilicate materials has proven its ability to improve the barrier properties of polymer materials according to a tortuous path model in which a small amount of platelet particles significantly reduces the permeability of gases through the nanocomposite. The key for nanocomposite technology is exfoliation of the clay into its individual platelets, thereby achieving the greatest barrier improvement and lowest haze. There are many ways in which nanocomposite technology can be used to improve the barrier properties of PET and PET containers. These technologies include the incorporation of clays into PET for monolayer containers and the formation of multilayer structures in which the clay is exfoliated into one or more internal layers of a PET or nylon nanocomposite. External barrier layer coatings, containing clays, are also under development for barrier applications with PET.

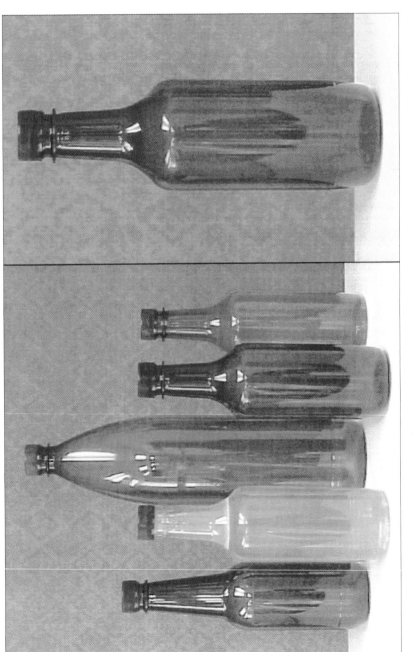

Figure 11 Photographs of examples of containers of various sizes, shapes, and colors containing an internal layer of a nylon nanocomposite

7 ACKNOWLEDGEMENTS

The authors wish to acknowledge the many Eastman Chemical Company scientists, technicians, and support personnel who contributed to the body of work discussed here, the management of Eastman Chemical Company for supporting this work and its publication, Southern Clay Products for providing refined sodium montmorillonite and organomontmorillonites, and Nanocor, Inc., for providing polymer-intercalated clay–ethylene glycol gels and Nanomers®.

8 REFERENCES

1. Barbee, R. B. and Wicker, T. US Pat. 4,440,922 (to Eastman Kodak Company) (4/3/84).
2. Davis, T. and Barbee, R. B. US Pat. 4,560,741 (to Eastman Kodak Company) (12/24/85).
3. Harada, M. *Plast. Eng.*, **44**(1), 27 (1988).
4. Messermith, P. B. and Giannelis, E.P. *J. Polym. Sci., Part A, Polym. Chem.*, **33**, 1049 (1995).
5. Neilson, L. E. *J. Macromol. Sci. (Chem.)*, **A1**(5), 929–942 (1967).
6. Frisk, P. and Laurent J. US Pat. 5,876,812 (to Tetra Laval Holdings and Finance, SA) (3/2/99).
7. Usuki, A., Mixutami, T., Fukushima, Y., Fujimoto, M., Fukumori, K., Kojima, Y., Sata, N., Kurauchi, T. and Kamigaito, O. US Pat. 4,889,885 (to Toyota) (12/26/89).
8. Ke, Y., Long, C., Ke, Y. and Oi, Z. *J. Appl. Polym. Sci.*, **71**, 1139 (1999).
9. Beall, G. W., Tsipursky, S., Sorokin, A. and Goldman, A. US Pat. 5,578,672 (to Amcol International Corporation) (11/26/96).
10. Tsipursky, S., Beall, G. W., Sorokin, A. and Goldman, A. US Pat. 5,721,306 (to Amcol International Corporation) (2/24/98).
11. Matayabas Jr, J. C., Turner, S. R., Sublett, B. J., Connell, G. W. and Barbee, R. B. PCT Int. Appl. WO 98/29499 (to Eastman Chemical Company) (7/9/98).
12. Barbee, R. B., Matayabas Jr, J. C. and Gilmer, J. W. PCT Int. Appl. WO 99/02593 (to Eastman Chemical Company) (1/21/99).
13. Vaia, R. A. and Giannelis, E. P. *Macromolecules*, **30**, 7990 (1997).
14. Vaia, R. A. and Giannelis, E. P. *Macromolecules*, **30**, 8000 (1997).
15. Hackett, S. E., Manias, E. and Giannelis, E. P. *J. Chem. Phys.*, **108**(17), 7410 (1998).
16. Lyatskaya, Y. and Balazs, A. C. *Macromolecules*, **31**, 6676 (1998).
17. Goettler, L.A. and Recktenwald, O. W. in Proceedings of *Additives '98*, 17 February, 1998.
18. Maxfield, M., Shacklette, L. W., Baughman, R. H., Christiani, B. R. and Eberly, D. E. PCT Int. Appl. WO 93/04118 (to Allied-Signal, Inc.) (1993).
19. Barbee, R. B., Matayabas Jr, J. C., Trexler Jr, J. W. and Piner, R. L. PCT Int. Appl. WO 99/32403 (to Eastman Chemical Company) (7/1/99).
20. Trexler Jr, J. W., Piner, R. L., Turner, S. R. and Barbee, R. B. PCT Int. Appl. WO 99/03914 (to Eastman Chemical Company) (1/28/99).
21. Okada, A., Usuki, A., Kurauchi, T. and Kamigaito, O. inMark, J. E., Lee, C. Y.-C. and Biancon, P. A. (Eds), *Hybrid Organic–Inorganic Composites*, ACS Symposium Series 585, 1995, Ch. 6, pp. 55–65.
22. Okada, A., Fukushima, Y., Kawasumi, M., Inagaki, S., Usuki, A., Sugiyama, S., Kurauchi, T. and Kamigaito, O. US Pat. 4,739,007 (to Toyota) (1998).
23. Lan, T., Kaviratna, P. D. and Pinnavaia, T. J. *Chem. Mater.*, **7**, 2144 (1995).
24. Wang, Z., Lan, T. and Pinnavaia, T. J. *Chem. Mater.*, **8**, 2200 (1996).

25. Biasci, L., Aglietto, M., Ruggeri, G. and D'Alessio, A. *Polym. Advd. Technol.*, **6**, 662 (1995).
26. Bagrodia, S., Germinario, L. T., Piner, R. L. and Trexler Jr, J. W. PCT Int. Appl. WO 99/44825 (to Eastman Chemical Company) (1999).
27. Matayabas Jr, J. C. in Proceedings of *NovaPak Europe '99*, 23 September, 1999.
28. Harrison, A. G., Meredith, W. N. E. and Higgins, D. E. US Pat. 5,571,614 (to Imperial Chemical Industries plc) (11/5/96).
29. Carblom, L. H. and Seiner, J. A. PCT Int. Appl. WO 98/24839 (to PPG Industries, Inc.) (6/11/98).

PART IV
Structure and Rheology

12

Structural Characterization of Polymer–Layered Silicate Nanocomposites

R. A. VAIA
Air Force Research Laboratory, Materials and Manufacturing Directorate,
WPAFB, OH, USA

1 INTRODUCTION

The distinguishing feature of polymer–layered silicate nanocomposites (PLSNs) and other polymer-based nanostructured materials is their morphology, in which the size of the inorganic particles is comparable to the spacing between particles, both of which are nanoscopic (1–100 nm). The conformation, dynamics and orientation distributions of the matrix polymers will depend on, as well as influence, the hierarchical inorganic morphology. Thus, detailed characterization of the nanoscale morphology of both the layered silicate and the polymer is critical to establish structure–property relationships for these materials. Conventionally, morphological characterization is dominated by wide-angle X-ray diffraction and transmission electron microscopy.

The objective of this chapter is to provide a summary of microscopy and scattering techniques to determine the morphology of PLSNs. Initially, silicate structure and PLSN fabrication techniques are summarized with regard to their impact on final PLSN morphology. Then, the breadth of potential morphologies is qualitatively discussed with respect to transmission electron microscopy studies, establishing parallels with other mesoscopic systems such as liquid crystalline polymers and block copolymers. Next, details associated with the Bragg–Brentano powder diffraction geometry are discussed, emphasizing the influence of layer order on the diffraction pattern, pitfalls associated with experimental procedures and

Polymer–clay nanocomposites Edited by T. J. Pinnavaia and G. W. Beall
© 2000 John Wiley & Sons Ltd

opportunities for detailed quantification of the structure of intercalated chains. Finally, small-angle scattering studies of layered silicate distribution and the structure of polymer crystallites in PLSNs are reviewed, as well as studies addressing the mesoscopic and microscopic arrangement of silicate layers in small molecule solutions.

2 BACKGROUND

Over the last decade, the utility of inorganic nanoparticles as additives to enhance polymer performance has been established [1]. Of particular interest are polymer–layered silicate nanocomposites (PLSNs) because of their demonstrated enhancements, relative to an unmodified resin, of a large number of physical properties including mechanical properties [2–6], barrier [5, 6], flammability resistance [7], ablation performance [8], environmental stability [9] and solvent uptake [10]. Furthermore, the nanoscale morphology affords opportunities to develop model systems consisting entirely of interfaces, and to study the structure and dynamics of confined and tethered chains using conventional bulk characterization techniques such as differential scanning calorimetry (DSC) [11–13], thermally stimulated current (TSC) [14], rheology [15], NMR [16] and spectroscopy [17].

In general, the greatest physical property enhancements for PLSNs are achieved with less than 4 vol % addition of a nanoscale dispersion of 1 nm thick silicate layers with diameters between 20 and 500 nm. These enhancements appear to be a general phenomenon related to the nanoscale dispersion of layers, but the degree of property enhancements is not universal for all polymers. The strength of interactions between the polymer and the silicate as well as the size and rigidity of the inorganic particles have been shown to influence the extent of enhancement and viscoelastic behavior associated with processing [3,18–20]. Dominant, though, is the dependence of property enhancements on layer distribution. Simply stated, the objective of fabrication and processing is to transform an initially heterogeneous, microscale morphology to a homogeneous morphology on the nanoscale. Homogenization of a system consisting of large-aspect-ratio (20–1000) layers, initially arranged in crystallite stacks (melt processing) or a percolative network (solution, *in situ*, emulsion or dispersion processing), must be expected to produce a hierarchical distribution of layers with a strong relationship to the specific processing history. The final morphology will consist not only of nanoscale layer separation but also meso- and microscale correlations related to fabrication approach, processing history and packing restrictions of the large-aspect-ratio layers.

Thus, given the potentially complex morphology, the fundamental mechanisms for property enhancements in PLSN, whether related to rigid-particulate reinforcement (e.g. interparticle connectivity and percolation) or to alterations of equilibrium chain conformations near the polymer–inorganic interface, are not fully understood. Note that issues associated with complex multiscale morphology are not entirely new or restricted to intercalated and dispersed layered silicates in polymers. Many parallels may be drawn with other mesoscopic systems, such as semicrystalline polymers, liquid crystalline polymers, block copolymers and colloidal suspensions, all providing characterization techniques and insight to facilitate establishment of structure–property relationships for PLSNs.

To begin to characterize and understand the possible complex hierarchical morphologies that occur in PLSNs, it is useful initially to review the structural details of layered silicates and the three general PLSN synthetic strategies with regard to their impact on final PLSN structure.

2.1 LAYERED SILICATES

Two general types of layered silicate (phyllosilicate), smectites and layered silicic acids, have been utilized for polymer nanocomposites. Conceptually, they consist of a weakly interacting stack of oxide layers. The oxide layers represent nanoscale units analogous to those in layered chalcogenides, double-layered hydroxides, graphite, deoxyribonucleic acid (DNA), imogolite and carbon nanotubes [21–23].

Most commonly utilized in PLSNs are the smectites which possess the same structural characteristics as the well-known minerals talc and mica [24–26]. Their crystal structure is characterized by a $0.96\,nm$ thick silicate layer consisting of two silica tetrahedral sheets fused to an edge-shared octahedral sheet of nominally alumina or magnesia (Figure 1). Depending on the particular silicate (synthetic or natural), the lateral dimensions of the layers vary from $20\,nm$ [laponite $Na_{0.7}(Mg_{5.4}Li_{0.4})Si_8O_{20}(OH)_4$] to greater than tens of microns [vermiculite $K_{1.8}(Mg_{4.4}Al_{1.0})(Si_{5.5}Al_{2.3}Fe_{0.1})O_{20}(OH)_4$]. Isomorphous substitution within the silicate layer generates a net negative charge on the layer, which is delocalized over surrounding oxygen atoms [27]. This layer charge is generally counterbalanced by hydrated alkali metal and alkali earth cations in the gallery. Exchange of the interlayer cations is relatively facile, allowing for the synthesis of numerous exchange ion derivatives of layered silicates. Typical cation exchange capacities (CEC) are in the range $60–120\,meq/100\,g$. Table 1 summarizes the classification, location of isomorphous substitution and nominal chemical composition of commonly used smectites in nanocomposites [28–35].

Layered silicic acids consist of nominally edge-shared silica tetrahedra. The family of layered silicic acids has at least five members [36]: kanemite ($NaHSi_2O_5.7H_2O$; $d_{001} = 1.03\,nm$), makatite ($Na_2Si_4O_9.5H_2O$; $d_{001} = 0.903\,nm$), octosilicate ($Na_2Si_8O_{17}.9H_2O$; $d_{001} = 1.10\,nm$), magadiite ($Na_2Si_{14}O_{29}.10H_2O$; $d_{001} = 1.56\,nm$) and kenyaite ($Na_2Si_{20}O_{41}.10H_2O$; $d_{001} = 1.20\,nm$). The homologous series exhibits varying layer thickness and stiffness, offering opportunities to examine the effect of these molecular level properties on the ultimate physical properties of PLSNs [19]. Only the crystal structure of a makatite layer is known and investigations are ongoing to determine the crystal structure of the remaining members of the family [37]. The net negative layer charge is balanced by exchange-

Table 1 Notional structure and chemistry of smectites commonly used in PLSNs

	Montmorillonite[a]	Hectorite[b]	Saponite[c]	Fluorohectorite[d]	Laponite[e]
Type	Dioctahedral	Trioctahedral	Trioctahedral	Trioctahedral	Trioctahedral
Substitution	Octahedral	Octahedral	Tetrahedral	Octahedral	Octahedral
Unit Cell Formula [26,34] ($x \sim 0.67$)	$M_x[Al_{4-x}Mg_x]$ $Si_8O_{20}(OH)_4$	$M_x[Mg_{6-x}Li_x]$ $Si_8O_{20}(OH)_4$	$M_y[Mg_{6-x}Fe_x^{3+}]$ $[Si_{8-y}Al_y]O_{20}(OH)_4$	$Li_{1.12}[Mg_{4.88}Li_{1.12}]$ $[Si_8O_{20}]F_4$	$Na_{0.7}[Mg_{5.4}Li_{0.4}]$ $Si_8O_{20}(OH)_4$

[a] For example: SWy1 and SWy2 (Source Clay Mineral Repository, Univ. Missouri): $M_{0.68}^+[Al_{3.32}Mg_{0.4}Fe_{0.26}^{3+}Fe_{0.12}^{2+}][Si_{7.84}Al_{0.16}]O_{20}(OH)_4$ [28, 29]; SAz-1 (Source Clay Mineral Repository, Univ. Missouri): $M_{0.56}^+[Al_{2.68}Mg_{0.12}Fe_{0.14}^{3+}]Si_8O_{20}(OH)_4$ [30]; Mnt (Nanocore, Inc.): $M_{1.08}^+[Al_{3.38}Mg_{0.54}][Si_{7.64}Al_{0.36}]O_{20}(OH)_4$ [31].
[b] For example: SapCa-1 (Source Clay Mineral Repository, Univ. Missouri) [29].
[c] For example: SHCa-1 (Source Clay Mineral Repository, Univ. Missouri) [32].
[d] Synthetic hectorite, Corning Inc. [33]. [e] Synthetic hectorite, Laporte Industries, Ltd UK, Laponite Technical Bulletin, L104/90/A.

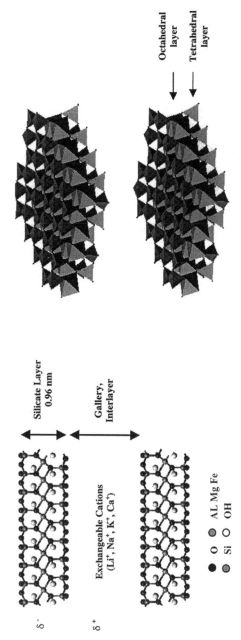

Silicate Layer 0.96 nm

Gallery, Interlayer

Exchangeable Cations (Li$^+$, Na$^+$, K$^+$, Ca$^+$)

δ^-

δ^+

AL Mg Fe

O

Si

OH

Octahedral layer

Tetrahedral layer

Figure 1 Crystal structure of 2:1 layered silicates (smectites). Van der Waals interlayer or gallery containing charge-compensating cations (y M$^+$) separates covalently bonded oxide layers 0.96 nm thick

able sodium cations in the interlayer, allowing for synthesis of numerous variations [38].

Smectites and silicic acids are hydrophilic, arising from the hydration of the interlayer alkali metal and alkali earth cations. Thus, they are incompatible with many polymer systems. However, ion exchange reactions with cationic surfactants, such as organic ammonium and phosphonium ions, render the layered silicate organophilic. These organophilic silicates are commonly referred to as organically modified layered silicates (OLSs). The equilibrium layer separation and conformation of the onium cations depend on charge per area of the layer (i.e. cation exchange capacity) and the size and molecular structure of the onium cations [39]. The large majority of OLSs used for PLSNs contain alkylammonium surfactants. The alkyl chains in the interlayer adopt conformations ranging from liquid-like to solid-like, depending on the chain length and exchange capacity (Figure 2) [17, 39]. The cations modify interlayer interactions by effectively lowering the interfacial free energy of the silicate. Furthermore, they can catalyze interlayer interactions, initiate polymerizations or serve as anchoring points for the matrix and thereby improve the strength of the interface between the polymer and inorganic. Industrially, OLSs are extensively used as gelling agents, thickeners and fillers [40].

The layered silicates of interest for PLSNs exhibit a hierarchical morphology whose habit depends on everything from processing history of the powders (freeze drying, air drying, coagulation) to size of the individual silicate layers. With regard to the distribution of the silicate layers in PLSNs, three general levels of structure can be identified: *crystallite* (*tactoid*), *primary particle* and *aggregates* [41, 42]. The crystallite is the fundamental unit of the powder morphology. High-resolution transmission electron microscopy (HRTEM) studies [43] show that layer stacking occurs to a much greater extent than lateral registry between unit cells (3–10 layers determined via breadth of *hkl* reflections). These layer stacks, often containing upwards of 100 layers, are commonly referred to as tactoids. The crystallites are the critical structure that must be disrupted to achieve uniform dispersion of the layers in the polymer. Polymer penetration is generally more facile within the primary particle (between the crystallites) than within the crystallites [41]. These weakly agglomerated primary particles, i.e. aggregates, form the powders that are mixed with polymers, solvent or monomers. The aggregates contain upwards of 3000 individual silicate layers and have potentially accessible surface areas of $700–800 \, \text{m}^2/\text{g}$. Finally, the chemical formula and layer charge in naturally occurring minerals represent an average over the entire sample because of compositional variations between layers and between crystallites.

2.2 POLYMER–LAYERED SILICATE FABRICATION

The first PLSN fabrication method is *solution mediated processing*. Here, a solvent is used to disperse the OLS as well as the polymer [21, 23, 44–46]. Upon removal of the solvent, uniform mixing of polymer and layered silicate is achieved. The final

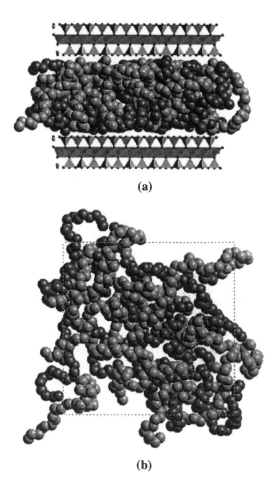

(a)

(b)

Figure 2   Molecular dynamic simulation of the disordered liquid-like interlayer structure of dioctadecylammonium-exchanged montmorillonite (CEC = 100 meq./100 g, $d_{001} = 2.4$ nm): (a) cross-section and (b) planar projection of the interlayer [39]. Gray and white chains correspond to molecules associated with the upper and lower surfaces; dashed lines correspond to the periodic boundary of the simulation. The ratio of *gauche* to *trans* conformers depends on packing density, chain length and temperature [17]

structure of the PLSN will depend on the thermodynamics of the three (or more)-component system and the extent of polymer adsorption to and bridging between layers. Detailed understanding of the influence of molecular factors (e.g. intermolecular interactions, surface energies) and kinetic factors (chain absorption, solvent evaporation rate and shear mixing) has yet to be examined. Recent variations of solution mediated processing include emulsion or suspension polymerizations, in which the silicate layers are suspended in an aqueous phase and a monomer is polymerized in a second phase within the suspension. Note that the silicate need not be organically modified for these variations. In general, the final morphologies, as well as the degree of polymer–OLS interaction, may be reminiscent of layer correlations in solution.

The second general fabrication approach is *in situ polymerization*, where the monomer is used directly as the solublizing agent for the OLS [2, 4, 6, 47, 48]. After combining the OLS and monomer, the monomer is polymerized in the presence of the layers. Functionalization of the OLS is chosen to control polymerization rates and initiation points so that layer separation occurs before or during polymerization. This is especially critical for thermoset systems where the extent of crosslinking reactions determines the gel point of the matrix and ultimately the extent of layer dispersion and final PLSN morphology.

Finally, *melt processing* is characterized by combining the OLS powder and the polymer in an uncrosslinked polymer melt [49, 50]. Polymer–OLS interactions determine the extent of layer separation and inclusion of polymer. Shear flows encountered in traditional polymer melt processing techniques facilitate homogenization of the PLSN morphology as well as global alignment of the layers. This approach is commercially advantageous, especially with regard to modification of commodity plastics.

3 NANOSCALE MORPHOLOGIES

Two idealized polymer–layered silicate morphologies are conventionally discussed, *intercalated* and *exfoliated*. Briefly, intercalated structures are analogous to traditional inclusion or guest–host compounds and result from polymer penetration into the interlayer and subsequent expansion to a thermodynamically defined equilibrium spacing [51, 52]. The expansion of the interlayer is finite, of the order of 1–4 nm, and is nominally associated with incorporation of individual polymer chains. In contrast, idealized exfoliated morphologies consist of individual nanometer-thick silicate layers suspended in a polymer matrix, and result from extensive penetration of the polymer within and delamination of the crystallites. The greatest property enhancements are observed for the exfoliated PLSNs. These two structures are thought to be thermodynamically stable and have been predicted by mean field [51] and self-consistent field models [52].

In practice, however, many systems fall between these idealized morphologies. Kinetic aspects associated with Brownian motion and shear alignment of the layers [20] as well as processing histories (melt processing and *in situ* polymerization) [53] will produce positional and orientational correlations between the plates. This is especially critical because of the relatively high viscosity of polymer melts. These kinetic factors will contribute to, and perhaps dominate, the development of a hierarchical morphology exhibiting nano- (1–100 nm), meso- (100–500 nm) and microlevel (500–10 000 nm) features. To develop a broader understanding of the potential PLSN morphologies, we initially consider the theoretical aspects of a collection of thin plates, followed by a discussion of microscopy of reported PLSNs.

3.1 MODELS

Owing to the large anisotropy of the silicate layer (diameter/thickness ~ 20–500), thermodynamically stable structures with positional (smectic) and orientational (nematic) ordering are anticipated for PLSNs. Although the silicate layers are not infinitely rigid, their phase behavior and dynamics can be understood using classical excluded-volume arguments. Analogous to rigid rod polymers [54], an ideal suspension of thin plates may be classified into three concentration regimes (dilute, semidilute and liquid crystalline), defined by the number of plates per volume, v, and the plate diameter, L. Dilute, isotropic solutions (no positional or orientational correlation exists between plates) occur where the volume per plate is greater than the excluded volume of a plate (defined as its volume of rotation and proportional to L^3). For these concentrations, the distance between the plates, $v^{-1/3}$, is larger than the lateral size of a plate, L. At higher concentrations ($v \geq 1/L^3 = v_1$), semidilute, isotropic solutions and liquid crystalline solutions occur, depending on the extent of intermolecular interactions between the plates. In semidilute isotropic solutions, the free rotation of a plate is hindered by neighboring plates, thus altering the system dynamics. However, plate–plate spacings are large enough for the effect on static properties to be negligible, resulting in an equilibrium orientation of a plate that is independent of its neighbors. At slightly higher concentrations ($v^* > v_1$) determined by the strength of intermolecular interactions, packing constraints will result in intermolecular correlation between plates, producing orientational ordering and macroscopic alignment at equilibrium, such as nematic-like ordering.

Thus, to a first approximation, the critical concentration for development of a superstructure, as characterized by plate–plate correlations, is of the order of v_1 $(1/L^3)$. Therefore, the critical volume fraction [volume of plate (L^2h)/excluded volume of plate (L^3)] will be proportional to α^{-1}, where the aspect ratio $\alpha = L/h$, and h is the thickness of the plate. Then, layer–layer correlations in PLSN $(20 < L < 500\,\text{nm}, \ h \sim 1\,\text{nm})$ should exist at concentrations as low as $0.2\,\text{vol}\%$ $(0.5\,\text{wt}\%)$, and thus should be anticipated in materials commonly examined. Furthermore, assuming parallel alignment of the layers at $v > v^*$ (nematic-like ordering), the layer–layer distance, d, will be approximately $h^*\phi^{-1}$ (where ϕ is the volume fraction of layers) and exhibit an inverse dependence on concentration.

Figure 3 shows the mean layer–layer separation for this idealized parallel alignment of layers that would result from various degrees of disruption of the crystallite stack. As the extent of exfoliation increases (thickness of plate, h, decreases), the mean distance between plates decreases at a given volume fraction of plates. For complete exfoliation of the individual layers ($h = 1\,\text{nm}$), the mean distance between layers is 100 and 25 nm for 1 and 4 vol% layered silicate, respectively. This nanoscale layer separation is comparable with the radius of

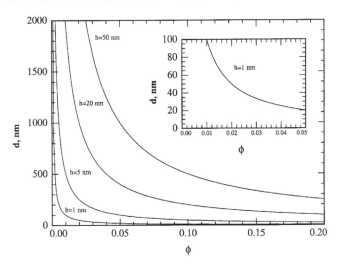

Figure 3   Equilibrium distance between uniformly aligned and dispersed plates of thickness h at various volume fractions of plates, ϕ. Distance between plates decreases as ϕ^{-1} ($h = 50$ nm corresponds to ~ 50 clay layers/crystallite; $h = 1$ nm is individual layers). For $h = 1$, h is comparable with the radius of gyration of high molecular weight polymers at volume fractions between 0.01 and 0.05 (inset)

gyration of high molecular weight polymers. Thus, perturbations of the equilibrium chain conformation of the matrix polymer, whether amorphous, crystalline or liquid crystalline, will occur, altering the fundamental structure–property relationship of the polymer relative to unperturbed bulk. Appreciation of the nanoscopic restrictions on chain conformations is critical and distinguishes PLSNs from conventional filled polymers containing micron-scale fillers in which the specific interfacial area of the filler is orders of magnitude less than the $\geqslant 700 \, m^2/g$ of the layered silicate. The entire matrix polymer within an exfoliated PLSN may be envisioned as interfacial in nature rather than bulk-like because of the large area and close proximity of the silicate surfaces.

Further refinements to the simple excluded volume argument have been proposed, which utilize an Onsager free-energy functional in combination with the Flory–Huggins mixing free energy and incorporate the anisotropic shape of the layers, entropic factors associated with a polymeric medium and intermolecular interactions between the various constituents [55]. These calculations show that the critical concentration for superstructure development depends on the molecular weight of the polymer and intermolecular interactions between the constituents of the system. Furthermore, higher-ordered superstructures, with increasing degrees of orientational and positional correlations between the layers (nematic, smectic A, columnar and crystal) are predicted at higher concentrations.

These higher-order structures, specifically nematic suspensions, have been experimentally observed in small molecule dispersions of layered particles, such as laponite [56] and vanadium pentoxide [57], and in poly(ε-caprolactam)/montmorillonite nanocomposites [58]. The presence of these ordered phases implies that domain boundaries and defect structures exist in dispersions of layered particles, paralleling liquid crystalline polymers and block copolymers. However, observation, characterization and determination of the importance of these defect structures to the physical properties of PLSNs are yet to be done.

3.2 TRANSMISSION ELECTRON MICROSCOPY

Experimental details relevant to transmission electron microscopy of layered silicates and PLSNs may be found in the literature [43]. Representative examples of PLSN morphologies are shown in Figures 4 to 8. Most obvious is the increased complexity of the observed morphology relative to the previously discussed idealized structures.

Figure 4 shows the morphology of a typical intercalated structure in which the volume fraction of polymer is much greater than that of layered silicate. The parallel

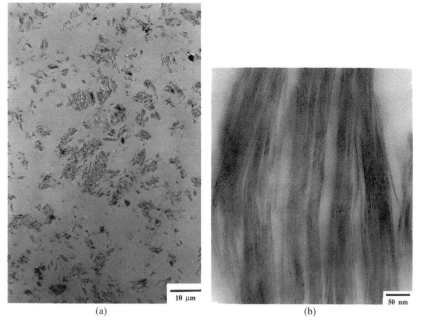

10 µm

(a)

50 nm

(b)

Figure 4 TEM bright-field images of an intercalated nanocomposite system (EPON 828/Jeffamine D-400 with 10 wt % OLS [48]: (a) micron-scale dispersion of primary particles; (b) intercalated crystallites (tactoids) of equally spaced silicate layers 0.05–0.5 µm in diameter. Note the flexibility of the silicate layers

dark lines in the micrographs are edges of individual silicate sheets. The original aggregates of the primary particles are disrupted, resulting in a micron-scale arrangement of primary particles in the polymer matrix. Within the primary particles, layer expansion swells the crystallites and the intercrystallite regions, preserving the highly parallel alignment of layers with nanoscopic separation.

Uniform expansion of the interlayers, as shown in Figure 4, is not always observed, however. Figure 5 shows a bright field electron micrograph of a poly(3-bromostyrene)–F12 (dodecylammonium-exchanged fluorohectorite) intercalate. As in Figure 4, micron-scale morphology is comprised of dispersed primary particles; however, layer expansion varies within the primary particle and crystallites. Smaller expansion predominately occurs in the interior, whereas layer expansion and individual layers are observed near the exterior of crystallites. These disordered intercalated structures occur during static melt processing, where homogenization of the layer stacks depends exclusively on Brownian motion and is impeded by a very high local concentration of layers, resulting in impingements. Generally, the smallest silicate layers predominately occur near the exterior of the crystallites and have been observed to separate more easily from the stack, resulting in intercalated crystallites surrounded by a distribution of uncorrelated layers [59].

Figure 5 Bright-field TEM of poly(3-bromostyrene)–F12 (dodecylammonium-exchanged fluorohectorite) [41]

In contrast to these intercalated PLSNs, morphologies containing both individual layers and structures reminiscent of the original crystallites occur, especially for thermoset systems and melt processing in the absence of high shear. Figure 6 shows a bright-field transmission electron micrograph of a thermoset nanocomposite containing 10 wt% Closite 30A in a diamine-cured (Jeffamine D-2000, Huntsman) epoxy (EPON 828, Shell). The low-magnification image shows submicron variation in the silicate content as well as packets containing parallel oriented layers with various degrees of swelling. Higher magnifications reveal individual layers. Similar microstructures have been reported for various thermoset systems [4, 6, 60] and cyclic oligomer synthesis [47]. For thermoset systems, these structures result from a competition between the rate of layer separation (Brownian motion and interlayer reactivity), the extent of layer–layer impingement and the rate of gelation of the surrounding matrix. As the uniformity of layer dispersion decreases, the local volume fraction of layers varies, implying that a mesoscopic superstructure consisting of compositional fluctuations exists. For example, contrast the micron-scale morphology in Figure 6 with that observed for ideal intercalants in Figure 4.

'Idealized' exfoliated structures have been reported for various thermoplastics and epoxies. However, in contrast to schematically depicted random distribution of layers, the orientation of the layers is at the minimum locally aligned and generally is globally aligned relative to processing history. Even at very low volume fraction of silicate (~ 0.01), preferential orientation of silicate layers is observed (Figure 7). Shear flows associated with processing have a substantial effect on macroscopic alignment of the layers. Figure 8 shows bright-field transmission electron micrographs of the mesostructure of:

(a) 5 wt% (NCH5) layered silicate/poly(caprolactam) nanocomposite synthesized by *in situ* polymerization of ε-caprolactam in the presence of montmorillonite organically modified by 12-aminolauric acid (Ube, Inc.) [2, 3];

(b) ~ 4 wt% (NLS4) layered silicate/poly(caprolactam) nanocomposite formed during melt processing of poly(caprolactam) (Capron B135WP, Allied Signal) and Cloisite 30A (organically modified montmorillonite, Southern Clay Products, Inc.) [8].

Both samples were preprocessed in an extruder and pelletized. Macroscopic orientation of the layers along the flow direction is observed, irrespective of the layer size. The average dimensions of the layers are approximately twice as large in NCH5 ($\sim 160 \pm 30$ nm) as in NLS4 ($\sim 70 \pm 20$ nm), relating to the different base montmorillonite used for the OLS. In general, the mean layer spacing is 35–45 nm in NCH5 and 40–60 nm in NLS4, in agreement with the excluded volume discussion (Figure 3). Note, however, that the layer distribution is not uniform in NLS4; a fraction of aggregates containing two or three silicate layers are observable. These partially exfoliated structures from melt processing may be associated with inhomogeneities (size and composition) inherent in natural smectites. Similar macro-

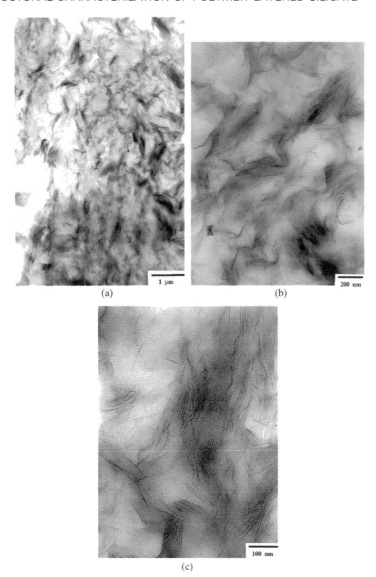

Figure 6 Bright-field TEM images of 10 wt% Closite 30A (Southern Clay Products, Inc.) in a diamine-cured (Jeffamine D-2000, Huntsman) epoxy (EPON 828, Shell) [48]: (a) submicron variation in silicate content as well as packets of parallel layers with various degrees of swelling; (b) regions with stacks of 25 layers with spacing of $\sim$5 nm (lower right) to regions with 3–5 layers separated by 10–20 nm (upper left); (c) individual silicate layers

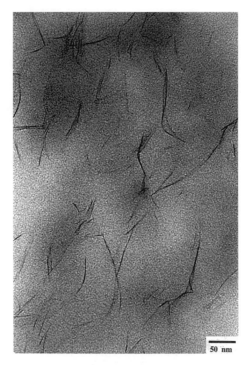

Figure 7 TEM bright-field image of exfoliated layers in an epoxy matrix [48]

scopic layer orientation has been observed in nylon 6 nanocomposite extrudate and films [2] as well as in solution processed polyimide (PMDA-ODA) films [5].

3.3 GENERAL CLASSIFICATION

The previous discussion indicates that a simple description of PLSN morphology as intercalated or exfoliated is far from adequate. Additional descriptors, such as *ordered*, *disordered* or *partial*, are helpful for accurate description of nanoscale morphologies. An expanded classification scheme is summarized in Figure 9. Disordered intercalated structures are characterized by weaker interlayer correlations sufficient to disrupt strong basal reflections observed in conventional wide-angle X-ray diffraction (discussed below). Furthermore, a minor fraction of smaller layers near the surfaces of the crystallites may be dispersed (e.g. Figure 5). On the other hand, PLSNs containing highly swollen interlayers still exhibiting registry associated with the original crystallite (e.g. Figure 6) or small stacks of 2–4 layers (e.g. Figure 8b) are *partially exfoliated*. Here, the layer correlation is not detectable by WAXD, and correct characterization requires microscopy. Finally, for true exfoliated layers, processing history and excluded volume (including intermolecular interactions) will

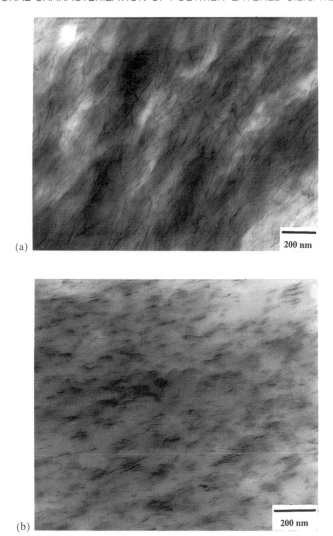

Figure 8 Bright-field TEM of extruded and pelletized poly(caprolactam) nano-composites: (a) *in situ* polymerized ε-caprolactam with 5 wt% OLS; (b) melt processed poly(caprolactam) with 4 wt% OLS

dictate ultralong-range (>10 nm) ordering of the layers above a critical concentration (e.g. Figures 7 and 8a). Maximum layer separation will depend on the volume fraction of silicate. Thus, depending on the degree of orientational (and translational) order and volume fraction, exfoliated PLSNs can further be described as *ordered* or *disordered*.

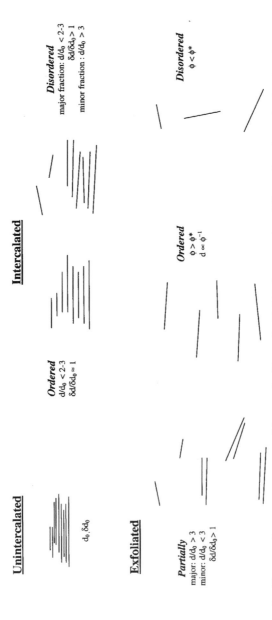

Figure 9 Nanoscale arrangements of layered silicate in PLSN that can be differentiated on the basis of: (1) the relative change in layer spacing and correlation (d and δd); (2) the relative volume fraction of single layers and layer stacks; (3) the dependence of single-layer separation on silicate volume fraction, ϕ, above a critical volume fraction for ordering, ϕ^*. Differentiation of morphologies based on $d/d_0 > 3$ is qualitatively associated with common lower-angle resolution in powder diffraction

Overall, the wealth of morphologies and heterogeneity of layer groupings indicates that the formation of intercalated and exfoliated PLSNs occurs by a more complex process than simple sequential swelling and separation of individual layers starting from the surface of the primary particle and crystallite [44, 61, 62]. Defect structures, local chemical inhomogeneities, electrostatic forces, viscoelastic properties of the polymer and stress fields arising from interlayer swelling will all contribute to mediate polymer transport and layer mobility and thus final morphology. These effects coupled with processing factors such as shear flows lead to more complex morphologies than those theoretically discussed to date. It is important to note that descriptors in Figure 9 only account for nanoscopic morphology, whereas the microscale arrangement of the crystallites or individual layers must also be determined and considered when establishing structure–property relationships. Previous efforts to ascertain this higher-level morphology for aqueous dispersion of smectite layers will be reviewed in Section 5.1.

To this point the discussion has exclusively focused on the distribution and arrangement of the silicate layers. However, the distribution of the layers will have an influence on, and be influenced by, the morphology of the polymer matrix. This is especially critical when layer separation is less than 2–3 times the radius of gyration of the polymer, where equilibrium chain conformations depend on interfacial interactions and extent of chain absorption at the silicate surface. Furthermore, the extent of chain–chain correlations such as liquid crystallinity and polymer crystallinity will be impacted by the 20–100 nm separation of layers. Mesoscopic structures inherent in polymers, such as crystal lamella and block copolymer ordering, are comparable in size to layer separation and thus present intriguing competition between various nanoscale morphologies. Very few detailed microscopy investigations of the polymer morphology in PLSNs have been reported.

4 POWDER DIFFRACTION

Microscopy is a useful tool to provide real-space information on the spatial distribution of phases and defect structures. However, it is not always feasible to use microscopy as a rapid characterization tool or for *in situ*, real-time studies. In contrast, diffraction and scattering rapidly provide globally averaged information, potentially over six or more orders of magnitude, and with the potential for *in situ*, real-time studies.

Traditionally, powder diffraction is used to characterize the structure of PLSNs. Most commonly used is the Bragg–Brentano reflection geometry (powder diffraction) in which the normal of the sample surface bisects the angle (2θ) between the incident and reflected beam. The extensive use of powder diffraction for PLSNs is largely based on the established procedures developed for identification and structural characterization of layered silicate minerals. Extensive discussion of the theoretical and practical aspects of powder diffraction relative to these minerals may be found in books by Brindely and Brown [25], Drits and Tchoubar [35], Bish and Post [63] and Moore and Reynolds [26].

4.1 EXPERIMENTAL FACTORS

Powder diffraction is used to monitor the position, full width at half maximum (fwhm) and intensity of the (001) basal reflection corresponding to the repeat distance perpendicular to the layers. Figure 10 summarizes general X-ray powder diffraction spectra, some from the PLSNs depicted in the micrographs. Figure 11 summarizes the general character of the X-ray spectra for various types of PLSN structures. For immiscible mixtures of polymer and OLS, the basal reflection does not change upon blending with the polymer. On the other hand, the finite layer expansion associated with intercalated structures results in a new basal reflection that corresponds to the larger gallery height. A decrease in the degree of coherent layer stacking (i.e. a more disordered system) results in peak broadening and intensity loss. In contrast, the extensive layer separation, beyond the resolution of Bragg–

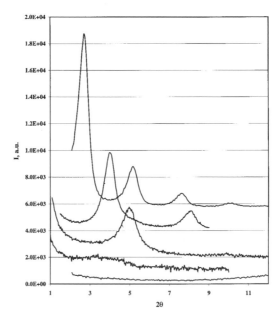

Figure 10   PLSN and OLS powder X-ray diffraction patterns, presented in top to bottom order, for: intercalated PLSN (EPON 828/Jeffamine D-400 with 10 wt% B34 [48]) shown in Figure 4; pristine quaternary ammonium-exchanged montmorillonite (B34, Rheox, Inc.); pristine fluorohectorite exchanged with dodecylammonium (F12); disordered intercalate (poly(3-bromostyrene)/F12 [41]) shown in Figure 5; partially exfoliated PLSN (EPON 828/Jeffamine D-2000 with 10 wt% Closite 30A [48]) shown in Figure 6

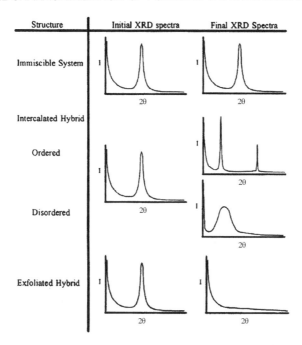

Figure 11 Schematic of powder X-ray diffraction spectra for various PLSN structures [51]

Brentano geometry, of exfoliated structures does not result in a new, observable, basal reflection, but leads to intensity loss and eventual disappearance of the unintercalated basal reflection. Real-time evolution of these scattering patterns provides detailed information on the dynamics of structural formation and intercalation, such as demonstrated by Vaia and coworkers examining the kinetics of polymer melt intercalation [64].

The large basal spacings ($d_{001} > 2.0\,\text{nm}$) and the absence of hkl reflections necessitates the collection of data at scattering angles, 2θ, of less than 10°. Because of their association with property enhancements, it is especially critical to determine the presence of exfoliated layers. However, layer disorder, silicate volume fractions less than 0.1 and experimental conditions all contribute to broadening and weakening of basal reflections. Thus, special considerations are necessary when preparing samples, performing experiments and interpreting diffraction spectra of PLSNs. Theoretical discussion and experimental details relative to sample preparation may be found in the literature [26, 63, 65, 66]. Powder diffraction spectra exhibiting the absence of basal reflections are often initially interpreted as evidence for an exfoliated PLSN. However, subsequent microscopy investigations show the presence of large packets of disordered layers separated by microns of polymer void of layers—far from the desired exfoliated morphology. For example, note the similarities in X-ray spectra between a disordered intercalate, a partially exfoliated PLSN and a fully exfoliated PLSN in Figure 10 and the corresponding micrographs.

Based on theoretical models of layer diffraction, Figure 12 shows the change in intensity relative to the mean number of layers per crystallite, $\langle N \rangle$, and the extent of layer disorder in the crystallite, α, for montmorillonite with a basal distance of $D = 3.0$ nm and a random crystallite distribution ($\sigma^* = 45°$) within the sample. As the crystallite size and internal disorder increase, the intensity decreases and the breadth of the basal reflections increases. In the absence of internal disorder, the breadths of the basal reflections are approximately constant at small 2θ. In contrast, internal disorder, α, broadens the higher-order basal reflections significantly more than (001). Additionally, the (001) reflection shifts to lower 2θ even though the mean layer spacing is still 3.0 nm. This arises from the product of a relatively weak interference peak with a steeply decreasing single-layer structure factor. Furthermore, the increase in featureless scattering at angles less than $2\theta_{001}$ arises from the increase in the interference function at these angles, resulting from the increased internal disorder. Additional shifts of the (001) reflection will occur for partially oriented systems, effectively broadening the basal reflection more, merging it with a monotonically decreasing signal at low 2θ.

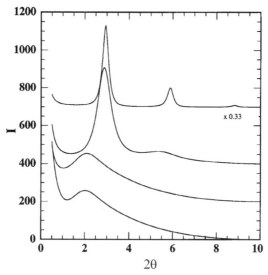

Figure 12   Calculated diffraction profiles for montmorillonite layers with basal distance of $D = 3.0$ nm and random crystalline distribution ($\sigma^* = 45°$) [65, 66]. From top to bottom, $\langle N \rangle = 5$, 5, 5 and 2 and $\alpha = 0.0$, 0.5, 1.5 and 1.5 nm, respectively, where $\langle N \rangle$ is the mean number of layers per crystallite and α is the extent of layer disorder. The effective magnitudes ($I_{max} - I_{back}$) of the first basal reflections are 1200, 450, 100 and 75, respectively. The intensity of the top curve is multiplied by 0.33 and all curves are vertically offset for clarity

In addition to these considerations, numerous experimental factors such as the θ-dependence of beam size, the θ-dependence of penetration depth, surface roughness, composition and sample alignment will also modify intensity at low angle and add to the uncertainty in resolving basal reflections from weakly scattering systems [26, 63, 65, 66]. The interlayer height derived from the maximum of the (001) reflection is approximate, since the maximum in the interference function is modulated by angular dependent factors that shift the position of the maximum. If numerous basal reflections are observable, it is more accurate to extrapolate to $\theta = 0$ the calculated interlayer distance from the recorded (00*l*)-order reflections relative to $\cos\theta_{(00l)}$ [35, 67].

Detailed modeling of the scattering profile can yield additional morphological data, such as crystallite size, extent of disorder and interlayer composition [65, 66]. For example, although determination of the particle thickness, L_z, using the Scherrer equation ($L_z = 0.9\lambda/b\cos\theta$) and the full width at half maximum, b, of the (001) reflection is rapid, it is only a *lower*-bound estimate because of the contribution to the peak breadth of internal disorder. Again, if available, the use of multiple-order reflections allows separation of these effects by extrapolation of the peak breadth at low 2θ to $\theta = 0$. The intercept will be proportional to the finite crystal size, whereas the slope will be proportional to the degree of internal disorder [35].

In summary, detailed analysis of powder diffraction data carefully corrected for experimental conditions, such as by complete profile fitting or using mineralogical analysis programs (e.g. MUDMASTER [68]), may yield new insights, such as the change in L_z and internal disorder upon polymer intercalation. Applying data analysis developed for mixed layered smectites [25, 26, 63] may yield additional information on staging (preferential intercalation of neighboring interlayers), especially in the early portions of polymer melt intercalation.

4.2 DENSITY PROJECTION: ELECTRON (X-RAY) AND NUCLEUS (NEUTRON)

For a highly ordered lamellar system, the correlation function is a series of narrow peaks with the same magnitude, and its product with the square of the structure factor essentially yields the magnitude of the one-dimensional structure factor $F^2(\theta)$ at integral multiples of the basal distance. After removal of experimental factors and Lorentz polarization [65, 66], observed scattering intensity is essentially a Fourier transform of the electron (X-ray) or nucleus (neutron) density distributions [67]. An inverse Fourier transform will provide detailed information of the atomic arrangements. However, this inverse problem is plagued by a loss of phase information arising from the dependence of scattered intensity on the square of the scattering amplitude $[F^2(\theta)]$ [35].

This issue may be circumvented by calculating the sign of $F(\theta_{00l})$ for various types of interlayer arrangements of intercalants. In the case of X-ray diffraction from smectites containing organic and polymeric intercalants, the interlayer species only moderately changes the magnitude of $F(\theta_{00l})$ and not the sign, which is determined by the structure of the smectite layer. Thus, the established structure of the smectite layers may be used to determine the relative sign of $F(\theta_{00l})$ determined from experimental data. Franzen [69] and Adams [70] used this approach to calculate one-dimensional density distributions of hexadecylammonium montmorillonite (X-ray) and tetrahydropyran (X-ray), 1,4-dioxane (X-ray) and pyridine (neutron) intercalants in Na^+-montmorillonite, respectively. The electron density projections for hexylde-cylammonium montmorillonite, originally interpreted with regard to a tilted paraffinic arrangements of alkyl chains, also agrees with disordered multilayer structures recently determined by Hackett and Manias via molecular dynamic simulations [39].

Numerous models, including a monolayer, bilayer and helical arrangement for the conformation of polyethylene oxide (PEO) within smectite interlayers, have been proposed based on X-ray diffraction and FTIR spectroscopy [71, 72]. Figure 13 shows the powder X-ray diffraction data for a highly ordered solution-cast film of PEO intercalated in Li^+-montmorillonite (SWy-2). The corresponding one-dimensional projection of electron scattering density normal to (001) is shown in Figure 14 along with modeled distributions for an ideal bilayer and helical conformation of intercalated PEO in montmorillonite (Table 1). The location and relative magnitude of the electron scattering density maximum for the aluminate octahedra and silicate tetrahedral layers derived from the experimental data agree well with those modeled

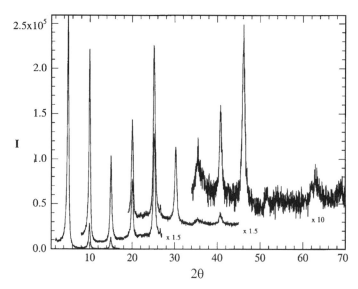

Figure 13 Powder diffraction pattern of a solution-cast film of intercalated polyethylene oxide in Li^+-montmorillonite (SWy-2). Data in the four overlapping regions were collected with increasing source slits and corrected for Lorentz polarization (highly oriented). Bragg spacing is 1.77 nm

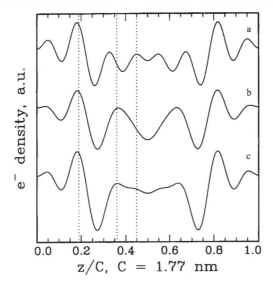

Figure 14 One-dimensional projection of electron scattering density perpendicular to the layer normal (001) for (a) an ideal helix and (b) an ideal bilayer configuration of intercalated PEO, as well as (c) the Fourier transform of experimental data (Figure 13). z/C is referenced with respect to the center (alumina tetrahedra) of the silicate layer. The interlayer spans from 0.28 to 0.72

from the crystallographic data [70]. The interlayer structure, however, does not agree with either the ideal bilayer or helical arrangements, but exhibits features reminiscent of each. The relatively featureless distribution of scattering length density alternatively may represent an interlayer chain conformation that is not rigidly restricted to a crystallographic lattice but is dynamic and disordered. This disordered chain conformation with dynamic mobility is in agreement with previous NMR [16] and thermally stimulated current measurements [14] of intercalated PEO montmorillonite and agrees with various molecular dynamic simulations of intercalated and near surface polymers [73, 74].

5 SMALL ANGLE SCATTERING: NEUTRON AND X-RAY

As the dimensions of the morphological features increase, absorption and complete external reflection prohibits the use of reflection geometries, such as Bragg–Brentano, to probe the bulk structure. However, small-angle scattering geometries, such as Laue transmission, provide resolution of the order of a few hundred nanometers with X-ray and neutron radiation. Even larger structures are accessible with Boson–Hardt cameras or six-circle goniometers. At these length scales, the atomic level details become lost as the dimension of the scattering entity increases.

Thus, the scattered intensity can be envisioned as arising from the coherent interaction of incident radiation scattered from a uniform medium with a varying electron (X-ray) or nuclear (neutron) composition. The structure is represented by a continuous function describing the density of scattering entities, $\rho(\mathbf{r})$, which is devoid of atomic or lattice details. Theoretical development and experimental details of scattering, both elastic and inelastic, by X-rays and neutrons are provided by Glater and Kratky [75], Wignall [76] and Feigin [77], respectively. Relative to the extensive use of powder diffraction, few examples of small angle scattering studies of PLSNs are reported.

5.1 DISPERSION OF LAYERED SILICATES

The dispersion of layered silicates in small molecules strongly parallels the process and potential morphologies of thermoset PLSNs before crosslinking. Overall, these smectite gels are idealized models for polymer–layered silicate dispersions, providing insight into possible higher-level layer–layer correlations and their impact on rheological and physical properties. Traditionally, microstructural models of these gels consisted of microflocculation of smectite platelets promoted by attraction between oppositely charged edges (arising from dissociation of amphoteric $-SiOH$, $-LiOH$, $-MgOH$ and $-AlOH$ on the terminus of the layer) and lateral surfaces of the layers and reinforced by *Van der Waals* interactions. This yielded the so-called *house-of-cards structure* (Figure 15).

More recent studies indicate that this mechanism for gelation is only favored at high ionic strengths (suspensions containing electrolytes) when the repulsive double layer potential between layers is screened [78]. Small-angle X-ray and neutron scattering studies imply that a more complex multiscale, multiphase structure exists that depends on plate size, plate shape, concentration, ionic strength and processing history of the medium [79]. Figure 15 also summarizes these biphasic models. Other studies of the concentration dependence of the structure of laponite [56] and vanadium pentoxide [57] solutions via scattering, rheology, osmotic and birefringence data indicate that discrete phases (flocculates, isotropic liquid, isotropic gel and nematic gel) as well as coexistence regions exist, depending on the volume fraction of layers and the composition of the medium (in this instance, ionic strength and pH).

For laponite gels ($2R \sim 20$ nm), above the volume fraction of layers necessary for gelation, ϕ_v^*, two critical length scales are present in these gels, as summarized in Figure 16 [80, 81]. A fundamental unit, consisting of swollen stacks of plates with a mass fractal dimension of 3 and a scale of tens of nanometers, forms aggregates on the scale of microns. At intermediate volume fractions ($\phi_v^* < \phi < \phi_{vc}$ where ϕ_{vc} is the critical volume fraction for formation of a two-phase suspension), alignment of the micron-size domains leads to the formation of a weak fibrous structure with a fractal dimension of around 1. At $\phi > \phi_{vc}$, two domains (high and low particle density) form, resulting in a mechanically stronger gel with a fractal dimension that

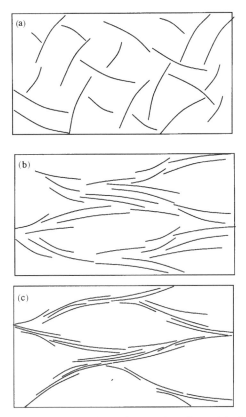

Figure 15   Models for the association of smectite plates (dark lines, 1 nm thick) in aqueous gels: (a) traditional house-of-cards structure which probably only forms at high ionic strength; (b) and (c) nematic ordered structures, surrounding or dispersed in regions of almost pure solvent, which may be highly swollen [(b) $d_{001} > 4$ nm, no diffraction peak] or collapsed [(c) $d_{001} \sim 3$–4 nm, finite diffraction peak] [79]

increases with volume fraction of laponite. Disruption of this larger length scale structure (i.e. fractal network) gives rise to yielding and thixotropic behavior of the gels, which is in contrast to the traditional view that disruption of layer–layer contacts is responsible.

Gelation of smectites with a larger layer size (e.g. montmorillonite), has yet to be adequately examined to validate the generality of these hierarchical structures. Preliminary investigation by Muzney and coworkers [82] has applied dynamic light scattering to monitor the dispersion of laponite and the influence of organic surfactants on layer association. Additionally, they examined the superstructure for dispersion of suitably modified laponite after polymerization of acrylamide. These

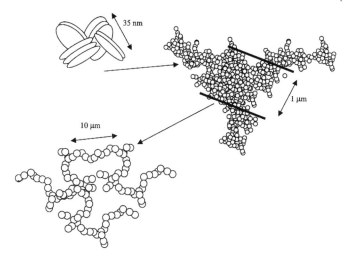

Figure 16   Proposed structure of aqueous laponite suspensions at rest. Packets of oriented smectite layers define a fundamental length scale of the order of a few nanometers. Tight aggregation of an isotropic distribution of subunits defines a second length scale of the order of microns. Loose connections of these microscale aggregates gives rise to an isotropic equilibrium structure [80]

studies indicated that a homogeneous, single-layer dispersion was achievable only when a large excess (5 times the CEC of laponite) of organic cationic surfactant was used. Jannai and coworkers [83] have used small-angle neutron scattering to study mixtures of organically modified vermiculite (n-butylammonium) with poly(vinyl methyl ether) in the presence of n-butylammonium chloride and deuterated water. They concluded that introduction of the polymer causes the silicate layers to become more strongly aligned with a more regular, but generally decreasing, interlayer spacing in the gel phase. However, inclusion of the polymer had a minimal effect on the phase transition temperature between the tactoid and gel phases.

As for investigation of layered silicate superstructure in polymer melts, molten polycaprolactam–layered silicate nanocomposites [2, 3, 8] provide excellent model systems because of the accessibility of various morphologies through different fabrication routes and modifications of the OLS. Vaia and coworkers have examined the small-angle scattering of two *in situ* polymerized PLSNs [NCH5 (5 wt % layered silicate), NCH2 (2 wt % layered silicate)] and a melt processed PLSN [NLS4 (4 wt % layered silicate)] [58]. Figure 8 shows the transmission electron micrographs of NCH5 and NLS4. In NCH2 and NCH5, approximately 30 and 50 % respectively of the poly(caprolactam) chains are end-tethered via an ionic interaction to the silicate surface [2], whereas polymer–silicate interactions in NLS4 are expected to be dominated by weak secondary interactions, such as *Van der Waals* forces [8].

Figure 17 summarizes the background-corrected small-angle scattering data of the molten nanocomposite at 250 °C. The absence of a strong basal reflection

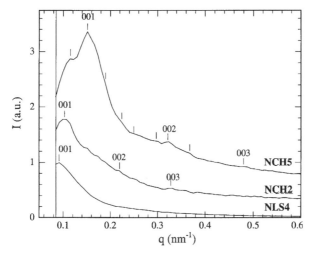

Figure 17 Small-angle X-ray scattering of molten poly(ε-caprolactam)/ montmorillonite nanocomposites formed via *in situ* polymerization (NCH5, 5 wt% layered silicate and NCH2, 2 wt% layered silicate) and melt processing (NLS4, 4 wt% layered silicate). Prominent reflections [NCH5: $q = 0.158$ nm^{-1} ($n = 1$), $q = 0.322$ nm^{-1} ($n = 2$), $q = 0.480$ nm^{-1} ($n = 3$); NCH2: $q = 0.105$ nm^{-1} ($n = 1$), $q = 0.220$ nm^{-1} ($n = 2$), $q = 0.320$ nm^{-1} ($n = 3$); NLS4: $q = 0.091$ nm^{-1}] indicate that superstructures exist (39.8, 59.8 and 69.0 nm, respectively). Minor peaks (e. g. NCH5, $q = 0.114$ nm^{-1}) indicate that additional microstructural features exist

($q \sim 2$ nm^{-1}) indicates a disruption of the original stacking of the silicate layers. However, additional Bragg reflections are observable in both the raw and Lorentz corrected (Iq^2) data. These results are consistent with previous TEM micrographs (NCH5: 35–45 nm; NLS4: 40–60 nm) [8] and correspond to the theoretical estimation of mean layer–layer spacing approximately 45 nm for a uniform nematic-like arrangement of layers at 2.2 vol%. Additionally, since the scattering regime is near $qH \sim 1$, analysis of the high-q data by high-q approximation to the form factor of a plate indicates that the effective layer thickness for NCH5 and NLS4 is 2.7 and 2.3 nm, respectively [35, 66, 67, 84]. This is close to the thickness anticipated for a silicate layer (0.96 nm) coated with alkylammonium surfactants (0.7–0.9 nm). Finally, the relative magnitude of the Bragg reflections and total scattered intensity (invariant) was observed to vary with sample processing history and even beam location (beam size ~ 300 µm) on the sample. For example, the invariant of NLS4 was twice as large as that for NCH5 even though there is less silicate in NLS4. These observations indicate that micron or larger domains exist with varying concentration or orientation of the layers parallel to the sample surface (and perpendicular to the incident X-rays). Determination of the translational and rotational diffusion rates of the layers will be necessary to understand relaxation and homogenization of these structures in the melt.

Detailed studies of the relationship between microstructure and rheological response of plate-like particles have been limited. This contrasts with the extensive characterization of microstructure and its relation to the flow of suspensions of rigid rod-like particles, such as liquid crystalline polymers. For aqueous suspensions of smectites, Ramsey and Lindner [85] and Hanley *et al.* [86] examined the structural response of montmorillonite and laponite gels *in situ* using two-dimensional small-angle neutron scattering and a Couette-type shear cell. Orientational correlations were found to be more extensive for the larger, more anisotropic montomorillonite layers, and preferential alignment may persist at distances of the order of the plate size after suspension of shear and re-equilibration. Later, Pignon and coworkers used *in situ* two-dimensional light, neutron and X-ray scattering to understand the deaggregation and recovery process of the micron-scale fractal network in laponite gels [80, 81]. A butterfly light scattering pattern was attributed to the formation of rollers within the dispersion, which on average aligned perpendicular to the shear direction. Two time scales were identified, corresponding to the rapid relaxation of particle orientation and a slow reaggregation process.

As for PLSNs, the presence of extended polymer chains, chemically or physically adsorbed on the layer surface, will drastically alter the equilibrium structure and subsequent response to dynamic forces relative to the small-molecule gels previously examined. Preliminary small-angle neutron scattering measurements on samples after large-amplitude oscillatory shear confirm the presence of global alignment of silicate layers as observed in injection molded nylon 6 hybrids (for example, see the micrographs in Figure 8) [87]. Preliminary real-time X-ray scattering studies of shear alignment and relaxation of the layered silicate structure in the poly(ε-caprolactam) melt indicate that relaxation occurs within minutes at $240\,^\circ$C (Figure 18) [88].

5.2 POLYMER STRUCTURE IN PLSN

The nanoscale and mesoscale organization of the smectite layers result in a bulk material in which the medium is no more than a few tens of nanometers from an interface (Figure 3). The proliferation of interfaces results in a material consisting entirely of interfacially modified polymers with a chain conformation [15] or network structure (thermosets) fundamentally different to that in the bulk.

For intercalated systems, the confinement is extreme, 1–3 nm, which is many times smaller than the radius of gyration of the polymer. Determination of the localized structure (distribution of monomer units within the interlayer) using powder diffraction (as well as computer simulations and NMR) was discussed in Section 4.2. In contrast, data on the global chain conformation and the degree of entanglements in these extremely confined environments are minimal. Using contrast matching techniques, Krishnamoorti and coworkers used blends of polystyrene (PS) and deuterated polystyrene (dPS) to remove small-angle neutron

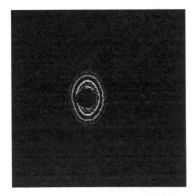
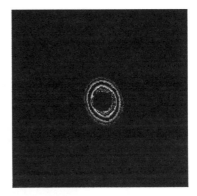
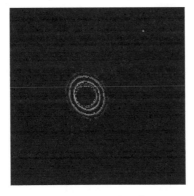

Figure 18   Temporal series (0, 10 and 20 min after cessation of shear) at 240 °C of small-angle X-ray scattering patterns (X-ray beam perpendicular to shear plate) recording the relaxation of shear-induced order of NCH5 (order prepared by steady state shear with parallel plates (direction vertical), 10 min, at 40 s⁻¹ and 240 °C)

scattering of the silicate layers to reveal the partial form factor (excess scattering) associated with the intercalated polystyrene chains. Figure 19 shows the excess small-angle neutron scattering for three intercalated PS–dPS blends with narrow polydispersity and different molecular weights [87]. The magnitude of the scattering increases with molecular weight, supporting the supposition that the excess scattering is associated with the intercalated polymer and not the silicate layers. The absence of a Guinier regime ($qR_g \sim 1$) and an intensity dependence that approaches q^{-2} within the measurement range imply highly distorted and extended coils with minimal entanglements. This study concludes that the intercalated polymer, and, by extrapolation, flexible chains confined to slits of the order of the chain persistence length, adopt an isolated, pseudo-two-dimensional conformation with minimal overlap/engagements with its neighbors relative to that present in the bulk. The

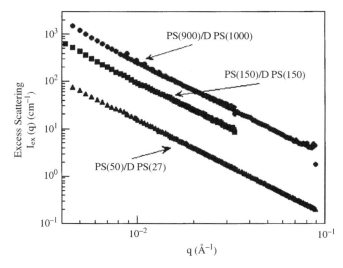

Figure 19   Excess small-angle neutron scattering for intercalated polystyrene/ deuterated polystyrene blends in dimethyldioctadecylammonium-exchanged montmorillonite (SWy-2)

change in entanglement density and conformations results in a polymer that is fundamentally different to the entangled bulk.

For exfoliated systems, layer separation is 20–50 nm and of the order of many mesoscopic features of polymer systems, such as crystal lamellae. The silicate spacing along with polymer–silicate interactions will impact the equilibrium crystal phase and arrangement of crystal lamellae in semicrystalline PLSNs. Figure 20 shows wide-angle and small-angle X-ray diffraction for liquid nitrogen-quenched samples of various poly(ε-caprolactam) nanocomposites [58, 89]. Whereas pure poly(ε-caprolactam) crystallizes predominantly in the α-crystalline phase, the presence of the exfoliated layers results in preferential crystallization in the metastable γ-phase. Furthermore, the presence of the layers disrupts the lamellae stacking, nominally present in pure poly(ε-caprolactam), resulting in a highly disordered, crystalline phase. The Toyota group has examined the influence of the layers on the morphology of the crystalline polymer. In addition to the possible presence of a new higher-temperature melting phase [90], the global orientation of the silicate layers, which arises from flow alignment (injection molding and film extrusion), dictates the orientation of the crystal phase. Depending on processing conditions, wide-angle X-ray diffraction indicated that the chain axis is parallel to the normal of the silicate layer in the interior of molded components, whereas it is parallel to the layer in the near-surface region of molded components and extruded films [53]. In addition to the crystal phase, evidence has been found that the layers also alter the formation of spherulites [91].

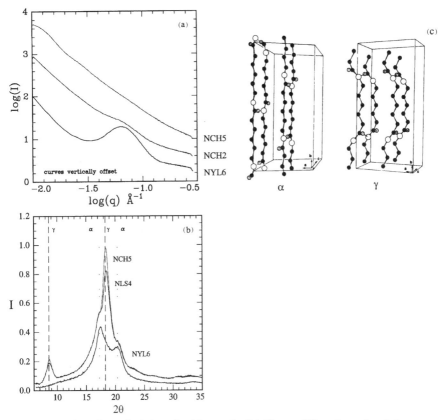

Figure 20   Small-angle (a) and wide-angle (b) X-ray diffraction of poly(ε-capro-lactam) nanocomposites and a schematic (c) of poly(ε-caprolactam) α and γ crystal phases [89]. Sample notation is the same as in Figure 17

Understanding the development of the crystal morphology in the presence of the layers is as important as understanding the development of the distribution of the layers with regard to establishing structure–processing relationships. Figure 21 shows the excess small-angle scattering associated with the crystalline polymer component of the nanocomposites, obtained as the difference between the profiles at an isothermal crystallization temperature, in this case $T_x = 215°\,C$, and the isotropic melt ($250\,°C$) after isothermal crystallization for 90 min. The excess small-angle scattering may be analyzed to yield a normalized correlation function (Figure 22). Correlation and interface function analysis [92, 93] yield the invariant, Q, the Porod constant, K, the Bragg length, L_B and the thicknesses of the constituent phases, l_1 and l_2, corresponding to the crystalline and amorphous layer thickness. Simultaneous wide-angle X-ray scattering (WAXS) gives complementary information on the

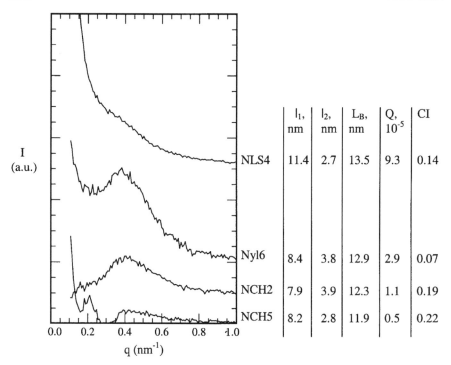

	l_1, nm	l_2, nm	L_B, nm	Q, 10^{-5}	CI
NLS4	11.4	2.7	13.5	9.3	0.14
Nyl6	8.4	3.8	12.9	2.9	0.07
NCH2	7.9	3.9	12.3	1.1	0.19
NCH5	8.2	2.8	11.9	0.5	0.22

Figure 21 Excess small-angle scattering associated with the crystalline component of the nanocomposites, obtained as the difference between the profiles at an isothermal crystallization temperature, $T_x = 215\,°C$, and the isotropic melt (250 °C) after isothermal crystallization for 90 min. Sample notation is the same as in Figure 17. The corresponding table summarizes the results from the correlation analysis of the scattering

fractional crystallinity index (CI) and crystal phase. These studies indicate major differences in morphology and development of morphology of the crystalline regions between pure poly(caprolactam) and the nanocomposites. When the chains are strongly associated with the silicate surface and at high packing density (~ 0.5–$0.6\,nm^2$/chain), such as for the tethered polymer chains in NCH2 and NCH5, the crystallizing polymer is not arranged in ordered stacks but is disordered as indicated by decreases in the invariant, lamella spacing and scattering intensity ($q = 0.42\,nm^{-1}$). Opposite trends are observed for relatively weak polymer–silicate interactions, such as in the melt processed NLS4 in which the crystalline regions (l_1) substantially increase in thickness. Irrespective of interfacial interactions, the addition of silicate increases the crystallization rate (both nucleation and growth) as well as the quantity of crystallized polymer, as indicated by the increase in CI at 90 min. Finally, the mean chain distance in the crystalline polymer regions of the

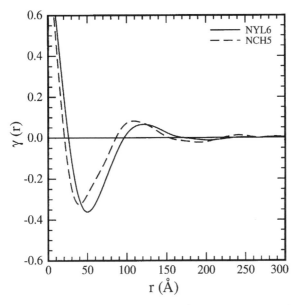

Figure 22 Normalized correlation function for excess scattering from poly-(ε-caprolactam) and a 5 wt% montmorillonite/poly(ε-caprolactam) nanocomposite. Sample notation is the same as in Figure 17

nanocomposite are greater than in the pure polymer at these temperatures $(215\,^{\circ}\text{C} > T_{\text{Brill}})$.

Overall, these observations are consistent with crystallization of the polymer near an interface with induced orientation and impeded mobility, resulting in poor chain packing and the incorporation of defects [94]. Additional isothermal and non-isothermal crystallization studies indicate that the degree of interaction between the silicate and the polymer has a marked effect on the nucleation, growth, secondary crystallization and Brill transition of the polymer crystal phase [95]. The presence of nanoscale distribution of inorganic–polymer interfaces is expected to alter the thermodynamic stability of other mesoscopically ordered systems, such as block copolymer phase behavior or polymer blend stability.

6 CONCLUSION

The utility of nanoscopic fillers to enhance polymer properties has been aptly demonstrated, especially with regard to the addition of layered silicates. Detailed characterization of both the hierarchical morphology of the nanofiller as well as the concomitant changes in the morphology of the polymer is critical to establishing structure–processing property relationships. In addition to the technological implications, these systems offer unique possibilities for use as models to study the structure and dynamics of confined polymers, brushes and blends using conventional analytical techniques (such as scattering and microscopy) on a wide range of different polymers with different surface interactions.

Microscopy is invaluable in visualizing the complex structures and establishing qualitative morphological groupings. The simple intercalated and exfoliated descriptions conventionally used are inadequate in fully describing the complex morphology of PLSNs as well as its dependence on fabrication and processing history. The nanoscale structures are more adequately conveyed using additional descriptors such as disordered, ordered and partial along with the traditional intercalated and exfoliated. Additionally, the large aspect ratio of the plates mandate layer–layer correlations at moderate and low volume fraction of layered silicate, leading to mesoscale organization which evokes possible connections with other mesoscopic systems such as block copolymers and liquid crystalline polymers in which the layers parallel a rigid block or a discotic mesogen, respectively.

Diffraction and scattering (and, although not extensively discussed herein, NMR, computer simulations and spectroscopy) provide qualitative structural details for these systems. In addition to rapid characterization of layer separation, powder diffraction may provide additional information on layer order, size distributions and atomic arrangements of constituents as well as a flexible method for *in situ* monitoring of structural changes, providing an insight into kinetic aspects of these systems. However, because of the strong θ-dependence of theoretical and experimental factors associated with powder diffraction, careful sample preparation and data evaluation is necessary and should be complemented by microscopic observations, especially for PLSNs with low volume fractions of OLS that are suspected to have exfoliated morphologies. Small-angle scattering provides complementary details to powder diffraction on larger length scales of both the polymer and silicate. For example, superstructures consisting of layer–layer correlations on 20–60 nm exist in melt processed PLSNs. Complementary small-angle scattering studies of small molecule–smectite suspensions imply that multiphase and micron-scale structures may exist in PLSNs and are critical for many bulk properties. With regard to polymer morphologies, previous studies indicate that intercalated chains adopt pseudo-two-dimensional configurations with minimal entanglements with their neighbors relative to those in the bulk. In addition, the presence of multiple constituents with dominant length scales on the nanolevel alter the phase behavior, as demonstrated for crystallization of semicrystalline polymers in the presence of exfoliated silicate layers.

In conclusion, complete structural characterization of polymer systems containing nanoscopic fillers is an evolving and challenging area with broad implications ranging from the establishment of structure–processing property relationships for these systems to the fundamental understanding of the physics of phase behavior, chain conformations and dynamics of confined polymer systems.

7 ACKNOWLEDGEMENTS

Invaluable discussion and input were kindly provided by numerous colleagues. R. Bharadwaj (Figure 1), E. Manias and E. Hacket (Figure 2), J. Brown (Figures 4, 6, 8 and 10), W. Ragland (Figures 4, 6 and 10), F.J.M. Rodriguez (Figure 18) and D. Lincoln (Figures 17 and 22) provided figures. S. Krumm kindly provided a shareware version of WinStruct used to calculate powder diffraction patterns (Figure 12). Finally, R. Krishanmoorti is gratefully acknowledged for discussions and Figure 19. This work was partially funded by the Air Force Office of Scientific Research and the Air Force Research Laboratory, Materials and Manufacturing Directorate.

8 REFERENCES

1. For example: *Global Assessment of R&D Status and Trends in Nanoparticles, Nanostructured Materials, and Nanodevices*, International Technology Research Institute, Loyola College in Maryland, Baltimore, MD, 1999; Godovski, D. Y. *Adv. Polym. Sci.*, **119**, 79 (1995); Edelstein, A. S. and Cammarata, R. C. *Nanomaterials: Synthesis, Properties and Applications*, Institute of Physics, Philadelphia, PA, 1996; Giannelis, E. P. *Adv. Mater.*, **8**, 29 (1996); Vaia, R. A., Lee, J.-W., Wang, C.-S., Click, W. and Price, G. *Chem. Mater.*, **10**, 2030 (1998); Winiarz, J. G., Zhang, L., Lal, M., Friend, C. S. and Prasad, P. N. *J. Am. Chem. Soc.*, **121**, 5287 (1999); Jin, L., Bower, C. and Zhou, O. *Appl. Phys. Lett.*, **73**(9), 1197 (1998); Pileni, M. P. *Langmuir*, **13**, 3266 (1997).
2. Usuki, A., Kawasumi, M., Kujima, Y. and Okada, A. *J. Mater. Res.*, **8**, 1174 (1993); Usuki, A., Kojima, A., Kawasumi, M., Okada, A., Fukushima, Y., Kurauchi, T. and Kamigatio, O. *J. Mater. Res.*, **8**, 1179 (1993).
3. Okada, A. and Usuki, A. *Mater Sci. Eng. C*, **3**, 109 (1995).
4. Lan, T. and Pinnavaia, T. J. *Chem. Mater.*, **6**, 2216 (1994); Wang, Z. and Pinnavaia, T. J. *Chem. Mater.*, **10**, 1820 (1998).
5. Yano, K., Usuki, A., Okada, A., Kurauchi, T. and Kamigaito, O. *J. Polym. Sci., Part A, Polym. Chem.*, **31**, 2493 (1993).
6. Messersmith, P. B. and Giannelis, E. P. *J. Polym. Sci., Part A, Polym. Chem.*, **33**, 1047 (1995); Messersmith, P. B. and Giannelis, E. P. *Chem. Mater.*, **6**, 1719 (1994).
7. Gilman, J. W. *Appl. Clay Sci.*, **15**, 31 (1999).
8. Vaia, R. A., Price, G., Ruth, P. N., Nguyen, H. T. and Lichtenhan, J. *Appl. Clay Sci.*, **15**, 67 (1999).
9. Goldman, A. Y., Montes, J. A., Barajas, A., Beall, G. W. and Eisenhour, D. D. *Ann. Tech. Soc. Plast. Eng.*, **56**(2), 2415 (1998).
10. Burnside, S. D. and Giannelis, E. P. *Chem. Mater.*, **7**, 1597 (1995).
11. Vaia, R. A., Ishii, H. and Giannelis, E. P. *Chem. Mater.*, **5**, 1694 (1993).
12. Vaia, R. A., Vasudevan, S., Krawiec, W., Scanlon, L. G. and Giannelis, E. P. *Adv. Mater.*, **7**, 154 (1995).
13. Vaia, R. A. and Giannelis, E. P. *Polym. Comm.* accepted (2000).
14. Vaia, R. A., Sauer, B. B., Tse, O. and Giannelis, E. P. *J. Polym. Sci., Part B, Polym. Phys.*, **35**, 59 (1997).
15. Giannelis, E. P., Krishnamorti, R. and Manias, E. *Adv. Polym. Sci.*, **138**, 108 (1999) and references therein.

16. Yang, D.-K. and Zax, D. B. *J. Chem. Phys.*, **110**(11), 5325 (1999); Wong, S., Vasudevan, S., Vaia, R. A., Giannelis, E. P. and Zax, D. B. *J. Am. Chem. Soc.*, **117**, 7568 (1995); Wong, S., Vaia, R. A., Giannelis, E. P. and Zax, D. B. *Solid State Ionics*, **86–88**, 547 (1996); Wong, S. and Zax, D. B. *Electrochim. Acta*, **42**, 3513 (1997); Harris, D. J., Bonagamba, T. J. and Schmidt-Rohr, K. *Macromolecules*, **32**, 6718 (1999).

17. Vaia, R. A., Teukolsky, R. K. and Giannelis, E. P. *Chem. Mater.*, **6**, 1017 (1994).

18. Yano, K., Usuki, A. and Okada, A. *J. Polym. Sci, Part A, Polym. Chem.*, **35**, 2289 (1997).

19. Katsoulis, D. E., Chao, T. C.-S., McQuiston, E. A., Chen, C. and Kenney, M. *Mater. Res. Soc. Symp. Proc (Organic/Inorganic Hybrid Materials)*, **519**, 320 (1998); Katsoulis, D. E., Chao, T. C.-S., Mcquiston, E. A., Chen, C. and Kenney, M. E. *Polym. Prepr. (Am. Chem. Soc., Div. Polym. Chem.)*, **39**(1), 514 (1998).

20. Krishnamoorti, R. and Giannelis, E. P. *Macromolecules*, **30**, 4097 (1997).

21. Whittingham, M. S. and Jacobson, A. J. (Eds), *Intercalation Chemistry*, Academic Press, New York, 1992.

22. Pinnavaia, T. J. *Science*, **220**, 365 (1983).

23. Pinnavaia, T. J. in Legrand, A. P. and Flandrois, S. (Eds), *Chemical Physics of Intercalation*, Plenum, New York, 1987.

24. Güven, N. in Bailey S. W. (Ed.), *Hydrous Phyllosilicates, Reviews in Mineralogy*, Mineralogical Society of America, Washington, DC, 1988, Vol. 19, pp. 497–560.

25. Brindely, G. W. and Brown, G. *Crystal Structure of Clay Minerals an their X-ray Identification*, Mineralogical Society, London, 1980.

26. Moore, D. M. and Reynolds, R. C. *X-ray Diffraction and the Identification and Analysis of Clay Minerals*, Oxford University Press, New York, 1997.

27. Bleam, W. F. and Hoffman, R. *Inorg. Chem.*, **27**, 3180 (1988). Bleam, W. F. and Hoffman, R. *Phys. Chem. Miner.*, **15**, 398 (1988).

28. Knechtel, M. M. and Patterson, S. H. *U. S. Geol. Surv. Bull.*, **1082-M**, 957 (1962).

29. van Olphen, H. and Fripiat, J. J. *Data Handbook for Clay Minerals and Other Nonmetallic Materials*, Pergamon, New York, 1979.

30. Janek, M. and Smrčok, L. *Clays Clay Miner.*, **47**(2), 113, (1999).

31. Pinnavaia, T. J. *et al. Chem. Mater.*, **10**, 3769 (1998).

32. Post, J. L. *Clays Clay Miner.*, **32**, 147 (1984).

33. Barrer, R. R. and Jones, D. L. *J. Chem. Soc. A.*, 1531 (1970).

34. Kloprogge, J. T., *J. Porous Mater.*, **5**, 5 (1998).

35. Drits, V. A. and Tehoubar, C. *X-ray Diffraction by Disordered Lamellar Structures*, Springer-Verlag, New York, 1990.

36. For example: Brandt, A., Schwieger, W. and Berbk, K. H. *Rev. Chim. Miner.*, **5**, 564 (1987); Lagaly, G., Beneke, K. and Weiss, A. *Am. Miner.*, **60**, 642 (1975); Beneke, K. and Lagaly, G. *Am. Miner.*, **62**, 763 (1977); Beneke, K. and Lagaly, G. *Am. Miner.*, **68**, 818 (1983); Borbely, G., Beyer, H. K., Karge, H. G., Schwieger, W., Brandt, A. and Bergk, K. H. *Clays Clay Miner.*, **39**, 490 (1991).

37. For example: Almond, G. G., Harris, R. K. and Franklin, K. R. *J. Mater. Chem.*, **7**, 681 (1997); Hanaya, M. and Harris, R. K. *J. Mater. Chem.*, **8**, 1073 (1998); Huang, Y., Jiang, Z. and Schwieger, W. *Chem. Mater.*, **11**, 1210 (1999).

38. Wang, Z., Lan, T. and Pinnavaia, T. J. *J. Chem. Mater.*, **8**, 2200 (1996).

39. Hackett, E., Manias, E. and Giannelis, E. P. *J. Chem. Phys.*, **108**, 7410 (1998).

40. Schoonheydt, R. A. in van Bekkum, H. Flanigen, E. M. and Jansen, C. J. (Eds), *Introduction to Zeolite Science and Practice*, Elsevier, Amsterdam, 1991, p. 201.

41. Vaia, R. A., Jandt, K. D., Kramer, E. J. and Giannelis, E. P. *Chem Mater.*, **8**, 2628 (1996).

42. Grim, R. E. *Clay Miner.*, McGraw-Hill, New York, 1953.

43. For examples of transmission electron microscopy studies of smectites see: Klimentidis, R. E. and Mackinnon, I. D. R. *Clays Clay Miner.*, **34**, 155 (1986); Lee, J. H. and Peacor, D. R.

Clays Clay Miner., **34**, 69 (1986); Ahn, H. H. and Peacor, D. R. *Clays Clay Miner.*, **34**, 180 (1986); Vali, H. and Köster, H. M. *Clay Miner.*, **21**, 827 (1986); Vali, H. and Hesse, R. *Am. Miner.*, **75**, 1443 (1990); Vali, H., Hesse, R. and Kohler, E. E. *Am. Miner.*, **76**, 1973 (1991); Marcks, C. H., Waschsmuth, H. and Reichenbach, H. G. V. *Clay Miner.*, **34**, 155 (1986); Vali, H. and Hesse, R. *Clays Clay Miner.*, **40**, 240 (1992).
44. Schöllhorn, R. *Physica*, **99B**, 89 (1980).
45. Solomon, D. H. and Hawthorne, D. G. *Chemistry of Pigments and Fillers*, Krieger, Malabar, FL, 1991.
46. Theng, B. K. G. *The Chemistry of Clay Organic Reactions*, John Wiley & Sons, New York, 1974; Theng, B. K. G. *Formation and Properties of Clay Polymer Complexes*, Elsevier, New York, 1979.
47. Huang, X., Lewis, S., Brittian, W. and Vaia, R. A. *Macromolecules*, **33**, (2000).
48. Brown, J. M., Curliss, D. B. and Vaia, R. A. PMSE preprint, American Chemical Society, 2000.
49. Vaia, R. A., Ishii, H. and Giannelis, E. P. *Chem. Mater.*, **5**, 1694 (1993).
50. Kato, M., Okamoto, H., Hasegwa, N., Usuki, A. and Sato, N. in Proceedings of 6th Japanese International SAPME Symposium, 1999, Vol. 2, p. 693.
51. Vaia, R. A. and Giannelis, E. P. *Macromolecules*, **30**, 7990 (1997); Vaia, R. A. and Giannelis, E. P. *Macromolecules*, **30**, 8000 (1997).
52. Balazs, A. C., Singh, C., Zhulina, E. and Lyatskaya, Y. *Acc. Chem. Res.*, **32**, 651 (1999).
53. Kojima, Y., Usuki, A., Kawasumi, M., Okada, A., Kurauchi, T., Kamigaito, O. and Kaji, K. *J. Polym. Sci., Part B, Polym. Phys.*, **33**, 1039, (1995); Kojima, Y., Usuki, A., Kawasumi, M., Okada, A., Kurauchi, T., Kamigaito, O. and Kaji, K. *J. Polym. Sci., Part B, Polym. Phys.*, **32**, 625, (1994).
54. Doi, M. and Edwards, S. F. *The Theory of Polymer Dynamics*, Oxford University Press, 1986, p. 324–326.
55. Ginzburg, V. V. and Balazs, A. C. *Macromolecules*, **32**, 5681 (1999); Lyatskaya, Y. and Balazs, A. C. *Macromolecules*, **31**, 6676 (1998).
56. Mourchid, A., Lécolier, E., Van Damme, H. and Levitz, P. *Langmuir*, **14**, 4718 (1998); Mourchid, A., Delville, A., Lambard, J., Lécolier, E. and Levitz, P. *Langmuir*, **11**, 1942 (1995).
57. Davidson, P., Garreau, A. and Livage, J. *Liq. Crystallogr.*, **16**, 905 (1994); Davidson, P., Bourgaux, C., Schoutteten, L., Sergot, P., Williams, C. and Livage, J. *J. Phys. II*, **5**, 1577 (1995).
58. Vaia, R. A., Lincoln, D. M., Wang, Z.-G. and Hsiao, B. S. *Polymer*, accepted (2000).
59. Krishnamoorti, R. Unpublished results.
60. Lan, T., Kaviratna, P. D. and Pinnavaia, T. J. *Chem. Mater.*, **7**, 2144 (1995).
61. Parry, G. S. *Physica*, **105B**, 261 (1981).
62. Lagaly, G. *Phil. Trans. R. Soc. (Lond.) A*, **311**, 315 (1984).
63. Bish, D. L. and Post, J. E. (Eds), *Modern Powder Diffraction, Reviews in Mineralogy*, Mineralogical Society of America, Washington DC, 1989, Vol. 20.
64. Vaia, R. A., Jandt, K. D., Kramer E. J. and Giannelis, E. P. *Macromolecules*, **28**, 808 (1995).
65. Krumm, S. *Comput. Geosci.*, **25**, 501 (1999).
66. Vaia, R. A. (to be published).
67. Klug, H. D. and Alexander, L. E. *X-Ray Diffraction Procedures for Polycrystalline and Amorphous Materials*, John Wiley & Sons, New York, 1974.
68. Eberl, D. D., Dritis, V. A., Írodoñ, J. and Nüesch, R. US Geological Survey, open-file report 96-171, Boulder Colorado, 1996; Drits, V. A., Erberl, D. D. and Írodoñ, J. *Clays Clay Miner.*, **46**, 38 (1998).
69. Franzen, P. *Clay Miner. Bull.*, **2**, 223 (1954).

70. Adams, J. M. *J. Chem. Soc. Dalton Trans.*, **70**, 2286 (1974); Adams, J. M., Thomas, J. M. and Walters, M. J. *J. Chem. Soc. Dalton Trans.*, **71**, 1459 (1975).
71. Aranda, P. and Ruiz-Hitzky, E. *Appl. Clay Sci.*, **15**, 119 (1999); Ruiz-Hitzky, E. *Adv. Mater.*, **5**, 334 (1993); Aranda, P. and Ruiz-Hitzky, E. *Chem. Mater.*, **4**, 1395 (1992).
72. Lemmon, J., Wu, J., Oriakhi, C. and Lerner, M. *Electrochim. Acta*, **40**, 2245 (1995); Wu, J. and Lerner, M. M. *Chem. Mater.*, **5**, 835 (1993).
73. Lee, J. Y., Baljon, A. R. and Loring, R. F. *J. Chem. Phys.*, **111**, 9754 (1999); Baljon, A. R., Lee, J. Y. and Loring, R. F. *J. Chem. Phys.*, **111**, 9068 (1999).
74. Manias, E., Bitsanis, G., Hadziioannou, G. and Brinke, T. *Europhys. Lett.*, **33**, 371 (1996); Manias, E. and Giannelis, E. P. (to be published).
75. Glatter, O. and Kratky, O. *Small Angle X-ray Scattering*, Academic Press, New York, 1982.
76. Wignall, G. D. in *Encyclopedia of Polymer Science and Engineering*, Wiley, New York, 1987, Vol. 10.
77. Feigin, L. A. and Svergun, D. I. *Small Angle X-ray and Neutron Scattering*, Plenum Press, New York, 1987.
78. Rand, B., Pekenc, E., Goodwin, J. W., and Smith, R. W. *J. Chem. Soc. Faraday Trans. I*, **76**, 225 (1980).
79. Morvan, M., Espinat, D., Lambard, J. and Zemb, T. *Colloids Surf. A, Physiochem. Eng. Asp.*, **82**, 193 (1994).
80. Pignon, F., Magnin, A., Piau, J. M., Cabane, B., Linder, P. and Diat, O. *Phys. Rev. E*, **56**(3), 3281 (1997); Pignon, F., Piau, J. M. and Magnin, A. *Phys. Rev. Lett.*, **76**(25), 4857 (1996); Pignon, F., Magnin, A. and Piau, J. M. *Phys. Rev. Lett.*, **79**(23), 4689 (1997).
81. Pignon, F., Magnin, A. and Piau, J. M. *J. Rheol.* **42**(6), 1349 (1998).
82. Muzney, C. D., Butler, B. D., Hanley, H. J. M., Tsvetkov, F. and Peiffer, D. G. *Mater. Lett.*, **28**, 379 (1996).
83. Jannai, H., Smalley, M. V., Hasimoto, T. and Koizumi, S. *Langmuir*, **12**, 1199 (1996).
84. Bongiovanni, R., Ottewill, R. H. and Rennie, A. R. *Prog. Colloid Polym. Sci.*, **84**, 299 (1991).
85. Ramsey, J. D. F. and Lindner, P. *J. Chem. Soc. Faraday Trans.*, **89**(23) 4207 (1993); Ramsay, J. D. F., Swanton, S. W. and Bunce, J. *J. Chem. Soc. Faraday Trans.*, **86**(23) 3919 (1990).
86. Hanley, H. J. M., Straty, G. C. and Tsvetkov, F. *Langmuir*, **10**, 3362 (1994).
87. Krishnamoorti, R. and Vaia, R. A. Unpublished data.
88. Murthy, N. S., Curran, S. A., Aharoni, S. M. and Minor, H. *Macromolecules*, **24**, 3215 (1991).
89. Kojima, Y., Matsuoka, T., Takahasji, H. and Kurauchi, T. *J. Appl. Polym. Sci.*, **51**, 683 (1994).
90. Jimenez, G., Ogata, N., Kawai, H. and Ogihara, T. *J. Appl. Polym. Sci.*, **64** 2211 (1997).
91. Hsiao, B. S. and Verma, R. K. *J. Synchrotron Rad.*, **18**, 1343 (1990).
92. Verma, R. K., Marand, H. and Hsiao, B. S. *Macromolecules*, **29**, 7767 (1996).
93. Muratoglu, O. K., Argon, A. S. and Cohen, R. E. *Polymer*, **36**(11), 2143 (1995); Mukai, U., Cohen, R. E., Bellare, A. and Albalak, R. J. *J. Appl. Polym. Sci.*, **70**, 1985 (1998).
94. Vaia, R. A., Lincoln D. M., Hsiao, B. S. and Krishnamoorti, R. *Polymer*, submitted (2000).

13

New Conceptual Model for Interpreting Nanocomposite Behavior

G. W. BEALL

Missouri Baptist College, St. Louis, MO, USA

1 INTRODUCTION

The pioneering work conducted by Toyota Central Research on nanocomposites [1] has generated a large amount of interest and research over the past decade. The research efforts have begun to bear fruit with the introduction of new products both in thermoplastics and thermosets. Applications for nanocomposites can be divided into two broad categories that relate to engineering or barrier properties. It has been well established in the engineering applications that nanocomposites afford substantial increases in tensile strength, tensile and flexural modulus, heat distortion temperature, solvent resistance, and flame resistance [1–7]. These improvements are accompanied with little or no loss [5] in impact resistance in many cases. This gain in strength and stiffness without loss in impact is quite unusual. The improvement in barrier properties also has been well documented for many polymer systems [1, 4, 8]. In some cases the change in permeability to certain gases has been lowered by more than an order of magnitude. These gains in barrier properties are manifested at low loading of clay in the range of 1–5 wt% nanoparticles. These improvements in barrier are also normally gained without loss of clarity in the film.

The vast majority of work on nanocomposites has focused on the use of smectite-type clays as the nanoparticle. Smectite clays have been studied the most because they are naturally occurring minerals that are commercially available and exhibit platy morphology with a high aspect ratio, large specific surface areas, and substantial cation exchange capacities. Several models have been proposed for predicting the behavior of polymer composites with smectite clay. The most extensively cited model is for barrier properties [10], but some models also exist for physical properties. This paper will examine in some detail the existing model,

Polymer–clay nanocomposites Edited by T. J. Pinnavaia and G. W. Beall
© 2000 John Wiley & Sons Ltd

compare its predictions with experimental data, and finally present a new conceptual model for predicting nanocomposite behavior.

2 BARRIER MODEL

The traditional view of changes in barrier properties in polymer/clay nanocomposites has been one of increase in path length for permeant owing to the tortuous path presented by the high aspect ratio clay. This theory can be reduced to a fairly simple equation derived for a two-dimensional case by Nielsen [10]. The relevant equation is

$$P_f/P_u = V_p/1 + (L/2W)V_f$$

where P_f is the permeability of the composite, P_u is the permeability of the polymer, V_p is the volume fraction of the polymer, V_f is the volume fraction of the filler, L is the average length of the face of the filler, and W is the average thickness of the filler. The L/W ratio for montmorillonite is critical in contributing to the barrier in the model. The consensus from a number of published reports is that, for the clays normally employed for this application, the L/W value is in the range 100–500 [1, 3, 4, 6, 8, 9].

The Nielsen model is a simple two-dimensional model that assumes perfect registry of the clay plates parallel to the surface of the film. This assumption is supported by transmission electron microscopy on films of nanocomposites where the alignment of the clay particles is close to parallel to the film surface [11]. The model predicts that the relative permeability would only be a function of the aspect ratio at a given loading of clay for all permeants. The relative permeability should not be affected by parameters such as humidity or temperature.

The experimental data on a number of systems show that this simple model appears to predict the permeability behavior of systems reasonably well at very low loadings of clay (less than 1 wt%). At higher clay loadings the data deviates substantially from the model. The following examples will illustrate the types and magnitude of deviations that can be found in the literature.

Work conducted on the permeability of polyimide nanocomposites to O_2, CO_2, and water yielded curves that would predict aspect ratios of 87, 83, and 132 respectively using this model [4]. The measured aspect ratios of the montmorillonite are closer to 200. Secondly, the behavior of H_2O relative to O_2 and CO_2 yields substantial differences in relative permeability that the model cannot predict. In further tests on the permeability of ethyl acetate [12] on these same polyimide nanocomposites it was found that the relative permeability at 0 % relative humidity (RH) was 0.19 and at 50 % RH it was 0.09. These composites contained 2.5 vol % clay. In these experiments the permeability of the pure polyimide increased with increased relative humidity (from 1.24 to 2.89) while the nanocomposite went down

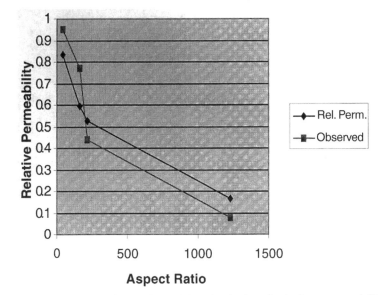

Figure 1 Comparison of experimental and calculated relative permeabilities for polyimide composites with various minerals

(from 0.19 to 0.09). This behavior can certainly not be explained by the simple tortuous path model.

In another study [13] a relationship between aspect ratio of a number of platy minerals and relative permeability to water was claimed. The minerals studied were hectorite, saponite, montmorillonite, and synthetic mica with aspect ratios of 46, 165, 220, and 1230. Figure 1 contains a comparison of the observed and calculated relative permeabilities using the model. It can be clearly seen that hectorite and saponite are well above the predicted values. In the composites these two minerals were observed not to exfoliate, and therefore one would not expect them to yield values that would fit the model. The values for montmorillonite and mica fall well below the predicted value. Others have reported in these studies that the montmorillonite also was not fully exfoliated and, therefore, could yield a relative permeability even further below the predicted curve and potentially closer to the high aspect ratio mica.

The water vapor transmission rate of high-density polyethylene (HDPE) [14] nanocomposites has been measured for a number of different surface-modified minerals. The minerals studied included montmorillonite treated with octadecylamine (M18), montmorillonite treated with dodecylpyrrodidone (M12), and a high aspect ratio fluoromica treated with octadecylamine (FM18). Figure 2 contains a comparison of the water vapor transmission rate for these three mineral nanocomposites as well as the pure HDPE. It is clear from the plot that the dodecylpyrro-

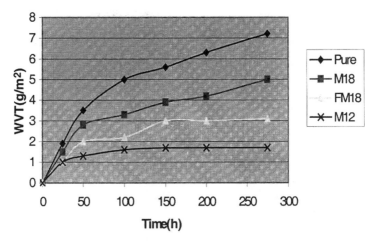

Figure 2 Water vapor transmission rate for HDPE and several nanocomposites

lidone-treated clay yields relative permeabilities well below the values for the same montmorillonite treated with octadecylamine and substantially below the higher aspect ratio fluoromica.

In the previous examples it is relatively easy to explain changes in relative permeability that are less than those predicted by invoking poor alignment of the clay plates or incomplete exfoliation. Deviations from the predicted relative permeability that exhibit lower values are more problematical to explain. The experimental observation of different relative permeabilities for different permeants on the same composite, effects of humidity upon relative permeability, and less than complete correlation between aspect ratio and relative permeability are impossible to reconcile with the simple tortuous path model.

3 CONCEPTUAL MODEL

The conceptual model for interpreting nanocomposite phenomena is based upon treating the clay/polymer interface as the dominating factor. In order fully to understand the model it is useful to review the fundamental properties of smectite clays. In particular the discussion will focus on montorillonite. A typical montmorillonite particle will have an aspect ratio of 250 and a total surface area of over $700\,m^2/g$. The plate will have lateral dimensions of 250 nm in two directions and a thickness of 1 nm. The individual plate will have a molecular weight of over 100 million and therefore should be treated as a true macromolecule.

In the conceptual model there are three regions around the clay plates that can be defined. The first is near the surface of the clay and is occupied by the surface modifier utilized to compatiblelize the clay to the polymer. This region is of the order of 1–2 nm and is generally well delineated. This region can be easily measured with X-ray diffraction on organoclays and is well documented since organically modified clays have been of commercial importance since their discovery in the 1940s. In the past few years a new method for surface modification of clays called ion-dipole treatment has been developed [15]. The size of the surface-modified region can also be directly measured by X-ray diffraction. The second region is a constrained polymer region. This region is less well defined and may extend 50–100 nm away from the surface of the clay. The size of this region is determined by a host of variables including the type of bonding between the surface modifier and the polymer, the strength of interaction between the polymer molecules, and the extent of nucleation imparted by the clay. The type of interaction of the surface modifier and the polymer can be divided into two broad categories. The first is direct chemical bonding between the surface modifier and the polymer. This is illustrated by the polymerization of caprolactam with 12-aminododecanoic acid montmorillonite. The second type of interaction would be a combination of Van der Waals attraction and hydrogen bonding. An example of this type of interaction is the direct compounding of octodecylamine montmorillonite into nylon 6. The strength of polymer/polymer interactions is important in determining the size of the constrained polymer region since, as the distance increases away from the surface, the amount of constraint on each successive layer of polymer will decrease proportionally to this polymer/polymer interaction. Therefore, a polymer that has strong intermolecular forces will have a larger constrained region than a polymer that exhibits weak intermolecular forces. Specifically, owing to strong hydrogen bonding, the constrained region in nylon 6 would be expected to be much greater than in polypropylene where the main interaction is due to Van der Waals attractions. The final factor, nucleation, may be the most important one. The ability of the clay to impart a nanophase structure to the composite will have a profound effect on both the permeability and physical properties of the polymer. The phenomena of nucleation in nanocomposites has been reported by a number of workers.

The third and final region is the unconstrained polymer which is not directly affected by the clay. The properties of this third region will be largely those of the pristine polymer. The physical properties of the other two regions will most likely be different from those of the pure polymer and most likely from each other. With respect to permeability, the diffusion coefficients in these three regions and the relative volume fractions will be the important parameters. If we label these diffusion coefficients as D_p, D_c, and D_s for the pure polymer, the constrained region, and the surface-modified region respectively, then there are four cases where the permeability of a composite could differ significantly from the pure polymer. These possible cases are as follows: D_c and D_s are greater than D_p; $D_c > D_p$ and $D_s < D_p$; $D_c < D_p$ and $D_s > D_p$; and D_c and D_s are both less than D_p. The two cases where both constrained and surface-modified regions have values greater than or less than the pure polymer would yield the extreme cases, and the other two cases

Table 1 Values for various parameters in model nanocomposites

Wt% clay	Volume fraction of clay	Number of clay molecules	Volume fraction of constrained polymer
1	0.004	7.86×10^{13}	0.077
2.5	0.01	1.97×10^{14}	0.193
5.0	0.02	3.94×10^{14}	0.387
7.5	0.03	5.91×10^{14}	0.580

would fall somewhere between them. In the case where both the constrained region and surface-modified region have diffusion coefficients greater than the pure polymer, the net effect would be to negate some if not all of the high aspect ratio of the clay. This case will be typified by negative deviations from the simple tortuous path model. In the following discussion, a disk shape will be assumed for the constrained polymer region with a thickness of 20 nm and a radius of 125 nm. A clay plate of this size would have a molecular weight of 76.6 million. Table 1 contains the number of plates per gram of composite at various wt% clay, as well as corresponding volume fractions for the clay and for the clay plus constrained polymer surface-modified region.

The effect of having D_c and D_s greater than D_p on relative permeability can be divided into two components. Since the constrained and surface-modified regions are intimately associated with the clay surface, the change in path length owing to the clay aspect ratio will all occur in the volume fraction of the surface modifier. In this region the effect of the higher diffusion coefficient of the surface modifier will be to negate some of the increase in path length that is due to the clay. This can be calculated by simply multiplying the $L/2W$ by the ratio of the diffusion coefficients D_p/D_s and employing the simple tortuous path equation. This value must be further corrected for the second effect that occurs in the constrained polymer volume fraction. In this region, $D_c > D_p$ and therefore will have the effect of lowering the apparent path length. This factor can be calculated by summing the contribution to diffusion in the volume fractions of constrained and unconstrained polymer regions. By assuming the diffusion coefficient D_p to be 1, the equation becomes

Correction factor $= V_p + [V_c/D_p/D_c)]$
where

V_p = volume fraction of the pure polymer
V_c = volume fraction of the constrained polymer
D_p = diffusion coefficient of the pure polymer
D_c = diffusion coefficient of the constrained polymer

This factor is then multiplied by the previously calculated relative diffusion coefficient. Figure 3 contains a comparison of the relative diffusion coefficient calculated for the simple tortuous path model, assuming an aspect ratio of 250, and

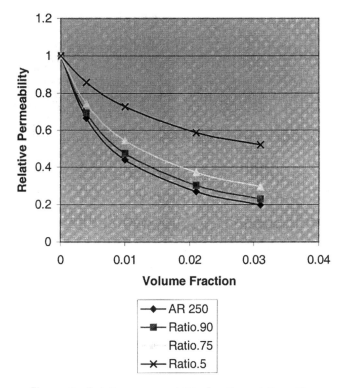

Figure 3 Relative permeability for D_c and $D_s > D_p$

several cases where the D_c and D_s are greater than D_p. It can be seen that very significant negative deviations from the tortuous path model occur at relatively small differences in diffusion coefficients.

The case where the constrained region has a diffusion coefficient greater than the polymer and the surface modifier a diffusion coefficient less than the polymer reduces to the tortuous path model with an apparent aspect ratio that is lower than reality. This case would yield similar negative deviations from the tortuous path model to the previous case where D_c and D_s are both greater than D_p.

The case where significantly lower relative coefficients can be observed is one where D_c and D_s are significantly lower than D_p. In this case the constrained region would not contribute to diffusion in any appreciable way and in effect would become a barrier. Utilizing the previously described model for the constrained polymer region, relative permeabilities can be calculated. Figure 4 contains a comparison of the relative permeability of the tortuous path model for a plate with an aspect ratio of 250 and a modified model where the constrained polymer region is 20 nm thick, which yields an aspect ratio of 12.5.

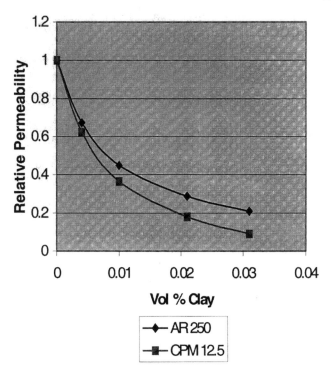

Figure 4   Relative permeability for the tortuous path and constrained polymer models

It can be seen that at low loadings the two curves are in reasonable agreement, but at loadings above 1 wt% they deviate significantly, with the modified model predicting significantly lower permeabilities. At about 7.5 wt% clay, this model would predict a volume fraction of 42% for the pure polymer. This volume is below the theoretical value for a cubic close packed sphere of 47.6%. Clay loading of 5–7.5 wt% will result in significant packing constraints, although rhombohedral close packing can yield values as low as 26%. In systems where packing begins to become a constraining factor, an empirical formula has been derived called the Kozeny relationship. This relationship has been measured mainly for the permeability of various porous rocks. The relationship is

$$K = P^3/CS^2$$

where P is the pore volume of the rock, S is the specific surface area of a unit volume of the rock, C is a constant characteristic of the rock, and K is the permeability of the rock.

In the case of a nanocomposite, P would be equivalent to the volume fraction of the pure polymer phase and S would be the specific surface area of the constrained polymer phase. Application of this model to nanocomposites would require utilizing C as a normalizing factor to fit it to the curve at low clay loadings. At loading above 6–7 wt %, this model would predict large decreases in permeability, far exceeding that predicted by the simple tortuous path model. The last case, where $D_c < D_p$ and $D_s > D_p$, would yield relative permeabilities similar to the case just discussed, but with the magnitude of deviation smaller.

It has been generally accepted that there is a correlation between aspect ratio and the observed change in relative permeability. The constrained polymer model also yields this same relationship. However, the effect is somewhat different from that predicted by the simple tortuous path model. Figure 5 contains a comparison of the relative permeabilities of several clay minerals with different aspect ratios, utilizing the tortuous path model and the constrained polymer model. The constrained polymer model predicts a much larger effect at low aspect ratio and a smaller difference between high aspect ratio materials such as montmorillonite and mica.

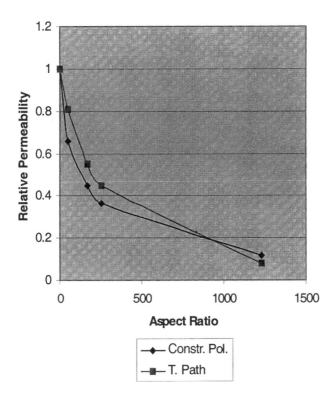

Figure 5 Relation between aspect ratio and relative permeability for the two models

There is substantial experimental evidence that the constrained polymer region exists. In the case of nylon 6 nanocomposites, both reactor polymerized and compounded composites exhibit profound changes in crystallinity and type of crystal phases present in the composites. In pure nylon 6 the predominant crystalline form is the alpha form and in nanocomposites it is the gamma form [1]. This change in predominant phase has been well established by a number of papers. In work on compounded nylon 6 [16] nanocomposites, the crystallinity of injection molded parts was measured. In the case of pure nylon the crystallinity was approximately 18 % but for the 4 % clay loaded nanocomposite it was 50 %. In both cases the predominant crystalline phase in very thin parts was the gamma form. For a constrained polymer region 30 nm thick and a 4 wt % loading of clay, the modified tortuous path model gives a calculated value of 44 % for the volume fraction of polymer in the constrained polymer region. Unfortunately, the relative permeability of this composite was not reported. It was further reported in this work that the crystallinity was very non-spherelitic and strongly influenced by the presence of the clay in the nanocomposite. Theoretical work done by Balaz predicts that for smectites with aspect ratios of 30 a nematic phase will predominate above 20 vol %. For commercially available smectites that exhibit aspect ratios more in the range 200–250, the nematic phase would appear much earlier and most likely below 5 wt %, since at 5 wt % the distance between plates is about 50 nm which is small compared with the lateral dimensions of the plates. In work conducted on polydimethyl siloxanes, a bound rubber phase was measured by solvent uptake on nanocomposites loaded with different levels of clay (Figure 6). It can be seen that the graph levels off at approximately 5 wt % and has a constrained polymer region of roughly 50 %. The constrained polymer model also predicts many of the changes in physical properties as a function of clay loading. An example is the change in heat distortion temperature (HDT) for nylon 6 nanocomposites [1]. Figure 7 is a plot of HDT as a function of clay loading. Again, the curve levels off at approximately 5 wt %. These examples all point to the fact that at about 5 wt % in most polymers the constrained polymer region is a large enough portion of the total volume for packing considerations to become dominant.

In the previous discussion an assumption was made about the relative diffusion coefficients of the constrained polymer region and the bulk polymer. The model allows the permeability of a given nanocomposite for different permeants to vary since the relative diffusion coefficient for the two phases can change depending upon the permeant.

4 CONCLUSION

A fairly simple model for predicting the behavior of nanocomposites has been proposed. The model is a modification of the very simple tortuous path model that

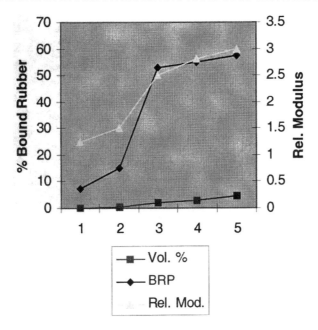

Figure 6   Bound rubber phase and relative modulus for PDMS as a function of clay loading

incorporates a constrained polymer and surface modifier region. The model predicts that:

(a) The observed permeabilities can be lower or higher than that predicted by the simple tortuous path model, depending on the relative diffusion coefficients of the phases.
(b) The permeability of a given nanocomposite can be quite different for different permeants.
(c) Nematic phases will predominate at or above 5 wt% loading of clay in most polymers.
(d) The aspect ratio of the mineral only may not always predict the permeability correctly for a given polymer.

The model is based upon the assumption of the existence of four distinct phases in a nanocomposite, including a clay phase, a surface-modified phase, a constrained polymer phase, and a polymer phase similar to the pure polymer. There is much experimental evidence for the existence of these phases, both direct and inferred. Future research should be directed towards a more complete understanding of these phases in nanocomposites.

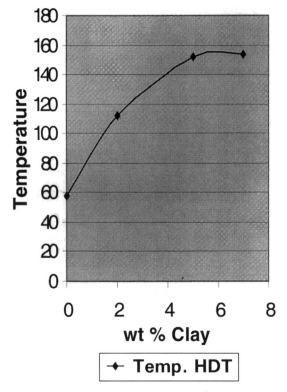

Figure 7   Heat distortion temperature as a function of clay loading

5   REFERENCES

1. Okada, A., Kawasumi, M., Usuki, A., Kojimo, Y., Kurauchi, T. and Kamigaito, O. in Schaefer, D. W. and Markes, J. E. (Eds), Materials Research Society Proceedings, Vol. 171, Polymer Based Molecular Composites, 1990.
2. Kojima, Y., Usuki, A., Kawasumi, M., Okada, A., Fukushima, Y., Kurauchi, T. and Kamigaito, O. *J. Mater. Res.*, **8**, 1185–1189 (1993).
3. Lan, T. and Pinnavaia, T.J. *Chem. Mater.*, **6**, 2216–2219 (1994).
4. Lan, T., Kaviratna, P. D. and Pinnavaia, T. J. *Chem. Mater.*, **6**, 573–575 (1994).
5. Massam, J. and Pinnavaia, T. J. *Mater. Res. Soc. Symp. Proc.*, **520**, 223–232 (1998).
6. Kawasumi, M., Hasegawa, N., Kato, M., Usuki, A. and Okada, A. *Macromolecules*, **30**, 6333–6338 (1997).
7. Hasegawa, N., Kawasumi, M., Kato, M., Usuki, A. and Okada, A. *J. Appl. Polym. Sci.*, **67**, 87–92 (1998).
8. Kojima, Y., Fukumori, K., Usuki, A., Okada, A. and Kurauchi, T. *J. Mater. Sci. Lett.*, **12**, 889–890 (1993).
9. Messersmith, P. B. and Giannelis, E. P. *J. Polym. Sci., Part A, Polym. Chem.*, **33**, 1047–1057 (1995).

10. Neilsen, L. E. *J. Macromol. Sci. (Chem.)*, **A1**(5), 929–942 (1967).
11. Okade, A. Personal communication, of Toyota Central Research Laboratory.
12. Gu, J. Master of Science thesis, Michigan State University, 1997.
13. Yano, K., Usuki, A. and Okada, A. *J. Polym. Sci., Part A, Polym. Chem.*, **35**, 2289–2294 (1997).
14. Geannelis, E. P. Personal communication, Cornell University.
15. Beall, G. W., Tsipursky, S., Sorokin, A. and Goldman, A. US Pat. 5,698,624, 16 December 1997.
16. Akkapeddi, M. K. in Proceedings of *Antec '99*, New York, 2–5 May 1999.
17. Balazs, A., Singh, C., Zhulina, E. and Lyotskoya, Y. *Acc. Chem. Res.*, **32**, 651–657 (1999).
18. Burnside, S. D. and Giannelis, E. P. *Chem. Mater.*, **7**, 1597–1600 (1995).

14

Modeling the Phase Behavior of Polymer–Clay Nanocomposites

A. BALAZS, V. V. GINZBURG, Y. LYATSKAYA, C. SINGH
AND E. ZHULINA
Chemical Engineering Department, University of Pittsburgh, Pittsburgh, PA,
USA

1 INTRODUCTION

One of the major gaps in our understanding of the properties of polymer/clay nanocomposites [1–8] concerns the equilibrium phase behavior of the mixture. Whether or not the system forms a uniform dispersion depends not only on processing but also, ultimately, on thermodynamic considerations. Even if well-separated, isolated clay sheets are intermixed with polymers, these high aspect ratio platelets can form ordered or crystalline structures, or can phase-separate from the matrix material. The task of describing the stability and morphology of polymer–clay mixtures can be divided into two parts: (i) determining the polymer-mediated interactions between the clay sheets, and (ii) calculating phase diagrams as a function of the composition of the mixture once these interactions are known. In effect, the problem involves analyzing the system on two different length scales. To obtain information about the nanoscale interactions among the organic modifiers, the underlying clay sheets and the polymer chains, one can treat the sheets as infinite, planar surfaces. To probe the macroscale behavior of the entire system, the clay disks can be treated as rigid particles that are dispersed in an incompressible fluid and interact via excluded-volume and effective long-range potentials. The first problem deals with the properties of polymers confined between the clay surfaces, while the second problem concerns the thermodynamics of anisotropic colloidal particles in a melt or solution. For accurate description of the equilibrium behavior of polymer/ clay mixtures, we ultimately must address both these issues.

Polymer–clay nanocomposites Edited by T. J. Pinnavaia and G. W. Beall
© 2000 John Wiley & Sons Ltd

The behavior of polymer melts in confined geometries (between two parallel plates) has been studied using different theoretical (self-consistent field, Monte Carlo) and experimental (neutron scattering, X-ray diffraction, FTIR) methods. Recently, Vaia and Giannelis [7, 8] used a lattice model and Balazs et al. [9–12] used both numerical and analytical self-consistent field (SCF) theories to calculate free energy profiles as two surfaces are pulled apart within a bath of molten polymers. These calculations indicate whether the mixture will form an intercalated or exfoliated composite, or a phase-separated system. In an intercalated system, the polymers penetrate the galleries and enhance the separation between the plates by a fixed amount; in an exfoliated composite, the sheets are effectively separated from each other and dispersed within the polymer melt. Calculating the complete phase diagram for the large-scale system, however, requires a detailed balance of all the contributions to the free energy, including the translational and orientational entropy of the clay particles, as well as their excluded-volume interactions. To account properly for these factors, one needs to turn to theoretical methods that describe the macroscopic phase behavior of colloidal systems.

The macroscopic phase behavior of polymer–colloid mixtures has been the subject of many theoretical, computational and experimental studies [13–18]. In most of the theoretical and computational investigations, the colloidal particles were assumed to be spherical in shape and the Asakura–Oosawa (AO) [13] model was adopted to describe the polymer–colloid mixture. In the AO model, the polymers are represented by small spheres, and the only interaction is due to excluded-volume effects. Thus, the problem of characterizing the complex blend was reduced to describing the properties of a binary hard sphere mixture. The phase diagram of binary hard sphere mixtures can be calculated to obtain liquid and solid (colloidal crystal) phases. Calculated phase diagrams can then be tested against experimental data, and good qualitative agreement is generally found between theory and experiment for mixtures of spherical colloids and polymers [17].

In the case of polymer–clay mixtures, however, a complicating factor is introduced by the strong anisotropy in the shape of the clay sheets. As noted above, because of their high aspect ratio, clay sheets experience orientational ordering even at low volume fractions and can form liquid crystalline phases (nematic, smectic, columnar or plastic solid), in addition to traditional liquid and solid phases (see Figure 1). Although there have been a few numerical and experimental studies of the thermodynamics of fluids containing anisotropic colloidal particles [19–24], the majority of these investigations were focused on rod-like colloids rather than discotic particles. Dijkstra et al. [23] studied the phase behavior of clay particles (modeled as infinitely thin disks with quadrupolar potentials), and Bates and Luckhurst [24] modeled a system of Gay–Berne [25] disks. Both of these simulations, however, considered only disks in the absence of solvent or polymer.

In recent studies [26, 27], we developed simple models that describe the liquid crystalline ordering in the polymer–platelet systems. The former theory [26] is based on the Onsager model [28] of the nematic ordering in hard-body fluids, while the

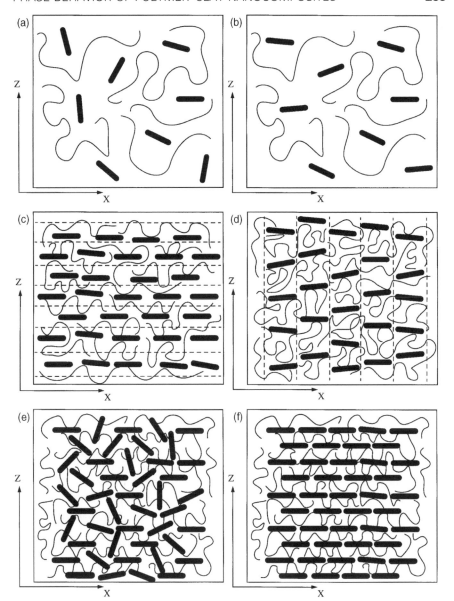

Figure 1   Mesophases of oblate uniaxial particles dispersed in a polymer: (a) isotropic, (b) nematic, (c) smectic A, (d) columnar, (e) plastic solid and (f) crystal. The nematic director **n** in ordered phases is aligned along the Z axis, and the disks lie in the XY plane. Dashed lines show smectic layers (c) and columns (d)

latter study [27] uses density functional theory to incorporate the positional (smectic and crystal) ordering of the platelets. By explicitly including excluded-volume interactions in these models, one can make direct correlations between the phase behavior of the system and the geometric properties of the colloidal particles, unlike the more phenomenological Landau-type [29] or Maier–Saupe-type [30, 31] theories.

The above theories [26, 27] take into account the long-range interactions between the clay sheets in only a rather approximate manner. An important task is to incorporate a more accurate description of the polymer-mediated clay–clay interactions into the density functional approach and thus to obtain the direct dependence of the phase behavior on chemical and molecular characteristics of the mixture. In a recent study [32], we combined the density functional formalism for the ordering of the clay sheets with the SCF-calculated potentials for the effective clay–clay interactions within a polymer melt. We incorporated this potential into the Somoza–Tarazona [33, 34] free energy functional for anisotropic oblate ellipsoids of revolution. The model was used to describe the phase diagram for several systems of clay disks with grafted organic modifiers, or 'surfactants', dispersed in a polymer melt.

In this chapter, we review our studies on the phase behavior of polymer/clay mixtures. We begin by discussing our SCF studies where we model neighboring sheets as two infinite, planar surfaces and determine the nanoscale interactions between the polymers and these interfaces. We then present a more macroscopic theory that allows us to describe the phase behavior of the polymer/clay mixture in terms of isotropic and nematic phases, as well as the two-phase coexistence regions. Through this model, we determine the role that shape of the clay sheet and polymer chain length play in the overall phase diagram. We then describe our use of density functional theory in conjunction with the SCF calculations to obtain the phase diagrams for several values of the surfactant grafting density, ρ_{gr}, and length, N_{gr}. The resulting phase diagrams not only confirm our earlier qualitative predictions [9, 12] but also offer quantitative estimates of the volume fractions of clay that correspond to the different phases. Finally, we present a simple kinetic argument that, coupled with our equilibrium analysis, provides additional insight into the factors that affect the morphology of the composite [26].

As discussed below, we investigate the properties of three distinct classes of polymer/clay mixtures. Firstly, we discuss the case of untreated clay particles in a melt of non-functionalized polymers. The second case involves a mixture of polymers with pretreated clay particles (where surfactants are grafted on each clay sheet). Finally, we consider dispersing clay particles in a melt of functionalized macromolecules. In such a system, some or all of the polymers have functionalized groups that are strongly attracted to the clay surfaces. This interaction can promote the exfoliation of clay sheets and thus yield homogeneous nanocomposites.

2 POLYMERS PENETRATING THE CLOSELY SPACED SHEETS

2.1 SELF-CONSISTENT FIELD CALCULATIONS

To investigate the nanoscale interactions between a polymer melt and the closely spaced sheets, we use the self-consistent field (SCF) theory derived by Scheutjens and Fleer [35]. In this treatment, the phase behavior of polymer systems is modeled by combining Markov chain statistics with a mean field approximation for the free energy. The equations in this lattice model are solved numerically and self-consistently. The self-consistent potential is a function of the polymer segment density distribution and the Flory–Huggins interaction parameter, χ, between the different components. (Note that χ is typically inversely proportional to temperature, and thus variations in χ are comparable with variations in temperature.) Since the method is thoroughly described in reference [35], we refer the reader to that text for a more detailed discussion. While such SCF calculations do not necessarily yield quantitative predictions, the results indicate how to tailor the system to modify the stability and morphology of the mixture.

Using this method, we consider two infinite, parallel plates immersed in a bath of molten polymer. To model organically modified clays, where the sheets are coated with linear alkyl chains, we introduce terminally anchored short-chain surfactants on to the plates. These surfaces lie parallel to each other in the xy plane and we investigate the effect of increasing the separation between the surfaces in the z direction. As the surface separation is increased, polymer from the surrounding bath penetrates the gap between these walls and the calculations yield the corresponding change in the free energy, ΔF, of the system. By systematically increasing the surface separation, H, we obtain the ΔF versus H curves discussed below. In previous studies [36] we used a similar approach to determine $\Delta F(H)$ for two polymer-coated surfaces in various solvents.

We note that the different components in these systems can contain charged species. In the following calculations we do not consider electrostatic interactions. Nonetheless, the strong attraction between oppositely charged sites can be modeled through a large, negative value for the relevant χ. Below, we determine how the features of the surfactants, melt and substrate affect the ability of the polymers to penetrate the gallery [9].

2.1.1 Tailoring the Surfactants

To provide a basis of comparison, we first consider the behavior of the polymer melt in the presence of the bare surfaces. The interaction parameter between the polymers and the substrate is given by $\chi_{surf} = 0$ and the length of the polymers is fixed at $N = 100$ monomers. In the reference state, the surfaces are in intimate contact. As the surfaces are pulled apart, polymers occupy all the space between these walls (because of the incompressibility of the system). When the chains within the gallery come into contact with the solid surfaces, they lose conformational entropy against the impenetrable interfaces. As a result, ΔF becomes positive as H is increased (see Figure 2) and the intermixing of these components is unfavorable. In other words, at $N = 100$, the polymers and neutral sheets are immiscible and the mixture would phase-separate.

In the case of the organically modified clay surfaces, favorable enthalpic interactions between the tethered surfactants and the polymers can overwhelm the entropic losses and lead to effective intermixing of the polymer and clay. Here, the reference state (corresponding to $\Delta F = 0$) is taken to be the state where the tethered surfactants form a melt that completely saturates the space between the two walls [37]. Our aim is to isolate conditions where ΔF is reduced relative to the reference state, that is, $\Delta F < 0$. For such values, the intermixing of the polymers and modified sheets is favorable relative to the unmixed state.

We first vary χ, the polymer–surfactant interaction parameter, fixing the length of the surfactants at $N_{gr} = 25$ and the grafting density at $\rho_{gr} = 0.04$. [The average distance between the surfactants is $(1/\rho_{gr})^{1/2}$]. Figure 3 shows the $\Delta F(H)$ profiles for various values of χ. For $\chi > 0$ [cases (a) and (b)], $\Delta F > 0$ and consequently the corresponding mixture would be immiscible. For $\chi \cong 0$ [cases (c) and (d)], the plots show distinct local minima for $\Delta F < 0$. Such local minima indicate that the mixture forms an 'intercalated' structure [7]. In particular, the lowest free energy state is one where the polymers have penetrated the gallery and enhanced the separation between the plates by a fixed amount. For $\chi < 0$ [case (e)], the plot indicates that there is a

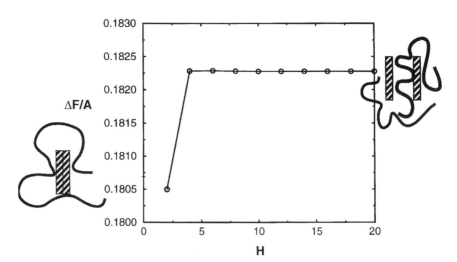

Figure 2   Free energy per unit area, $\Delta F/A$, versus H for the bare surfaces in the melt. Here, $\chi_{surf} = 0$ and polymer length is $N = 100$. The illustration on the left shows the surfaces in intimate contact, and on the right the sketch shows the polymers localized between these substrates

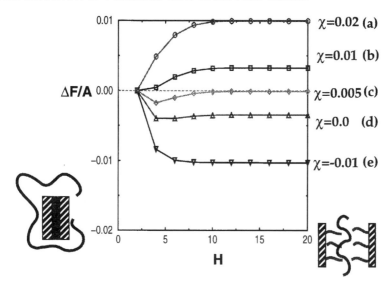

Figure 3 $\Delta F/A$ versus H for various χ values. The parameters are: $N_{gr} = 25$, $\rho = 0.04$, $N = 100$ and $\chi_{surf} = 0$. The illustration on the left shows the reference state, where the grafted chains form a melt between the surfaces, and in the sketch on the right the surfaces are separated by the intervening polymers

global minimum at large (infinite) separations. Such plots point to an 'exfoliated' structure [7], where the sheets are effectively separated from each other and dispersed within the melt. Thus, from a purely thermodynamic argument, we see that increasing the attraction between the polymers and surfactants promotes the formation of stable composites and could result in the creation of exfoliated structures.

Increasing the length of the polymers, N, increases the disparity in the lengths of the free and grafted chains and increases the values of ΔF relative to those in Figure 3. Thus, increasing N promotes phase separation and requires more attractive χ values in order for the system to be miscible. One way of decreasing the disparity between the free and tethered chains is to increase the length of the surfactants. Figure 4 shows the free energy profiles [38] at various surfactant lengths ($N_{gr} = 25$, 50 and 100) for $\chi = 0$, 0.01 and 0.02. The grafting density is fixed at $\rho_{gr} = 0.04$. For the $\chi = 0$ case, increasing the length of the tethered chain alters the structure of the composite from intercalated to exfoliated. More dramatically, for $\chi = 0.01$, increasing N_{gr} drives an immiscible mixture to form a thermodynamically stable inter-calated hybrid.

The interactions between the short surfactants and long polymers are characterized by a sharp, thin interface and little interpenetration between the different chains [9, 39]. In contrast, when the polymer and surfactant are of the same length

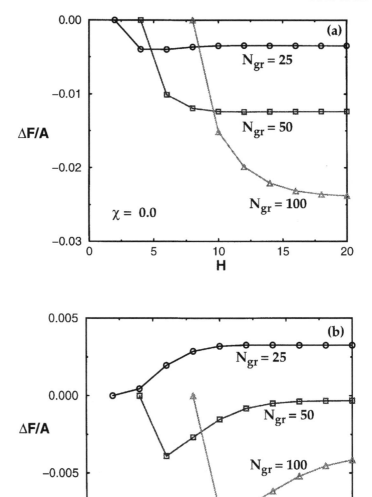

Figure 4 $\Delta F/A$ versus H for different surfactant lengths. Here, $\rho = 0.04$, $N = 100$ and $\chi_{surf} = 0$. The figures are for the following values of the interaction parameter: (a) $\chi = 0.0$, (b) $\chi = 0.01$ and (c) $\chi = 0.02$

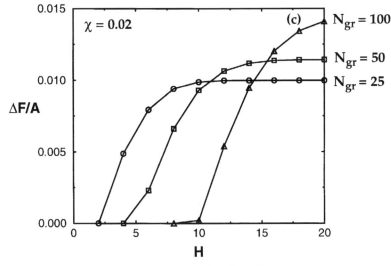

Figure 4   (*Continued*)

($N_{gr} = N = 100$), the interaction between the species leads to a broad interface, or 'interphase', which allows the polymers more conformational degrees of freedom and therefore is more entropically favorable. As a result, the values of ΔF are lower for the longer surfactants.

Note that these entropic effects dominate even at values of χ ($= 0.01$) where the energetic interactions are unfavorable. If, however, χ is increased beyond a critical value, energetic effects dominate and increasing the surfactant length increases the free energy. For $\chi = 0.02$, increasing the length of the surfactant increases the number of unfavorable polymer–surfactant contacts and thus increases the value of ΔF. Thus, within a finite range of polymer–surfactant interactions, the miscibility and morphology of the composite can be tailored by increasing the surfactant length.

These observations agree with recent experiments on the fabrication of nanocomposites with organically modified clays and polystyrene–polybutadiene block copolymers [40]. The researchers found that increasing the length of the alkyl chains on the clay promoted insertion of the copolymer, whereas no insertion was observed with the shorter alkyl chains. They also noted that, the longer the alkyl chain, the greater is the compatibility with the polystyrene block and the greater is the effective strength of the material.

Finally, we note that increasing ρ_{gr} to 0.12 shifts ΔF to higher values and destabilizes the mixture (see Figure 5). As the surfactant layer becomes more dense, it becomes harder for the free chains to penetrate and intermix with the tethered species. At high ρ_{gr}, more attractive (negative) values of χ are needed to promote polymer penetration into the interlayer. Comparing Figures 2, 3 and 5, we see that a

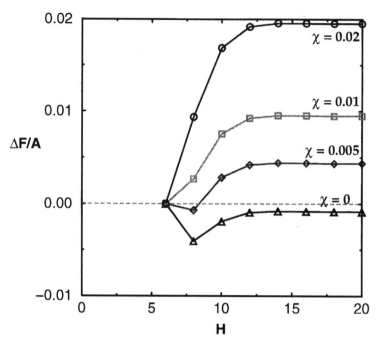

Figure 5 $\Delta F/A$ versus H for $\rho = 0.12$ and various χ values. Other parameters are the same as in Figure 3

bare surface and a surface with too many tethered surfactants are both unfavorable for forming the hybrids. The comparison points to the fact that there is an optimal grafting density for forming polymer/clay composites.

2.1.2 Tailoring the Polymer Melt

In the above discussions, we considered systems where χ is varied from 0.0 to 0.02. For cases where the specified polymers are highly incompatible with the alkyl chains or underlying substrate, an alternative means of intermixing the components must be developed. The ideal scheme would be applicable to a broader range of macro-molecules and would not involve the challenging task of optimizing the surfactant grafting density. Here, we introduce the notion of tailoring the melt by adding a fraction of functionalized polymers. These chains contain 'stickers' that are highly attracted to the surface; otherwise, they are chemically identical to the remainder of the chains in the bulk. The functionalized species will bind to the bare clay sheets and thus constitute long-chained surfactants. The free polymer can readily penetrate this thick grafted layer and form a broad interphase, which promotes the formation of exfoliated hybrids. While the experimental results will depend on the kinetics of the process, the SCF calculations can indicate, for example, how the fraction of functionalized chains in the mixture affects the thermodynamic stability of the product. To test the feasibility of this approach, we varied the fraction of functionalized polymers within a melt of chains of comparable length.

The functionalized polymers contain a sticker site at one end of the chain and are slightly shorter ($N = 75$) than the non-functionalized species ($N = 100$). (In a physical experiment, decreasing the length of the functionalized species by a small degree will permit these chains to diffuse to the surface faster than the longer chains and, once at the surface, provide an interfacial layer for the longer polymers.) The interaction between the reactive stickers and the surface [41] is characterized by the Flory parameter χ_{surf}; we set $\chi_{surf} = -75$. The interaction parameters between the stickers and all other species are set equal to 0. (Thus, the stickers do not react with themselves or other monomers.) For the other monomers in the system (the non-stickers), $\chi'_{surf} = 0$.

The results in Figure 6 imply that this scheme is successful in exfoliating the sheets. Note that having just a small fraction of such functionalized species in the melt promotes the dispersion of the particles. As a basis of comparison, recall that the interaction between the bare surface and non-functionalized polymers leads to an immiscible mixture (Figure 2). At low volume fractions of functionalized polymers, these chains do indeed act as high molecular weight surfactants, forming an interlayer that the polymer can readily penetrate. On the other hand, as the fraction of the functionalized chains is increased, the polymers are driven to coat the substrates and consequently the surfaces are pushed apart by the adsorbing chains. Note that increasing the fraction of functionalized chains beyond 30% has little effect on decreasing ΔF. In effect, once the surfaces have been coated with the 'sticky' polymers, increasing the volume fraction of these species does not lead to further decreases in this free energy. For that matter, since all curves in Figure 6 point to an exfoliated system, the most cost-effective treatment could be the 5% example, where only a small fraction of functionalized species is required for the process. These results are in excellent agreement with the findings of our analytical SCF model of this system [9]. Using this analytical model, we again find that only $\sim 5\%$ of functionalized polymers are needed to exfoliate the sheets.

The above results point to another route for forming stable polymer/clay dispersions. Instead of the functionalized polymers, a small fraction of AB diblock copolymers could be added to a melt of B polymers or a melt composed entirely of AB diblocks could be employed. Here, a short hydrophilic A block will anchor the chain to the bare (non-modified) clay sheets. The large organophilic B block will extend away from the surface and could drive the separation and dispersion of these sheets.

2.1.3   Tailoring the Clay Substrate

The plots in Figure 7 show the effect of introducing a relatively large attraction ($\chi_{surf} = -1$) between the polymers and the underlying substrate. By comparing Figures 7a and b, we see that a highly attractive surface interaction could promote

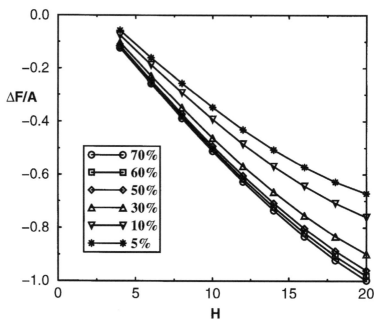

Figure 6 $\Delta F/A$ versus H. The melt contains two types of polymer: a functionalized chain ($N = 75$) and a slightly longer ($N = 100$) non-functionalized species. The functionalized polymer has a sticker at one end of the chain. The interaction between the sticker and surface is given by $\chi_{surf} = -75$. The percentages refer to the fraction of functionalized polymers. The other parameters are the same as in Figure 3

the formation of exfoliated composites from previously intercalated structures or immiscible components. Even when the interaction between the polymer and surfactants is relatively unfavorable ($\chi = 0.05$), the presence of the attractive substrate drives the system to form an intercalated hybrid. This effect is similar to the mechanism of intercalation proposed by Vaia and Giannelis [7, 8]. As indicated above, the effect of a large, negative χ_{surf} is comparable with the effect of a strong electrostatic attraction between the polymer and surface. Thus, the equilibrium results indicate that tailoring the relative charge or polarity of the substrate provides a means of controlling the phase behavior and structure of the mixture. As we caution below, the kinetics of polymer interpenetration can, however, lead to trapped states that affect the morphology of the final hybrid.

In the above calculations, we analyzed the thermodynamic behavior and isolated conditions that promote exfoliation by studying the free energy profiles of polymers confined between two surfaces. In order to determine the equilibrium morphology of a given system, it is important to consider other factors, such as the clay volume fraction and the geometric characteristics of the clay sheets. For this calculation, we

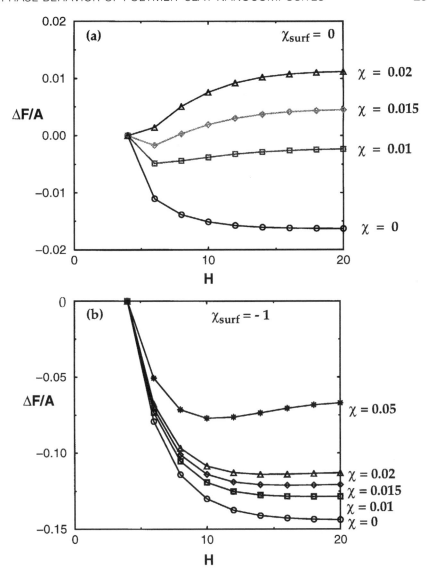

Figure 7 $\Delta F/A$ versus H in the presence of polymer–surface interactions. The other parameters are $N = 100$, $N_{gr} = 100$ and $\rho = 0.02$. The values of the polymer–surface interaction are: (a) $\chi_{surf} = 0$ and (b) $\chi_{surf} = -1$

must utilize more macroscopic models than the SCF; we describe these models below.

3 MODELING THE MACROSCOPIC PHASE BEHAVIOR

Our macroscopic approaches to calculating the phase behavior are based on the density functional theory (DFT) formalism. We first formulate a simple Onsager-type density functional and describe the orientational ordering of clay sheets as a function of their volume fraction, aspect ratio and the Flory–Huggins parameter between the polymers and clay. Next, the density functional is significantly modified and expanded to account for the formation of positionally ordered phases (smectic, columnar, 'house-of-cards' and crystal). Finally, we combine the DFT formalism with SCF calculations (using the SCF free energy profiles as pair potentials, which serve as input to the free energy functional) to obtain approximate phase diagrams for various polymer/clay mixtures. It is important to note that, in calculating such phase diagrams, we assume that the sheets are well separated and can undergo random mixing with the polymer chains.

3.1 ONSAGER-TYPE APPROACH: ORIENTATIONAL ORDERING OF CLAY SHEETS

The clay sheets are roughly oblong in shape and possess a high aspect ratio, being approximately 200 nm in diameter and 1 nm in width. To capture this physical asymmetry, we model the individual particles as broad, rigid disks of diameter D and thickness L. The homopolymers are modeled as flexible chains of length N. The volume fraction of disks in the incompressible blend is ϕ_d, and $\phi_p = 1 - \phi_d$ is the volume fraction of polymer. The interaction between a monomeric unit in the polymer chain and a surface site on the disk is characterized by χ.

In the mixed state, the broad disks can be either randomly oriented with respect to each other and form an isotropic phase, or be relatively aligned and form a nematic phase. A nematic ordering is especially beneficial in decreasing the permeability of gases through polymer/clay films [1]. To analyze the miscibility of the homopolymer/disk mixture and investigate the formation of the isotropic and nematic structures, we modify the Onsager model [28, 42, 43] for the nematic ordering of rigid rods. The free energy of the mixture has the following form [26]:

$$F = F_{conf} + F_{ster} + F_{int} + F_{transl} \tag{1}$$

The term F_{conf} describes the conformational losses due to the alignment of the disks, F_{ster} accounts for the steric interactions between the disks, F_{int} describes the enthalpic contributions due to polymer–disks contacts and F_{transl} takes into account the translational entropy of disks. Using this approach, we obtain the following expression for the free energy [26]:

$$F = n_d(\text{const} + \ln\phi_d + \sigma - \ln(1 - \phi_d)(b/v_d)(\rho - 1) - \chi v_d\phi_d) + n_p \ln\phi_p \quad (2)$$

where n_p is the number of polymer molecules, n_d is the number of disks per unit volume and $v_d = (\pi D^2 L)/4$ is the volume of an individual disk. The parameters σ and ρ allow us to distinguish between the isotropic and nematic phases; σ effectively describes the orientation of the disks with respect to the nematic direction, and $\rho = (4/\pi)\langle|\sin\gamma|\rangle$, where γ is the angle between two disks. In the isotropic phase, $\sigma = 0$ and $\rho = 1$, while in the nematic phase, $\sigma_n = 2\ln[-(2/\sqrt{\pi})(v_d/b)\ln(1 - \phi_d)] - 1$ and $\rho_n = -2(v_d/b)/\ln(1 - \phi_d)$.

To construct the equilibrium phase diagram, we equate the chemical potentials of the disks and polymer in the isotropic and nematic phases: $\mu_d^i = \mu_d^n$ and $\mu_p^i = \mu_p^n$, where $\mu_k^i = (\delta F_i)/(\delta n_k)|_{n_{j\neq k}} = \text{const}$. We solve this system of equations numerically and present the results as plots that highlight the role of χ, ϕ_d and N in the phase behavior of the system.

3.1.1 Isotropic–Nematic Phase Diagrams: The Role of Shape Anisotropy and Polymer Chain Length

Figure 8 shows the phase diagrams for the polymer/disk mixtures in the χ versus ϕ_d coordinate frame for various homopolymer lengths, N. The diameter of the disks is fixed at $D = 30$ and the thickness L is set at 1. The solid curves mark the phase boundaries for the system. The area between the respective boundaries encompasses the phase-separated region. The two regions below the boundaries pinpoint the regions of miscibility: the mixture forms an isotropic phase (i) at low concentrations of disks and exhibits a nematic phase (n) at high ϕ_d.

As can be seen from the figure, the region of miscibility decreases with an increase in N. The plots further reveal that, for relatively high N ($N \geq 100$), phase separation occurs even at negative values of χ, where there is an attractive interaction between the two components. In other words, at $N \geq 100$, the polymer and disks are totally immiscible even if there is no repulsion between these components, indicating an entropically driven phase separation. Note that this result agrees with our finding from the SCF calculations; Figure 2 shows that, for $N = 100$, the polymer and bare surface are immiscible even at $\chi_{surf} = 0$. The SCF results also show that increasing N promotes phase separation between the sheets and polymer.

The analytical model also indicates the effect of varying the size of the clay sheets. In particular, increasing the diameter of the disks promotes the nematic ordering of disks at lower values of ϕ_d. This behavior is shown in Figure 9. Note that in the two limiting cases of low and high D there is the possibility of isotropic–isotropic (i–i) phase coexistence in the former case and nematic–nematic (n–n) phase coexistence in the latter case. Our main interest is in the isotropic–nematic transition, and therefore we do not consider instances of i–i and n–n coexistence.

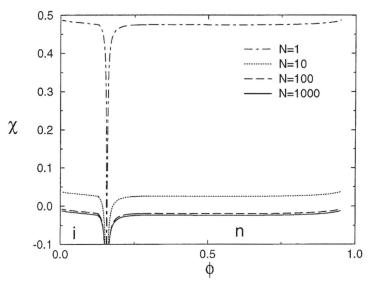

Figure 8 Phase diagrams for disks and polymers of various *N*, plotted as a function of χ versus ϕ_d, where ϕ_d is the volume fraction of the disks. The diameter of the disks is $D = 30$, and their width is $L = 1$. The area above the phase boundaries encompasses the region of immiscibility. The miscible regions lie below the boundaries. Within the regions of miscibility, i marks the isotropic phase and n indicates the nematic phase. The plots reveal that increasing *N* increases the region of immiscibility

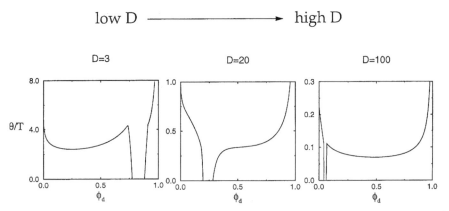

Figure 9 Phase diagrams illustrating the effect of increasing the disk diameter

We now modify the free energy density functional to include: (i) the positional ordering of the clay sheets at high volume fractions and (ii) a more detailed description of the interactions (attractive or repulsive) between the clay sheets. The first task is performed by replacing the Onsager-type free energy with the Somoza–Tarazona [34] functional, while the second one is implemented by inserting the SCF-calculated potentials (obtained from the free energy profiles) into the interaction term, F_{int}. Below, we discuss this combined DFT–SCF approach and present calculated phase diagrams for various mixtures of polymers and organically modified clay.

3.2 COMBINING DENSITY FUNCTIONAL AND SELF-CONSISTENT FIELD THEORY

We again consider a system of clay disks with a thickness $L = 1$ nm and diameter $D = 30$ nm (values similar to the experimental system of Muzny *et al.* [44]) dispersed in a polymer melt where the length of the polymers is given by $N = 300$. Surfactants are grafted to the surface of each clay platelet. Both the polymers and surfactants are taken to be flexible Gaussian chains. A monomer in either the polymer or surfactant is modeled by a sphere of radius $a = 1$ nm. The polymer–polymer interactions and the surfactant–surfactant interactions are restricted to excluded-volume effects. On the other hand, between the polymers and surfactants there are excluded-volume and short-range enthalpic interactions. The latter interaction is described through the Flory–Huggins χ parameter. In carrying out these calculations, our primary aim is to test the above prediction that the presence of a long, compatible 'surfactant' can lead to the formation of stable, exfoliated composites. Thus, in the ensuing study, we set the χ for the polymer–surface and surfactant–surface interactions equal to zero and focus on varying the χ for the polymer–surfactant interactions, the surfactant length, N_{gr}, and the grafting density, ρ_{gr}.

To study the dependence of the phase behavior on the volume fraction of clay, ϕ_c, and χ, we use the Somoza–Tarazona [33, 34] formulation of the density functional theory (DFT). This formalism was first developed for monodisperse fluids of rigid anisotropic particles. To adapt it to binary or ternary systems (colloidal particles dispersed in polymer melt or solution), we follow ideas developed in references [26] and [45] and impose the incompressibility condition

$$\phi_c + \phi_{gr} + \phi_p = 1 \tag{3}$$

where ϕ_c is the volume fraction of clay sheets, ϕ_{gr} is the volume fraction of the grafted surfactant and ϕ_p is the volume fraction of the polymer. The volume fractions of clay and surfactant are related via

$$\phi_{gr} = \phi_c (2\rho_{gr} N_{gv} V_m / L) \tag{4}$$

where V_m is the monomer volume. Here, we consider the clay sheets to be thin, oblate ellipsoids of revolution, rather than oblate cylinders or cut spheres. This approximation allows us to use the Somoza–Tarazona free energy density functional without introducing additional adjustable parameters. (Although our specific choice for the shape of the clay can affect the phase behavior for large volume fractions of clay, the general features of the phase diagram should remain relatively insensitive to differences between these various geometries. We will explore this point further in future studies.)

We can describe the state of the polymer and organically modified clay mixture by only two variables, χ and ϕ_c (in the following, we will drop the subscript and use ϕ instead of ϕ_c). The orientational and positional ordering of the clay sheets is described by the single-particle distribution function (SDF) $\gamma(\mathbf{r}, \mathbf{n})$, where $\mathbf{r}$ represents the position of the center of mass of a particle, and $\mathbf{n}$ is the nematic director (which denotes the direction of the short axis of a disk). Following Kventsel *et al.* [46], we decouple the orientational and positional degrees of freedom:

$$\gamma(\mathbf{r}, \ \mathbf{n}) = \rho(\mathbf{r}) f(\mathbf{n}) \tag{5}$$

where $\rho(\mathbf{r})$ is the positional SDF (the local density of clay disks), and $f(\mathbf{n})$ is the orientational SDF. The free energy of the system is written as a functional of ρ and f.

It is important to note that such an approximation (neglecting the compressibility of the mixture and assuming that the free energy depends only on the clay SDF) is a significant simplification. Recent studies of binary hard sphere mixtures [18] showed that the 'one-density' approach is reasonable when the ratio of the volumes of large and small spheres is sufficiently high. In our system, the volume of a colloidal particle is significantly larger than the volume occupied by a polymer chain, thus justifying the use of a 'one-density' approach. The free energy contributions due to the translational and conformational degrees of freedom of the polymer are added only as new terms in the entropic and enthalpic parts of the overall free energy. While the calculation of the translational part of the polymer free energy is straightforward (we simply use a Flory–Huggins expression), the evaluation of the conformational part is more complicated. Indeed, the conformational entropy of a polymer is strongly dependent on the spatial distribution of dispersed clay particles. Polymers confined in a narrow gallery between two adjacent disks have much smaller entropy than those remaining in the melt. The conformational free energy of the confined polymers can be rewritten as an effective interaction between the disks. This effective interaction can be calculated by using a numerical self-consistent field (SCF) model [35], as we will describe in the following section.

We rewrite the free energy of the system in the form

$$\beta F = \beta F_{id} + \beta F_{ster} + \beta F_{int} \tag{6}$$

where F_{id} is the free energy of an 'ideal gas' of colloidal particles and polymers, F_{ster} is the contribution of the excluded-volume effects for the colloidal (clay) sheets, F_{int} represents the long-range (attractive or repulsive) interactions between clay sheets and $\beta = 1/kT$.

The first ('ideal') term of the free energy on the right-hand side of equation (6) can be written as the sum of two parts:

$$\beta F_{id} = \beta F_c + \beta F_p \tag{7}$$

In specifying the exact form of βF_c and βF_p, we follow the principles used, for example, by Gast et al. [14] and Lekkerkerker et al. [15] for mixtures of polymers and spherical colloidal particles. This approach is modified to include the orientational ordering of the clay particles. The 'ideal' free energy of clay particles, βF_c, consists of translational and orientational terms:

$$\beta F_c = \int dr \rho(\mathbf{r}) \ln \rho(\mathbf{r}) + \int d\mathbf{r}\, d\mathbf{n} \rho(\mathbf{r}) f(\mathbf{n}) \ln[4\pi f(\mathbf{n})] \tag{8}$$

The 'ideal' free energy of the polymer melt, βF_p, includes only the translational (Flory–Huggins) contribution:

$$\beta F_p = (V/Nv_m)\phi_p \ln(\phi_p) \tag{9}$$

where V is the total volume of the system, v_m is the monomer volume and ϕ_p is the volume fraction of the polymer.

We introduce a new parameter, the effective particle thickness L_{eff}:

$$L_{eff} = L + 2\rho_{gr}N_{gr}V_m \tag{10}$$

and the effective shape anisotropy parameter $\kappa = L_{eff}/D \leq 1$. The minimal volume occupied by an organically modified particle (if the grafted chains were to form a monolayer) is $V_{eff} = (\pi/6)D^2 L_{eff}$.

For a system of ellipsoidal particles, a semi-empirical steric interaction free energy can be written as [33, 34, 47, 48]

$$\beta F_{ster} = \int dr \rho(\mathbf{r}) \Psi_{hs}[\phi_c'(\mathbf{r})](V_{excl}[f]/V_{phe}) \tag{11}$$

where $\Psi_{hs}(x)$ is the semi-empirical Carnahan–Starling [49] function, which describes the excess (non-ideal) free energy density for hard spheres as a function of their packing fraction. The parameter $V_{excl}[f]$ is the average excluded volume per particle for a given orientational distribution, and $V_{phe} = V_{eff}$ is the excluded volume per particle for perfectly aligned ellipsoids. If we consider only spatially uniform phases (e.g. isotropic and nematic phases), equation (11) would reduce to the Onsager-type expression with the correct second virial coefficient. For non-uniform phases, it is a complicated functional of the density, because the function $\phi_c'(\mathbf{r})$ depends on $\rho(\mathbf{r})$ in a non-trivial manner (for more details see references [33], [34] and [48]). The role of this functional is to properly describe the short-range correlations due to the excluded-volume interactions.

The sum of the free energy terms in equations (7) and (11) describes an athermal dispersion of hard oblate ellipsoids in a polymer melt. It is known that such a system is capable of forming liquid crystalline (nematic) and crystalline phases [19, 21]. In order to obtain additional liquid crystalline phases (smectic or columnar), strongly anisotropic long-range interactions are required [50]. We suppose that the interaction free energy, βF_{int}, is 'small' compared with the ideal and steric free energy terms, that is, $|\beta F_{int}| < |\beta F_{id} + \beta F_{ster}|$. In this case, one can assume that the pair correlation function for the particles, $g(1,2)$, is mostly determined by the excluded-volume effects (not long-range interactions), and calculate the interaction free energy βF_{int} as

$$\beta F_{int} = \int d\mathbf{r}_1 \, d\mathbf{n}_1 \, d\mathbf{r}_2 \, d\mathbf{n}_2 \rho(\mathbf{r}_1) f(\mathbf{n}_1) \rho(\mathbf{r}_2) f(\mathbf{n}_2) \delta(1 - \mathbf{n}_1 \cdot \mathbf{n}_2) g(1,2) V(\mathbf{r}_1 - \mathbf{r}_2) \quad (12)$$

where the mean field pair correlation function $g(1,2) = 0$ if particles overlap, and is equal to 1 if they do not overlap. This form is similar to the traditional way of representing enthalpic interactions between anisotropic particles (see, for example, equation (3) in reference [48]). The delta function on the right-hand side of equation (12) allows only those configurations in which interacting disks are parallel. Such an approximation is reasonable for very anisotropic particles where side-by-side configurations have a significantly larger 'contact' area than either side-to-edge or edge-to-edge configurations. In addition, since this potential is highly anisotropic, it is likely to promote the formation of smectic and columnar phases [49]. The potential function $V(\mathbf{r})$ is assumed to have the following generic form:

$$V(\mathbf{r}) = (\pi/4)D^2(1 - [r_\perp/D]^2)U(z) \quad (13)$$

where $\mathbf{r} = (x, y, z)$ and $r_\perp = (x^2 + y^2)^{1/2}$. The form of the interaction potential per unit area, $U(z)$, and its dependence on such factors as the Flory–Huggins parameter, χ, the polymer chain length, N, the grafting density of the surfactant, ρ_{gr}, and the surfactant chain length, N_{gr}, is discussed in the next section.

When evaluating the integral over positions and orientations of particles [the right-hand side of equation (12)], we use the orientational lattice approximation (see, for example, reference [51]). In this approach, particle orientations are restricted to being along the x, y or z directions only. Such an approximation is necessary to reduce the number of degrees of freedom and thus make it possible to analyze the whole phase diagram within a reasonable time frame. (The errors introduced by using the orientational lattice cannot significantly alter the phase diagram since they contribute only to βF_{int} and not to the dominant term βF_{ster}; however, it is possible that the use of the orientational lattice approximation slightly affects the equilibrium value of the nematic order parameter in the ordered phases.)

In order to describe the thermodynamic behavior of the system, it is necessary to minimize the free energy for all possible phases (isotropic, nematic, smectic, columnar and crystal) for each value of ϕ_c, and find the lowest energy state. After that, coexistence regions can be found by means of applying the Maxwell rule, or, equivalently, by equating the chemical potentials of the particles and the polymer for the different phases. The minimization is done using a variational approach in which the SDF is parameterized by specific functions that reflect the symmetry of a given phase. We use the following parameterization:

For the orientational SDF

$$f(\mathbf{n}) = \left\{\alpha/[4\pi\sinh(\alpha)]\right\}\cosh[\alpha\cos(\mathbf{n}, \mathbf{z})] \tag{14a}$$

For positional SDF

$$\rho(\mathbf{r}) = \rho A \exp(L\cos[Q_z z]) \qquad\qquad \text{for the smectic phase}$$
$$\rho(\mathbf{r}) = \rho A \exp(L\cos[Q_x x]\cos[Q_y y]) \qquad\qquad \text{for the columnar phase}$$
$$\rho(\mathbf{r}) = \rho A \exp(L\cos[Q_x x]\cos[Q_y y]\cos[Q_z z]) \qquad \text{for the crystal phase} \tag{14b}$$

In each case, A is a normalization constant, while L describes the strength of the positional ordering. In order to minimize the number of free parameters, we assume that the columnar structure is hexagonal ($Q_x = \sqrt{3}Q_y$), and the crystal structure is 'stretched' hexagonal close packed ($Q_x = \sqrt{3}Q_y = \kappa\sqrt{3}Q_z$).

To complete the free energy functional, it is necessary to calculate the interaction potential per unit area, $U(z)$, for the clay sheets. As mentioned above, this interaction potential consists of two parts:

(a) the interaction between 'bare' clay particles, $U_1(z)$ (due to electrostatic and Van der Waals attraction), and

(b) the contribution from the polymer chains confined in the gallery between adjacent clay sheets, $U_2(z)$.

Both contributions are shown schematically in Figure 10. The former term depends on the chemical structure of the clay itself, and we will not attempt to calculate it from first principles (such a calculation is described, for example, in reference [52]). Instead, we simply assume that it is purely attractive and short ranged, and use the following form:

$$U_1(z) = E_0\left\{1 - \exp[-(z - L_{\text{eff}})]\right\}. \tag{15}$$

In order to evaluate the pair interaction between clay sheets owing to the presence of the polymers and grafted chains, we employ the results of the above SCF calculations. After solving the equations and finding the equilibrium polymer and surfactant density profiles, we calculate the free energy density $\Delta F/A$ for each value of the plate separation z. Repeating this calculation for different separations, one obtains the effective potential per unit area, $U_2(z)$. In this study, we have only one non-zero χ parameter, that between the polymer and the surfactant. When the interaction between the polymers and surfactants is repulsive (positive χ), the polymers do not penetrate the gallery between the clay sheets, and the effective interaction between the sheets is predominantly attractive. When χ is negative, the interaction between the polymers and surfactants is attractive, and the polymer is pulled into the gallery, leading to the intercalation or possible exfoliation of the clay platelets.

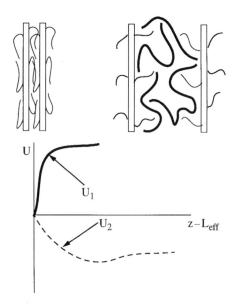

Figure 10 Schematic representation of the pair interaction between the two parallel plates with grafted organic modifiers. The solid line represents the contribution from the 'bare plate' attraction, U_1, and the dashed line depicts the effective contribution of the grafted 'surfactants' and the polymer chains, U_2. On the sketch, short thin lines are surfactant chains, and long thick lines are 'free' chains

The use of the self-consistent field approach to calculate interparticle potentials can be justified if the volume fraction of clay particles is sufficiently low for most polymer chains to interact with no more than two clay sheets at the same time. It can be shown that this requirement is satisfied if the clay volume fraction is not very high ($\Phi = \phi + \phi_{gr} < 0.5$) and the radius of gyration of a polymer chain, R_g, is less than the diameter of the clay particle, $R_g < D$. In the system we are considering, $N = 300$ and $R_g \approx \sqrt{N} \approx 17$, so this condition is satisfied.

To study the influence of the grafted chains on the equilibrium morphology of the mixtures, we consider the following systems:

(a) $\rho_{gr} = 0.2$, $N_{gr} = 5$;
(b) $\rho_{gr} = 0.04$, $N_{gr} = 25$;
(c) $\rho_{gr} = 0.02$, $N_{gr} = 50$;
(d) $\rho_{gr} = 0.04$, $N_{gr} = 50$;
(e) $\rho_{gr} = 0.02$, $N_{gr} = 100$.

The geometric characteristics of the clay sheets and the length of the polymers in the melt N are kept constant for all the systems: $D = 30$, $L = 1$, $N = 300$. We consider two values of the clay–clay attraction strength, E_0:

(a) $E_0 = 0$ (no long-range attraction between clay sheets);
(b) $E_0 = 0.1 \, kT/a^2$ (strong attraction between clay sheets).

The calculation of the phase diagrams is performed in two steps. Firstly, we use the SCF model to determine the free energy profiles $U_2(z)$ for each of the five systems, varying the Flory–Huggins parameter χ between 0.05 and -0.05. While this may appear to be a relatively small variation in χ, the SCF calculations described in the previous section indicate that the system can undergo significant changes in phase behavior (going, for example, from immiscible to intercalated or exfoliated) over this range of values. The calculated free energy represents the polymeric contribution to the clay–clay interaction potential. After adding the bare clay–clay interaction potential [described by equation (15)], we obtain the necessary potential energy profiles. To facilitate the subsequent discussion, and illustrate the connection between the potentials and the phase diagrams, we plot these potentials in Figures 11a to e. Recall that, for a fixed z, the polymer–clay interaction is more favorable for lower values of U. Comparison of the profiles allows us to make some qualitative conclusions about the role of grafted chains in determining the equilibrium phase behavior.

3.2.1 SCF Input Data

We first consider the case $E_0 = 0$. It can be seen, by comparing Figures 11a to c, that, for a fixed value of $\theta_t = \rho_{gr} N_{gr}$ (the total amount of grafted monomers), decreasing the grafting density and increasing the length of the grafted chains leads to decreases in the values of $U(z)$. This trend indicates better miscibility between the disks and polymer and hence a more repulsive interaction between the disks. The entropy of mixing is greater for polymers and a loose layer of long grafted chains than for polymers and a dense layer of short chains. Furthermore, the longer grafted chains are more effective at providing a steric repulsion between the surfaces. (The steric interaction is due to the entropic elasticity of the grafted chains, which creates an effective repulsion between the surfaces, until the surface separation becomes comparable with the radius of gyration of the grafted chains.)

For the range of grafting densities considered here, increasing ρ_{gr} while keeping the chain length constant (see Figures 11c and d) promotes greater mixing between the polymer and clay for $\chi \leq 0$. Now, the polymers interact with a 'soft' penetrable interface, rather than a hard wall. This observation holds for relatively low grafting densities. At higher grafting densities, the tethered chains form a dense surface layer and the intermixing between the polymers and clay is again unfavorable, as was discussed in the SCF section. Note that, for $\chi > 0$, increasing ρ shifts the curves to greater positive values. Here, increasing the number of chains increases the number of incompatible interactions and thus causes an upward shift in the free energy.

Finally, we keep the grafting density unchanged and vary the length of the grafted chains (see Figures 11b and d and Figures 11c and e). For $\chi \leq 0$, the strength of the

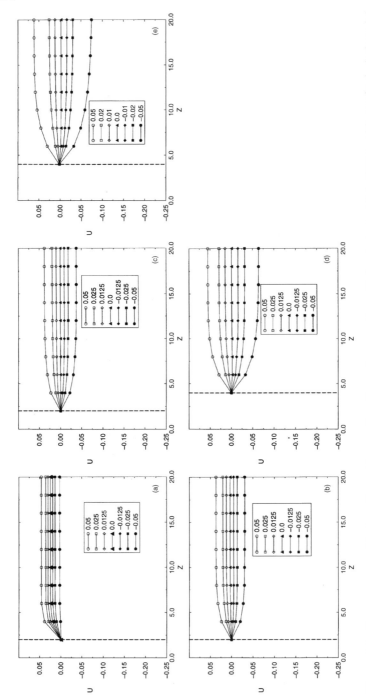

Figure 11 Free energy of interaction, U, as a function of separation, Z, for two parallel plates with grafted chains: (a) $\rho_{gr} = 0.2$, $N_{gr} = 5$; (b) $\rho_{gr} = 0.04$, $N_{gr} = 25$; (c) $\rho_{gr} = 0.02$, $N_{gr} = 50$; (d) $\rho_{gr} = 0.04$, $N_{gr} = 50$; (e) $\rho_{gr} = 0.02$, $N_{gr} = 100$. The dashed vertical lines on each plot denote the 'hard-wall' condition (Z cannot be smaller than $2\rho_{gr}N_{gr}$ because of the incompressibility of the polymer melt and the grafted chains). The attractive strength $E_0 = 0$

disk–disk repulsion increases as the chain length becomes longer. Thus, from considering all these different cases, one can conclude that for $\chi \leq 0$, increasing ρ_{gr} (up to an optimal value) and the length of grafted chains N_{gr} leads to better miscibility between the clay sheets and the polymer melt.

3.2.2 Phase Diagrams: the Role of ρ_{gr} and N_{gr}

We now turn our attention to the calculation of the (χ, ϕ) phase diagram for each of the five systems. To calculate phase diagrams, we vary the 'combined volume fraction', $\Phi = \phi + \phi_{gr}$, between 0.001 and 0.601 with a step of 0.01. This procedure is repeated for 13 values of χ in the range $[-0.05, 0.05]$. (Recall that positive values of χ correspond to the repulsion between the grafted chains and the polymer melt, and, therefore, to effective attraction between the clay disks; negative values of χ correspond to the attraction between the grafted chains and the polymer melt, causing effective repulsion between the clay disks.) For each point in the (χ, ϕ) plane, we calculate free energies of all possible phases. In terms of the morphologies in these diagrams, the isotropic or low-density nematic phase corresponds to an exfoliated composite. The crystal and columnar phases describe clay-rich crystal-lites, and the plastic solid phase represents the low-density structure known as a 'house-of-cards' (where clay disks have some positional ordering and prefer 'edge-to-face' configurations) [23, 53]. We also find two-phase regions where highly ordered clay-rich phases coexist with pure polymer. The phase diagrams for all five systems are shown in Figures 12a to e.

We will first comment on the general features of the phase diagrams. In the region where the χ parameter is positive, the systems exhibit a broad phase coexistence between the polymer-rich isotropic phase and the clay-rich crystal phase. Such behavior corresponds to a strong immiscibility between the organically modified clay and the polymer matrix; clay sheets, even if exfoliated during processing, will eventually reaggregate. As $\chi \to 0$, the two-phase region becomes narrower and splits into two (isotropic–nematic and nematic–crystal). The triple point (I–N–Cr) is dependent on the specific system parameters, as is the width of the nematic phase. For the systems under consideration, the smectic phase was always found to be metastable (within the broad isotropic–crystal coexistence). When χ becomes negative, the isotropic–nematic and nematic–crystal coexistence regions become narrower and shift towards higher values of ϕ. Finally, when χ is strongly negative, the new 'plastic solid' ('house-of-cards') and columnar phases appear. Indeed, at low clay loadings, the strong repulsion between neighboring disks forces them to adopt more energetically favorable 'edge-to-face' configurations. This leads to either gelation or crystallization. At higher clay volume fractions, the steric excluded volume effects dominate the long-range disk–disk repulsion, forcing the formation of columnar and crystal phases.

We now analyze the dependence of the phase behavior on the surfactant length and the grafting density. It can be easily seen that the phase diagrams plotted in

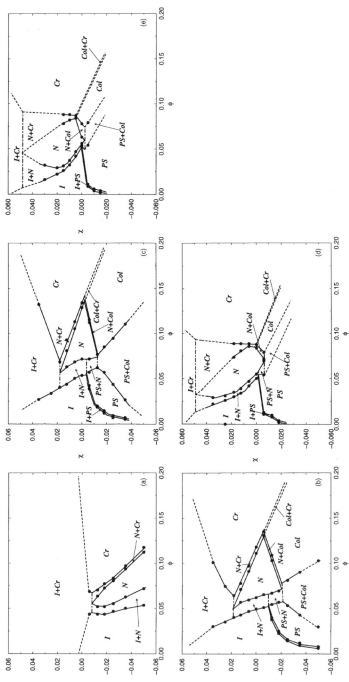

Figure 12 Phase diagrams of polymer–clay mixtures: (a) $\rho_{gr} = 0.2$, $N_{gr} = 5$; (b) $\rho_{gr} = 0.04$, $N_{gr} = 25$; (c) $\rho_{gr} = 0.02$, $N_{gr} = 50$; (d) $\rho_{gr} = 0.04$, $N_{gr} = 50$; (e) $\rho_{gr} = 0.02$, $N_{gr} = 100$. Phases: I = isotropic, N = nematic, Cr = crystal, PS = plastic solid, Col = columnar. Points represent calculated coexistence densities, lines serve as a guide to the eye; dashed lines represent approximate locations of phase transition boundaries (exact calculation was impossible because one or both coexistence points lie in the region $\Phi > 0.6$). ϕ is the clay volume fraction, χ is the Flory–Huggins parameter between the surfactant and the polymer

Figures 12d ($\rho_{gr} = 0.04$, $N_{gr} = 50$) and 12e ($\rho_{gr} = 0.02$, $N_{gr} = 100$) are very similar. The phase diagrams shown in Figures 12b ($\rho_{gr} = 0.04$, $N_{gr} = 25$) and 12c ($\rho_{gr} = 0.02$, $N_{gr} = 50$) are also very similar to each other. One can conclude, therefore, that the macroscopic phase behavior of the polymer–clay mixture is primarily determined by the total amount of surfactant θ_t. This statement holds for relatively long surfactants and relatively low grafting densities. For the case of short, densely grafted organic modifiers (Figure 12a), the immiscibility between the clay disks and polymer (see Figure 11a) dominates the phase behavior of the system.

To study the dependence of the phase behavior of the mixture on the surfactant length, N_{gr}, we compare the diagram plotted in Figures 12b ($\rho_{gr} = 0.04$, $N_{gr} = 25$) with that in 12d ($\rho_{gr} = 0.04$, $N_{gr} = 50$), and compare the diagram in Figure 12c ($\rho_{gr} = 0.02$, $N_{gr} = 50$) with the one in 12e ($\rho_{gr} = 0.02$, $N_{gr} = 100$). It can be seen that increasing N_{gr} moves the isotropic–nematic–crystal triple point upwards (for example, from $\chi \sim 0.02$ in Figure 12b to $\chi \sim 0.05$ in Figure 12d). The stability region of the nematic phase (and to a lesser extent, the isotropic phase) is extended to more positive values of χ. Increasing N_{gr} also narrows the two-phase isotropic–nematic region. All these effects are favorable in creating thermodynamically stable exfoliated composites. The only adverse factor from increasing N_{gr} is the narrowing of the nematic phase near $\chi = 0$. Indeed, with the increase in the total amount of surfactant, we simultaneously increase the effective excluded volume of the clay particles and decrease the shape anisotropy [see equation (10)]. As a result, the maximum 'inorganic volume fraction' where one can obtain an orientationally ordered exfoliated morphology is decreased dramatically (for example, from $\phi = 0.12$ in Figure 12b to $\phi = 0.07$ in Figure 12d).

Finally, we consider the case where the grafting density ρ_{gr} is changed while N_{gr} is held constant. By comparing Figures 12c ($\rho_{gr} = 0.02$, $N_{gr} = 50$) and 12d ($\rho_{gr} = 0.04$, $N_{gr} = 50$), we see that increasing ρ_{gr} shifts the isotropic–nematic–crystal triple point to higher values of χ, extends the stability region of the nematic region to more positive χ and reduces the size of the isotropic–nematic coexistence region. In effect, increasing ρ_{gr} (for the limited ranges examined here) improves the miscibility and enhances the thermodynamic stability of the dispersed composite morphologies (isotropic and nematic phases).

In the above analysis we assume that the clay particles have only excluded-volume interactions ($E_0 = 0$). In many real systems there is a strong Van der Waals interaction between the clay sheets that inhibits exfoliation. To model this effect and its influence on the phase behavior, we repeat all the calculations with $E_0 = 0.1$, mimicking a strong clay–clay attraction. We calculate the phase diagrams and see how the addition of the clay–clay attraction modifies the phase behavior.

For the systems with $\theta_t = 1$, the clay–clay attraction completely dominates the entropic contribution of the surfactant molecules. The phase diagrams (not plotted here) show the complete immiscibility between polymer and clay for all values of χ in the $[-0.05, 0.05]$ range. For the systems with $\theta_t = 2$ (Figures 13a and b), the miscibility is significantly diminished relative to the corresponding $E_0 = 0$ cases

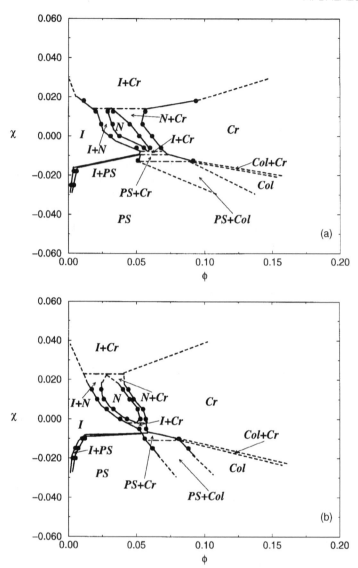

Figure 13 Phase diagrams of polymer–clay mixtures: (a) $\rho_{gr} = 0.04$, $N_{gr} = 50$; (b) $\rho_{gr} = 0.02$, $N_{gr} = 100$. $E_0 = 0.1$. The phase diagrams for the other three systems are not shown because, for all $-0.05 < \chi < 0.05$, the polymer and clay are completely immiscible

(see Figures 12d and e), as one would expect. The isotropic–nematic–crystal triple point is shifted downwards from $\chi \approx 0.05$ ($E_0 = 0$, Figure 12d) to $\chi \approx 0.02$ ($E_0 = 0.1$, Figure 13a). The total areas occupied by isotropic and nematic phases shrink dramatically, and the nematic–crystal transition occurs at lower values of ϕ.

The calculated phase diagrams suggest qualitative trends in the macroscopic phase behavior of polymer–clay nanocomposites. At the same time, quantitative comparison of phase diagrams of model and currently studied experimental systems is somewhat difficult. Typically, the surfactant molecules range in length between 10 and 20 CH_2-groups, and the grafting density is quite high (1–3 chains/nm^2) [7, 8]. More systematic experimental studies involving longer grafted chains are needed to test our predictions.

It is important to note that the calculated phase diagrams describe only the equilibrium properties of the polymer–clay mixtures. As suggested in reference [26], to understand exfoliation in polymer–clay nanocomposites, one must also consider the kinetic aspects of the penetration of the polymers into the gap between the clay sheets.

3.3 COMBINING THERMODYNAMICS WITH KINETICS

To calculate the phase diagrams, we assumed that the disks are well separated from each other. Thus, each particle could undergo random mixing with the polymer. In reality, however, this situation might not be reached. Consider the mixing of the polymer and clay agglomerate [5]. Within the agglomerate, the individual crystallites or sheets are arranged in the closely spaced parallel stacks. In order for the different components to intermix, the polymer must penetrate the gap from an outer edge and then diffuse towards the center of the gallery. When the polymer moves through the gap, two scenarios are possible, depending on the value of χ. In the first case, $\chi \geq 0$ and there is no attraction between the polymer and disks. Here, the polymer can separate the disks, as the chain tries to retain its coil-like conformation and gain entropy. In the second case, $\chi < 0$ and the polymer and disks experience an attractive interaction. Here, the polymer 'slides' through the gallery, maximizing contact with the two confining surfaces. The overall conformation of the polymer is rather 'flat', but the losses in conformational entropy are compensated for by enthalpic gains when the chain comes in contact with one of the disks. The final morphology of the mixture will obviously be different for the two scenarios. In the first case, the disks will be totally separated and dispersed in the polymer matrix, forming an exfoliated structure. In the second case, the disks are effectively glued together by the intervening polymer and the platelets remain parallel to each other, forming an intercalated structure. These trends are supported qualitatively by recent computer simulations of polymer flow into an attractive slit [54].

We can now combine the results of the equilibrium study with our kinetic arguments to analyze the structure of the final composite. As discussed above, for kinetic reasons, a true exfoliated structure, where the disks are totally separated, can only occur for $\chi \geq 0$. Figure 8, however, shows that, for $\chi \geq 0$, the homopolymers and disks will ultimately demix *unless* the polymer is relatively short, namely $N < 100$. Therefore, a stable exfoliated structure can only be attained by mixing disks with relatively short homopolymers.

Again from kinetic considerations, we observed that an intercalated structure is expected when $\chi < 0$. The equilibrium analysis (see Figures 8 and 12a) shows that χ must be negative for mixtures of long homopolymers and disks to be thermodynamically stable. Combining these arguments, we see that stable mixtures of high molecular weight polymers and discotic particles will exhibit an intercalated morphology.

The proposed scheme of using a mixture of functionalized and non-functionalized polymers for the melt could be a way around this problem, providing a means of creating composites with exfoliated morphologies. While the stickers at the chain ends are highly attracted to the surface, the remainder of polymer does not react with the substrate. Thus, as the polymers penetrate the sheets, the majority of the chain is not likely to glue the surfaces together. Similar behavior is expected to occur in melts containing diblock copolymers, where a short block is attracted to the substrate, but a longer hydrophobic block will separate the hydrophilic sheets.

4 CONCLUSIONS

We used two distinct theoretical methods to probe the interactions between polymers and clay sheets, or more generally, solid particles. Through the SCF models, we determined the free energy profiles as a polymer melt penetrates the gap between bare and organically modified sheets. The shape of the profiles reveal whether the polymer–clay interaction is favorable and, thus, if the mixture is miscible. For miscible mixtures, the curves indicate the structures (intercalated versus exfoliated) of the composite. Through these calculations, we could modify the characteristics of the surfactants, polymers and substrate and thus isolate factors that drive the polymers to permeate the clay galleries. This model will not, however, yield a complete phase diagram for the system.

To generate such a phase diagram, we developed density functional theories for a mixture of homopolymers and rigid disks. Through these models, we could take into account the possible orientational and positional ordering of the disks and pinpoint the boundaries for the isotropic, nematic, smectic, columnar and crystal phases, as well as the phase-separated regions of phase space. In this chapter, we used the Onsager-type model to show that an increase in N requires a decrease in χ for the mixture to be thermodynamically stable. Similar conclusions were obtained when we used our second DFT model to probe the effect of varying N [27]. These observations also agree with the SCF calculations [9].

By combining the DFT and SCF approaches, we could obtain additional insight into the equilibrium behavior of the system. In particular, by using the SCF-generated free energy curves as an input to the DFT free energy functional, we can connect the macroscopic phase behavior of the mixture with details of the polymer–clay interaction. We applied this multiscale technique to determine the phase behavior of organically modified clays in polymer melts and derived several important conclusions concerning the stability of such mixtures and the effect of varying the surfactant length, N_{gr}, and density, ρ_{gr}.

The actual phase behavior and morphology of the mixture can be affected by the kinetics of the polymers penetrating the gap. Combining the results of our equilibrium studies with simple kinetic arguments, we predict that, for large N and $\chi < 0$, a stable hybrid will only exhibit an intercalated morphology. Recent experimental studies [8] reveal that the melt mixing of organically modified clays and highly attractive polymers does in fact lead to intercalated hybrids.

The above results provide design criteria for synthesizing optimal exfoliating agents. For facile penetration into the gallery, the polymer must contain a fragment that is highly attracted to the surface. This fragment also promotes miscibility between the polymer and clay. In addition, this copolymer must contain a longer fragment that is not attracted to the sheets. The non-reactive block will attempt to gain entropy by pushing the sheets apart. Once the sheets are separated, these blocks will also sterically hinder the surfaces from coming into close contact. Our SCF results on melts containing functionalized polymers give credence to this scheme. The macroscopic phase diagrams also show that having a long chain (comparable with the length of the polymer) anchored to the surface of the clay promotes the stability of the isotropic and nematic phases. These results reveal that the optimal polymeric candidates for creating stable exfoliated composites are those that would constitute optimal steric stabilizers for colloidal suspensions.

5 ACKNOWLEDGEMENTS

This work was supported by the Dow Chemical Company, the Army Office of Research, the N.S.F. through grant number DMR-9709101 and ONR through grant N00014-91-J-1363. Part of the calculations in this study were carried out on the computers in the Laboratory of Molecular and Materials Simulations, University of Pittsburgh, funded by the N.S.F. and by a grant from IBM Corp. We thank Drs David Moll, Dmitri Kuznetsov and June Huh for useful discussions.

6 REFERENCES AND NOTES

1. Yano, K., Uzuki, A., Okada, A., Kurauchi, T. and Kamigaito, O. *J. Polym. Sci., Part A, Polym. Chem.*, **31**, 2493–2498 (1993).
2. Wang, Z. and Pinnavaia, T. J. *Chem. Mater.*, **10**, 1820–1826 (1998).

3. Vaia, R. A., Jandt, K. D., Kramer, E. J. and Giannelis, E. P. *Macromolecules*, **28**, 8080–8085 (1995).
4. Messersmith, P. B. and Stupp, S. I. *J. Mater. Res.*, **7**, 2599–2611 (1992).
5. Krishnamoorti, R., Vaia, R. A. and Giannelis, E. P. *Chem. Mater.*, **8**, 1728–1734 (1996).
6. Giannelis, E. P., Krishamoorti, R. and Manias, E. *Adv. Polym. Sci.*, **138**, 107–147 (1999) and references therein.
7. Vaia, R. A. and Giannelis, E. P. *Macromolecules*, **30**, 7990–7999 (1997).
8. Vaia, R. A. and Giannelis, E. P. *Macromolecules*, **30**, 8000–8009 (1997).
9. Balazs, A. C., Singh, C. and Zhulina, E. *Macromolecules*, **31**, 8370–8381 (1998).
10. Balazs, A. C., Singh, C. and Zhulina, E. in Wahl, K. J. and Tsukruk, V. (Eds), *Microstructure and Tribology of Polymer Surfaces*, Plenum Press, 1999, Ch. 23.
11. Zhulina, E., Singh, C. and Balazs, A. C. *Langmuir*, **15**, 3935–3943 (1999).
12. Balazs, A. C., Singh, C., Zhulina, E. and Lyatskaya, Y. *Acct. Chem. Res.*, **32**, 651–657 (1999).
13. Asakura, S. and Oosawa, F. *J. Polym. Sci.*, **33**, 183–192 (1958).
14. Gast, A. P., Hall, C. K. and Russel, W. B. *J. Colloid Interface Sci.*, **96**, 251–267 (1983).
15. Lekkerkerker, H. N. W., Poon, W. C. K., Pusey, P. N., Stroobants, A. and Warren, P. B. *Europhys. Lett.*, **20**, 559–564 (1992).
16. Meijer, E. J. and Frenkel, D. *J. Chem. Phys.*, **100**, 6873–6887 (1994).
17. Illett, S. M., Orrock, A., Poon, W. C. K. and Pusey, P. N. *Phys. Rev. E*, **51**, 1344–1352 (1995).
18. Dijkstra, M., van Roij, R. and Evans, R. *Phys. Rev. Lett.*, **82**, 117–120 (1999).
19. Vroege, G. J. and Lekkerkerker, H. N. W. *Rep. Prog. Phys.*, **55**, 1241–1309 (1992).
20. Bolhuis, P. G., Stroobants, A., Frenkel, D. and Lekkerkerker, H. N. W. *J. Chem. Phys.*, **107**, 1551–1564 (1997).
21. Adams, M., Dogic, Z., Keller, S. L. and Fraden, S. *Nature*, **393**, 349–351 (1998) and references therein.
22. Dijkstra, M. and van Roij, R. *Phys. Rev. E*, **56**, 5594–5602 (1997).
23. Dijkstra, M., Hansen, J. P. and Madden, P. A. *Phys. Rev. E*, **55**, 3044–3053 (1997).
24. Bates, M. and Luckhurst, G. R. *J. Chem. Phys.*, **104**, 6696–6709, (1996).
25. Gay, J. G. and Berne, B. J. *J. Chem. Phys.*, **74**, 3316–3320 (1981).
26. Lyatskaya, Y. and Balazs, A. C., *Macromolecules*, **31**, 6676–6680 (1998).
27. Ginzburg, V. V. and Balazs, A. C. *Macromolecules*, **32**, 5681–5688 (1999).
28. Onsager, L. *Ann. N. Y. Acad. Sci.*, **51**, 627–659 (1949).
29. Liu, A. J. and Fredrickson, G. *Macromolecules*, **26**, 2817–2824 (1993).
30. Brochard, F., Jouffroy, J. and Levinson, P. *J. Phys. (France)*, **45**, 1125–1136 (1984).
31. Kyu, T. and Chiu, H.-W. *Phys. Rev. E*, **53**, 3618–3622 (1996).
32. Ginzburg, V. V., Singh, C. and Balazs, A. C. *Macromolecules*, **33**, 1089–1099 (2000).
33. Tarazona, P. *Phys. Rev. A*, **31**, 2672–2679 (1985).
34. Somoza, A. M. and Tarazona, P. *J. Chem. Phys.*, **91**, 517–527 (1989).
35. Fleer, G., Cohen-Stuart, M. A., Scheutjens, J. M. H. M., Cosgrove, T. and Vincent, B. *Polymers at Interfaces*, Chapman and Hall, London, 1993.
36. Singh, C., Pickett, G., Zhulina, E. B. and Balazs, A. C. *J. Phys. Chem. B*, **101**, 10614–10624 (1997).
37. Our SCF calculations are based on an incompressible model. In the reference state for the organically modified clays, we consider the system to be composed of the two surfaces and the tethered surfactants; there are no void or solvent sites in the system.
38. In the reference state, the surfactants form a melt that completely fills the gap between the surfaces. Since the length of the surfactants is now being varied, the reference states for different values of N_{gr} ($= 25$, 50, 100) are not identical. Namely, the shorter chains will achieve this state at smaller surface separations than the longer surfactants.

39. Leibler, L. and Mourran, A. *Mater. Res. Soc. Bull.*, **22**, 33–37 (1997) and references therein.
40. Laus, M., Francescangeli, O. and Sandrolini, F. *J. Mater. Res.*, **12**, 3134–3139 (1997).
41. In the case of surface adsorption, we must divide χ_{surf} by the coordination number of the cubic lattice in order to relate this Flory–Huggins parameter to experimentally relevant values. Thus, our value of χ_{surf} is comparable with a binding energy of $(75/6) = 12.5$ kT.
42. Odijk, T. *Macromolecules*, **19**, 2313–2329 (1986).
43. (a) Khokhlov, A. R. and Semenov, A. N. *J. Stat. Phys.*, **38**, 161–182 (1985); (b) Khokhlov, A. R. and Semenov, A. N. *Macromolecules*, **19**, 373–378 (1986).
44. Muzny, C. D., Butler, B. D., Hanley, H. J. M., Tsvetkov, F. and Peiffer, D. G. *Mater. Lett.*, **28**, 379–384 (1996).
45. Schaink, H. M. and Smit, J. A. M. *Macromolecules*, **29**, 1711–1720 (1996).
46. Kventsel, G. F., Luckhurst, G. R. and Zewdie, H. B. *Mol. Phys.*, **56**, 589–610 (1985).
47. Velasco, E., Somoza, A. M. and Mederos, L. *J. Chem. Phys.*, **102**, 8107–8113 (1995).
48. Ginzburg, V. V., Glaser, M. A. and Clark, N. A. *Liq. Cryst.*, **23**, 227–234 (1997).
49. Carnahan, N. F. and Starling, K. E. *J. Chem. Phys.*, **51**, 635–636 (1969).
50. Luckhurst, G. R. and Simmonds, P. S. J. *Mol. Phys.*, **80**, 233–252 (1993).
51. Sokolova, E. P. and Vlasov, A. Yu. *J. Phys. Condens. Matter*, **9**, 4089–4101 (1997) and references therein.
52. Hackett, E., Manias, E. and Giannelis, E. P. *J. Chem. Phys.*, **108**, 7410–7418 (1998).
53. Frenkel, D. *Mol. Phys.*, **60**, 1–20 (1987).
54. Baljon, A. R. C., Lee, J. Y. and Loring, R. F. *J. Chem. Phys.*, **111**, 9068–9072 (1999).

15

Rheological Properties of Polymer–Layered Silicate Nanocomposites

R. KRISHNAMOORTI AND A. S. SILVA

Department of Chemical Engineering, University of Houston, Houston, TX, USA

1 INTRODUCTION

Understanding the rheological properties of polymer–layered silicate nanocomposites is crucial to gain a fundamental understanding of the processability and structure–property relations for these materials. Moreover, these nanocomposite materials have proved to be model systems for examining the underlying molecular level underpinnings of the structure and dynamic properties of confined polymer systems and polymer brushes using macroscopic characterization techniques [1]. In fact, some of these nanocomposites have proved to be very useful in describing the properties of molten polymer brushes, and have demonstrated the substantial difference in behavior between solution and melt brushes [2]. This review focuses mainly on the rheological properties of layered silicate-based polymer nanocomposites in the melt state. The linear and non-linear viscoelastic properties for these nanocomposites are correlated with their nanoscale and mesoscopic structures and comparisons are drawn with the rheological properties of layered silicates dispersed in low molecular weight solvents.

Dispersions of organically modified layered silicates in organic solvents and unmodified layered silicates in water have suggested [3–8] that the rheological properties of these dispersions are rich and intimately linked to the mesoscopic structure of the inorganic filler in the dispersions. For instance, in the case of layered silicates dispersed in low molecular weight solvents, negative thixotropy has frequently been observed and attributed to the breakdown and slow reformation of the "house-of-cards" structure owing to the application and subsequent removal of shear. Furthermore, on account of the mesoscopic structure under quiescent conditions, many of these systems are also known to exhibit finite yield stresses. While the

Polymer–clay nanocomposites Edited by T. J. Pinnavaia and G. W. Beall
© 2000 John Wiley & Sons Ltd

'house-of-cards' structure has been attributed to the electrostatic attraction between edges of positively charged and negatively charged faces, others have suggested that the layers are associated with each other in tactoids, and the face–face mediated interactions result in a granular mesoscopic structure. This mesoscopic arrangement and its relative fragility to applied shear have been confirmed by *in situ* small-angle neutron scattering measurements [3, 4].

It has been conjectured that, in the case of polymer–layered silicate nanocomposites, the melt rheological properties are dictated by a combination of the mesoscopic structure and the strength of the interaction between the polymer and the layered silicate. The mesoscopic structure would be crucially dependent not only on the strength of the polymer/layered silicate interaction but also on the inherent viscoelastic properties of the matrix in which the layers or collection of layers are dispersed. In this review, the melt state rheological properties of such nanocomposites are examined in the light of their mesoscopic structure as well as the strength of the polymer surface interactions. In particular, this review will focus on the rheological characterization of three systems: exfoliated end-tethered poly(ε-caprolactone) (PCLC)-based nanocomposites [9], nylon 6 (NCH)-based nanocomposites [9, 10] and a series of intercalated nanocomposites of a roughly symmetric disordered poly(styrene)–poly(isoprene) diblock copolymer (PSPI) [11]. The pseudo-solid-like behavior observed at long times in the linear viscoelastic response is attributed to the percolation of a three-dimensional filler network structure comprising a random orientation of grains consisting of locally correlated layers. The non-linear behavior is strongly dependent on the physical connectivity or interaction between the polymer and the layered silicate, with end-tethering resulting in strain hardening and weaker interactions leading to shear thinning.

2   LINEAR VISCOELASTIC PROPERTIES

The melt-state linear viscoelastic properties for the nanocomposites are typically examined in a constant strain rheometer in either a cone and plate or parallel plate geometry. A time-dependent oscillatory strain of the form

$$\gamma(t) = \gamma_0 \sin(\omega t) \tag{1}$$

is applied, where γ_0 is the strain amplitude, ω is the frequency (varied from 0.001 to 100 rad/s) and t is time. The resulting time-dependent stress measured is of the form

$$\sigma(t) = \gamma_0[G' \sin(\omega t) + G'' \cos(\omega t)] \tag{2}$$

where $\sigma(t)$ is the shear stress, G' is the storage or elastic modulus and G'' is the loss or viscous modulus. The linearity of the measurements is typically confirmed by repeating the measurements at higher and lower strain amplitudes and observing the independence of the measured storage and loss moduli. For the nanocomposites with low silicate loadings, linear viscoelasticity is observed for strain amplitudes below 5 %, while for the higher-silicate nanocomposites, linear viscoelastic response is observed over a more restricted strain amplitude range, with non-linear effects being observed in some cases for strain amplitudes as small as 1 % [9–11].

The frequency-dependent viscoelastic properties (G', G'' and $\tan \delta = G''/G'$) are obtained at several different temperatures and superposed by the application of the Boltzmann time–temperature superposition principle to prepare viscoelastic master curves [12]. Both frequency (horizontal) shift factors a_T and modulus (vertical) shift factors b_T are applied simultaneously to the data to create the viscoelastic master curve. Depending on the relative values of the glass transition temperature of the matrix polymer and the experimental temperatures, the frequency shift factors can be fitted to the WLF equation [12]:

$$\log a_T = \frac{-C_1(T - T_0)}{[C_2 + (T - T_0)]} \tag{3}$$

where C_1 and C_2 are constants and T_0 is the reference temperature, or to the Arrhenius equation given as

$$\log a_T = -\frac{E_a}{R}\left[\frac{1}{T} - \frac{1}{T_0}\right] \tag{4}$$

where E_a is the flow activation energy.

In the following sections, the linear viscoelasticity results for the exfoliated end-tethered nanocomposites and intercalated PSPI hybrids are presented, followed by a detailed analysis of the possible structural origins of the viscoelastic response.

2.1 END-TETHERED EXFOLIATED NANOCOMPOSITES

The rheological master curves based on linear viscoelastic measurements for a series of delaminated hybrids, poly(ε-caprolactone) on montmorillonite (PCLC), prepared by *in situ* polymerization [13, 14] with the polymer chains being end-tethered to the silicate surface via cationic surfactants, are shown in Figure 1. The storage modulus for the PCLC hybrids show a monotonic increase at all frequencies with increasing silicate content [15], with the exception of the 2 wt % nanocomposite which, at the highest frequencies, has a slightly lower value than the 1 wt % nanocomposite [9]. The loss moduli, on the other hand, show a somewhat non-monotonic dependence, with the value for the 1 wt % hybrid exceeding that for the 2 and 3 wt % nanocomposites. However, the trend for the 2–10 wt % nanocomposites suggests that G'' increases with increasing silicate loading. These slight deviations from the expected monotonic increase in G' and G'' with increasing silicate loading are attributed to the decreasing molecular weight of the PCL with increased silicate loading, with the largest decrease occurring for the 2 wt % hybrid in comparison with that of the 1 wt % nanocomposite.

The storage and loss moduli for the nylon 6–layered silicate end-tethered nanocomposites (NCH), prepared by Ube Chemical Company [16–18] and with the molecular weight of the matrix polymer roughly comparable for all three samples, are shown in Figure 2. Measurements were conducted at a single

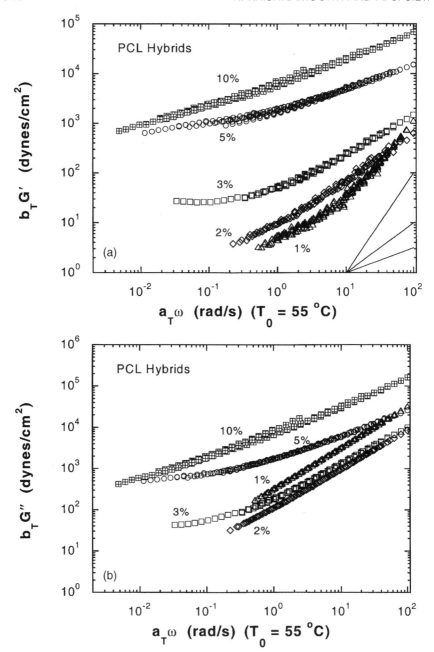

Figure 1 Time-temperature superposed linear viscoelastic dynamic moduli: (a) storage modulus G' and (b) loss modulus G'' for PCLC nanocomposites. Frequency scans from 55 to 160 °C were superposed using frequency and moduli shift factors

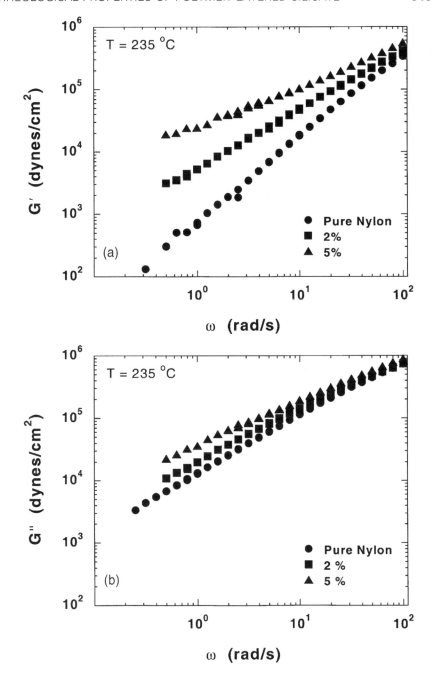

Figure 2   Linear viscoelastic dynamic moduli: (a) storage modulus G' and (b) loss modulus G'' for NCH nanocomposites at 235 °C

temperature of 235 °C [9]. These results clearly indicate that, in the absence of molecular weight differences, both G' and G'' increase monotonically with increasing silicate loading at all frequencies.

For both the PCLC and NCH systems described above, the molecular weights are low enough for the polymer chains to be fully relaxed at the frequencies and temperatures accessed, and, if present as homopolymers and in the absence of polydispersity effects, they would exhibit liquid-like flow behavior (i.e. $G' \propto \omega^2$ and $G'' \propto \omega$). While the polydispersity of the polymer chain lengths would affect the attainment of this limiting behavior, the effect is expected to be minimal in the dynamic regime probed in these experiments. Examination of Figures 1 and 2 indicates that the frequency dependence ($1 \leqslant a_T\omega \leqslant 100\,\text{rad/s}$ for the PCLC and $1 \leqslant \omega \leqslant 10$ for the NCH) of both G' and G'' decreases with increasing silicate content. Furthermore, the hybrids with 5 wt% or greater silicate content display a limiting power-law behavior for both G' and G'', with both scaling as $\omega^{0.5}$. The frequency dependence of G' and G'' in the intermediate frequency regime is summarized in Table 1. Additionally, at low frequencies (i.e. $a_T\omega \leqslant 0.1\,\text{rad/s}$), the PCL-based nanocomposites with silicate contents in excess of 3 wt% exhibit nearly frequency-independent G' and G'', suggesting the possibility of a pseudo-solid-like behavior at long times owing to the incomplete relaxation of the chains and the composite. This pseudo-solid-like behavior and its structural origins are discussed below.

In the development of the linear viscoelastic master curves for PCLC, both frequency and modulus shift factors, a_T and b_T respectively, were employed. For the low silicate content hybrids (i.e. with 1, 2 and 3 wt% silicate) only frequency shift factors were required and b_T was set to unity. On the other hand, for the high silicate content (i.e. 5 and 10 wt% hybrids), a small temperature-dependent b_T was employed ($0.9 \leqslant b_T \leqslant 1.0$) in order to obtain good time–temperature superposed master curves. The temperature dependence of the a_T values for the PCL-based hybrids is shown in Figure 3. These values are independent of the silicate loading

Table 1 Poly(ε-caprolactone)^a and nylon 6 nanocomposites

Sample^b	wt% OMTS	$\bar{M}_w$	$\bar{M}_n$	$\bar{M}_w/\bar{M}_n$	G'	G''
PCLC1	1	30900	17200	1.80	1.60	1.00
PCLC2	2	16900	10300	1.64	0.97	0.97
PCLC3	3	14500	9200	1.57	0.80	0.90
PCLC5	5	13600	8800	1.54	0.50	0.65
PCLC10	10	16000	9900	1.63	0.55	0.70
Nylon 6	0	21700			1.50	0.93
Nylon 6-2	2	22200			1.00	0.80
Nylon 6-5	5	19700			0.60	0.70

OMTS = organically modified mica-type silicate; CL = ε-caprolactone; PCL = poly(ε-caprolactone).
^{a} GPC performed on polymer recovered from PCLC samples using ion exchange reaction.
^{b} Number at end of acronym indicates wt% OMTS in the composite.

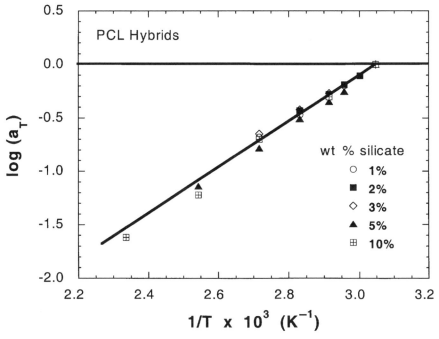

Figure 3 Temperature dependence of the frequency shift factors a_T for the PCLC hybrids studied in this work. The line shown in Figure 3 is the best fit Arrhenius equation [equation (4)] to the a_T values for all the hybrids and the pure polymer

and are consistent with the temperature dependence for PCL homopolymers as demonstrated in Figure 3. The flow activation energy, E_a, was obtained by fitting the data to the Arrhenius equation [equation (4)] and was found to be ~19 kJ/mol, consistent with that of PCL homopolymers [19, 20]. The non-dependence of the a_T values on the silicate loading and the near equivalence of the temperature dependence of a_T for the hybrids and PCL homopolymer suggest that the temperature-dependent relaxation being probed in these linear viscoelastic measurements is representative of the polymer. Since the silicate layers to which the polymer is tethered do not have a temperature-dependent relaxation, they do not contribute to the observed a_T values. However, the homopolymer-like temperature dependence of the observed relaxation does not imply that the relaxation behavior or the temperature dependence of the relaxation of the polymer near the surface of the layers is unaffected. The observed shift factors and the agreement with the principle of time–temperature superposition can easily be reconciled with the notion of two distinct populations of dynamical species: a matrix component that is unaffected and an adsorbed component whose relaxation processes are much slower than the time scale of the experiments. The experiments described here are incapable of probing, if indeed such a distinct bimodal population exists, and only offer the insight that, for all the populations whose relaxation time changes with temperature, the temperature dependence is identical to that of the pure polymer.

2.2 INTERCALATED PSPI NANOCOMPOSITES

The linear dynamic viscoelastic master curves for the unfilled disordered nearly symmetrical polystyrene-polyisoprene diblock copolymer PSPI18 and five hybrids prepared with a dimethyldioctadecylammonium-substituted montmorillonite (2C18M) (with 0.7, 2.1, 3.5, 6.7 and 9.5 wt% silicate) are shown in Figure 4 [11]. The master curves, generated by applying the principle of time–temperature superposition to isothermal frequency scans, were shifted to a common reference temperature, T_0, of 85 °C using both frequency shift factors, a_T, and modulus shift factors, b_T. The data for the 0.7 wt% silicate hybrid is almost identical to those obtained for the PSPI18 and not shown in Figure 4. Prolonged heating of the samples in the rheometer with flowing N_2 resulted in no change in the viscoelastic data. Additionally, gel permeation chromatography of the polymer after rheological testing exhibited no crosslinking or degradation.

At all frequencies, both G' (Figure 4a) and G'' (Figure 4b) for the nanocomposites increase monotonically with increasing silicate loading. The viscoelastic behavior at high frequencies ($a_T\omega > 10\,\text{rad/s}$, i.e. for frequencies where the polymer chains are not fully relaxed) is qualitatively unaffected by the addition of the layered silicate, with the exception of a monotonic increase in the modulus value. We note that the terminal relaxation time of the polymer chains is between 1.2 and 1.8 s at 85 °C [12, 21, 22], implying that below $a_T\omega$ of 10 rad/s, liquid-like flow behavior is expected for the polymer. Further, at low $a_T\omega$ (corresponding to a regime where the unfilled PSPI18 exhibits liquid-like behavior), both G' and G'' for the nanocomposites exhibit a diminished frequency dependence, which becomes weaker with increasing silicate content. For the 6.7 and 9.5 wt% nanocomposites in the low ω region, G' exceeds G'', with G' nearly independent of frequency, suggesting the possibility of pseudo-solid-like behavior for time scales at least of the order of 10^3–10^4 s.

The temperature dependence of the frequency and modulus shift factors, a_T and b_T, used to generate the master curves in Figure 4 are shown in Figure 5. The values of a_T are, within the errors of the experiment, independent of the silicate loading and well represented by a WLF relationship [equation (3)] with $C_1 = 11.2 \pm 0.4$ and $C_2 = 95 \pm 4\,°C$. The value of C_1 is consistent with that expected on the basis of the values of C_1 for PS and PI ($C_{1,\text{PS}} = 22.5$ and $C_{1,\text{PI}} = 2.8$ for $T_0 = 85\,°C$) [23, 24] and the known composition of the copolymer examined. The near non-dependence of the frequency shift factors on the silicate loading is similar to that observed for the PCL hybrids and suggests that the temperature-dependent relaxation processes observed in the viscoelastic measurements are essentially unaffected by the presence of the silicate layers. A substantial portion of the polymer is not intercalated between the layers of the silicate, and thus the temperature-dependent relaxation observed

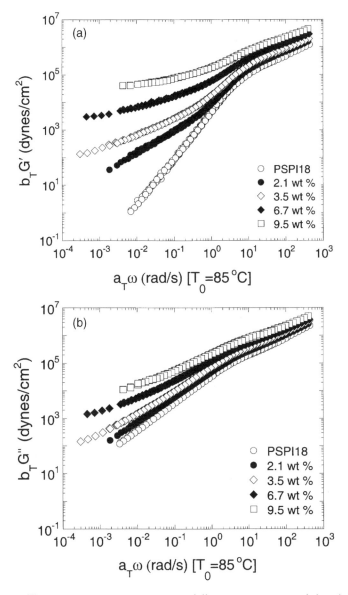

Figure 4 Time–temperature superposed linear storage modulus (a) and loss modulus (b) for the series of 2C18M-based PSPI18 intercalated hybrids. As expected, the moduli increase with increasing silicate loading at all frequencies. At high frequencies, the qualitative behavior of the storage and loss moduli are essentially unaffected. However, at low frequencies the frequency dependence of the moduli gradually changes from liquid-like to solid-like for nanocomposites with 6.7 and 9.5 wt % silicate

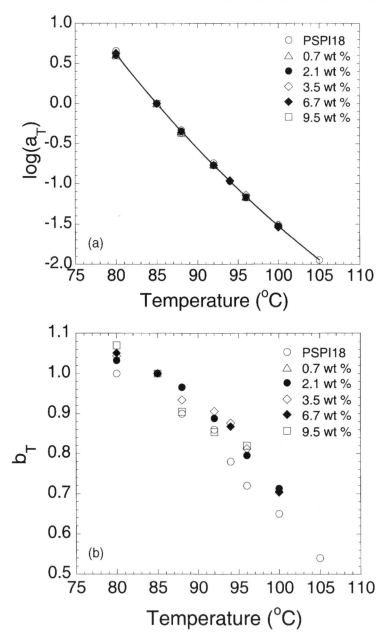

Figure 5 Frequency shift factors a_T (a) and moduli shift factor b_T (b) for PSPI18 and the different hybrids studied in this work. The line shown in Figure 5a is the best fit WLF equation [equation (3)] to the a_T values for all the hybrids and the pure polymer

could be attributed to that of the matrix polymer. Again, the possibility of a bimodal population of polymer relaxations, one unaffected and another considerably slowed down by the interaction with the silicate layer, cannot be ruled out on the basis of these measurements.

The b_T values required for time–temperature superpositioning of the linear viscoelastic data are large in magnitude and cannot be explained in terms of a simple density effect, i.e.

$$b_T = \frac{\rho_0 T_0}{\rho T} \tag{5}$$

where ρ_0 and ρ are the densities at T_0 and T respectively. Based on a simple density effect [equation (5)] [12], it would be expected that the values would not vary far from unity. These b_T values do not change significantly with silicate loading and are, within the errors of the experiments, comparable for all silicate loadings. One possible explanation for the substantial b_T values could be the block architecture of the polymer examined, and the relative proximity of the order–disorder temperature to the temperature of the measurements. Although the polymer is disordered at all experimental temperatures, there exist substantial repulsive thermodynamic interactions between the PS and PI blocks, leading to considerable concentration fluctuations in the disordered state, the magnitude of which diminishes with increasing temperature [25]. Larson and Fredrickson [26, 27] have suggested that, owing to the increase in concentration fluctuations while approaching the order–disorder transition from the disordered state, there can be additional contributions to the moduli and this might explain the magnitude of the vertical shift factors required. However, it is also expected that the principle of time–temperature superposition would fail if the magnitude of the corrections due to the concentration fluctuations to the moduli were substantial. It is possible that this failure of the principle of time–temperature superposition is subtle and thus not observable in our mechanical measurements.

To confirm the viscoelastic behavior observed at low frequencies in the dynamic measurements for the layered silicate-based hybrids, we have undertaken linear stress relaxation measurements, and the results of these are shown in Figure 6. A single-step strain γ_0 is applied at time $t = 0$, and the shear stress $\sigma(t)$ is measured as a function of time, with the modulus $G(t)$ obtained as $G(t) = \sigma(t)/\gamma_0$. The stress relaxation data reported in this paper were obtained at low strains and verified to be in the linear regime.

For any fixed time after the imposition of strain, the modulus increases with increasing silicate loading, similar to that observed in the dynamic viscoelastic measurements. Furthermore, at short times (i.e. for $t < 0.5$ s), the stress relaxation behavior is qualitatively similar for the hybrids and the unfilled polymer. At long times, however, the unfilled polymer relaxes like a liquid, while the hybrids with high silicate contents behave like a solid-like material for times as long as $\sim$2000 s (at 85 °C).

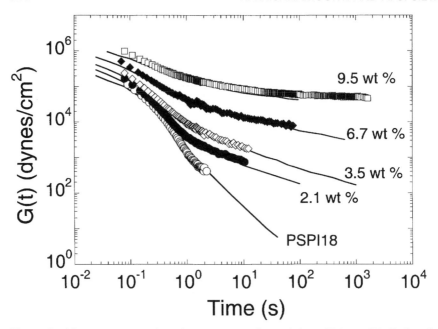

Figure 6 Linear stress relaxation measured modulus $G(t)$ at 85 °C for the unfilled PSPI18 and the hybrids. The weight fraction of the silicate in the hybrids is noted in the figure. Only 30% of the data collected are shown for all the datasets. The lines represent the predictions of equation (6) with the dynamic moduli presented in Figure 4. The agreement between the predictions and the measured data is excellent, suggesting the validity of equation (6) in describing the linear moduli for the hybrids

The linear stress relaxation modulus is related to the dynamic oscillatory shear moduli G' and G'' via the relaxation spectrum $H(\tau)$ as [12]

$$G'(\omega) - G(t)|_{t=1/\omega} = \int_{-\infty}^{\infty} \left[\frac{\omega^2 \tau^2}{1 + \omega^2 \tau^2} - e^{(-1/\omega\tau)} \right] H \, d(\ln \tau) \qquad (6)$$

Using a two-point collocation method, Ferry and coworkers [12, 28] have shown that the transient $G(t)$ can be related to the corresponding dynamic moduli G' and G'' by

$$G(t) = G'(\omega) - 0.40G''(0.40\omega) + 0.014G''(10\omega)|_{\omega=1/t} \qquad (7)$$

This approximate relation has proved useful in relating the dynamic and transient moduli for homopolymers [12]. Recently, the same relationship was successfully used to describe the viscoelastic properties of micellar systems dispersed in a homopolymer [29, 30]. Based on the dynamic viscoelastic response shown in Figure 4, and using equation (7), the modulus $G(t)$ was calculated and is shown in Figure 6. We note in passing that the time–temperature superposed viscoelastic master curves

(Figure 4) allow for a calculation of $G(t)$ over a much wider time scale than that measured experimentally and reported in Figure 6. The agreement between the measured and the calculated values of $G(t)$ is excellent, suggesting that indeed the measurements in both dynamic oscillatory and stress relaxation modes are linear and that the approximate relation suggested by equation (7) is also valid.

Based on both the dynamic oscillatory shear and the stress relaxation moduli, it is clear that the addition of layered silicate has a profound influence on the long-time relaxation of the hybrids. With increasing silicate loading, the liquid-like relaxation observed for the unfilled polymer gradually changes to solid-like (or pseudo-solid-like) behavior for hybrids with silicate loadings in excess of 6.7 wt%.

2.3 STRUCTURAL ORIGINS OF LINEAR VISCOELASTICITY

The two end-tethered exfoliated and the intercalated nanocomposites display extraordinary similarity in their linear viscoelastic response, with marked deviation from terminal flow behavior and suggestive of pseudo-solid-like behavior at extremely low concentrations of layered silicates. In particular for the end-tethered nanocomposites, non-terminal flow behavior precedes the pseudo-solid-like behavior. Similar non-terminal low-frequency rheological behavior has been observed in ordered block copolymers and smectic, liquid crystalline, small molecules [31–33]. Several hypotheses have been suggested to explain the observed rheological behavior in these systems [34–39]. It has been suggested that undulations and defects might contribute to the unusual low-frequency behavior in layered block copolymers. It has also been postulated that the domain structure of the ordered mesophases is responsible for this behaviour owing to the coupling of dynamic processes on the microscopic and mesoscopic length scales. Most commonly, however, the long-range domain structure and the presence of defects that might undergo annihilation and birth controlled by diffusion have been suggested to be responsible for the non-terminal flow behavior.

In the end-tethered exfoliated PCLC and NCH systems, we expect the highly anisotropic layers with lateral dimensions of the order of 0.1–1 µm to form domains where a distinct long-range order persists even in the melt state of the polymer [40, 41]. The local correlations result from the highly anisotropic nature of the layered silicates, whereby, beyond a critical concentration, which is extremely low, the layers are unable to be dispersed randomly and yet satisfy mutual impenetrability and maintain rigidity. Moreover, this long-range correlation and the domain structure are likely to be better defined for composites with higher silicate loadings, where the geometrically imposed mean distance between the layers becomes considerably smaller than the lateral dimensions of the layers. It would be expected that the local correlations, in terms of orientation and distance between adjacent layers, would be considerably polydisperse and under quiescent conditions not lead to a liquid crystalline-like order. Many such randomly oriented grains would be expected to make up the entire sample, leading to the presence of a generally disordered material. Thus, for these end-tethered nanocomposites it is postulated that there exists a granular structure with local correlations between individual silicate layers, intergranular boundaries and defects. It is most likely that this granular structure, along with the intimate contact, because of the tethering, between the polymer (soft phase) and the layered silicate (hard phase), would lead to the presence of non-terminal flow behavior at intermediate frequencies for the end-tethered nanocomposites. The intimate contact because of tethering appears to be important, as previous reports on the viscoelastic properties of free polymer-based exfoliated nanocomposites [10] did not display such non-terminal behavior preceding the pseudo-solid-like behavior.

On the other hand, in both the end-tethered exfoliated nanocomposites (PCLC) and the intercalated PSPI-based nanocomposites, a nearly frequency-independent storage and loss moduli are observed at low frequencies for the higher silicate content nanocomposites. In the case of the end-tethered nanocomposites, it is conceivable that the tethering of the polymer chains to the essentially immobile silicate layers leads to the pseudo-solid-like behavior. However, for the PCLC hybrids with less than 3 wt% silicate, the behavior at the lowest frequencies does not resemble that of a solid-like material. Thus the tethering, while altering the relaxation of the chains, does not by itself lead to a long-time pseudo-solid-like behavior.

In the case of the disordered PSPI-based nanocomposites, it is possible that the preferential attraction of the PS block to the silicate layers could cause the block copolymer to undergo an ordering transition, i.e. forming ordered microdomains. Previously, Green and coworkers [42] showed that thin films of a disordered PS–PMMA diblock copolymer can undergo an ordering transition when cast on silicon oxide substrates, a surface preferentially attractive to PMMA. It has also been previously shown that, for ordered block copolymers, the low-frequency response can be significantly diminished [33, 43–45]. In particular, for materials with microstructures possessing cubic symmetry (such as spheres or gyroids), the low-frequency dependence of G' is nearly independent of frequency. However, the viscoelastic properties of a PS homopolymer ($M_w = 30\,000$, $M_w/M_n < 1.06$) + 6.7 wt% 2C18M, shown in Figure 7 [46], clearly demonstrate that the low-frequency viscoelastic behavior is identical to that of the PSPI + 6.7 wt% 2C18M hybrid. Thus, the pseudo-solid-like behavior of the PSPI-based nanocomposites cannot be attributed to a surface-induced ordering phenomenon, as the homopolymer-based hybrid is incapable of forming an ordered structure.

Alternatively, the low-frequency viscoelastic response can be explained in terms of a physical jamming of the dispersed layered silicates owing to their highly anisotropic nature. The frequency dependence of the low-frequency G' and G'', where a pseudo-solid-like behavior is observed, is similar for the PSPI hybrids with 6.7 and 9.5 wt% silicate and for the PCLC hybrids with 3, 5 and 10 wt%. We suggest, just as in the case of the PCLC-based hybrids, that in the PSPI-based intercalated nanocomposites the mesoscopic structure consists of silicate tactoids of

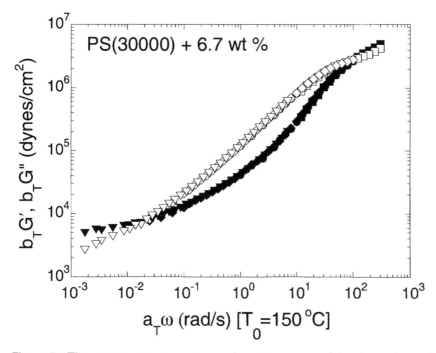

Figure 7 Time–temperature superposed master curves of the dynamic moduli for the PS (30 000) + 2C18M-based 6.7 wt % silicate hybrid. The data were shifted to 150 °C. Open symbols correspond to $b_T G''$ while the filled symbols correspond to $b_T G'$. The remarkable similarity in the low-frequency behavior of this hybrid to that of the PSPI18 + 2C18M hybrid (6.7 wt % silicate) shown in Figure 4 suggests that the unusual viscoelastic properties observed for the PSPI18-based hybrids are not a result of a surface-induced ordering

several tens of well-ordered layers and the occasional presence of individual layers removed from the tactoids. Transmission electron micrographs of a hybrid formed with the same organically modified layered silicate and a polystyrene–poly(ethylene-co-butene-1) (PS–PEB) block copolymer revealed a mesoscopic structure for the PSPI-based nanocomposites to be consistent with the above picture [47]. On the basis of this mesoscopic structure and at low silicate concentrations, we suggest that, beyond a critical volume fraction, the tactoids and the individual layers are incapable of freely rotating and when subjected to shear are prevented from relaxing completely. This incomplete relaxation due to the physical jamming or percolation of the nanoscopic fillers leads to the presence of the pseudo-solid-like behavior observed in both the intercalated and exfoliated hybrids.

To verify that the percolation of the stacks of layers (i.e. tactoids) is a plausible explanation for the intercalated PSPI nanocomposites, we estimate the percolation threshold for these hybrids parameterized in terms of the stack size. For this we consider a hypothetical hydrodynamic sphere surrounding each tactoid, and consider the percolation of these hydrodynamic spheres to signify the onset of incomplete relaxation and the presence of pseudo-solid behavior in the hybrids (Figure 8). The percolation of random spheres in three dimensions has been calculated to occur for a volume fraction, ϕ_{per}, of ~ 0.30 [48], and is only slightly changed by the inclusion of excluded-volume interactions. A simple volume filling calculation, assuming a uniform distribution of identically sized tactoids, yields a relationship between the critical tactoid size required, n_{per} (the number of silicate layers per tactoid), and the layered silicate weight fraction at percolation, $w_{sil,per}$, as

$$n_{per} = \frac{4}{3\phi_{per}} \left[\frac{w_{sil,per}\rho_{org}}{w_{sil,per}\rho_{org} + (1 - w_{sil,per})\rho_{sil}} \right] \frac{R_h}{h_{sil}} \quad (8)$$

where R_h is the radius of the hydrodynamic volume (in this case equivalent to the radius of the disk-like layered silicates), h_{sil} is the thickness of the silicate layers and ρ_{org} and ρ_{sil} are the densities of the organic component and layered silicate respectively. Assuming that the silicate layers can be adequately represented by

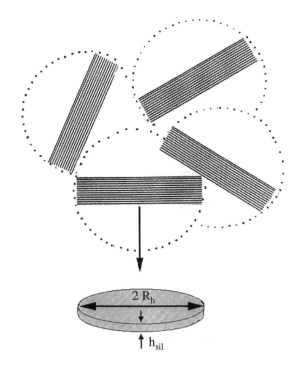

Figure 8 A schematic representation of the tactoids of layered silicates and their interaction with each other, resulting in incomplete relaxation of the hybrids

uniform disks of $0.5\,\mu m$ diameter $(2R_h)$ and $1\,nm$ thickness, h_{sil}, and assigning the weight fraction of silicate at the percolation threshold, $w_{sil,per}$, to be 0.067 (pseudo-solid behavior is observed for the hybrids with $\geqslant 6.7\,wt\%$ layered silicate), we obtain the average tactoid size to be $\sim\!30$ layers. This average tactoid size is not unreasonable in the light of the presence of several higher order X-ray diffraction peaks and evidence from TEM measurements on hybrids of PS–PEB prepared with the same layered silicate [47] that, in addition to a few delaminated layers, the tactoid size can range from 10 layers to over 50 layers. Furthermore, this tactoid size implies that the effective anisotropy associated with the filler is $\sim\!5$–10. It is this anisotropy, along with the random relative arrangement of the tactoids, that leads to the observation of the percolation phenomenon at extremely low loadings of the silicate.

We can use equation (8) to estimate the percolation threshold for the exfoliated nanocomposites. Assuming the geometrical parameters to be roughly the same, percolation would be expected for truly uncorrelated hybrids with $\sim 0.1\,wt\%$ layered silicate. It is clear from the data presented that in the PCLC hybrids, percolation, as observed by rheological measurements, occurs at $\sim 2.5 \pm 0.5\,wt\%$ silicate. These suggest that, even in the case of exfoliated nanocomposites, there exists considerable local orientational order which has enormous consequences for the physical and mechanical properties of the hybrids. This orientational order along with the end-tethering of the chains is in fact responsible for the non-terminal intermediate frequency response in the PCLC and NCH hybrids.

3 NON-LINEAR VISCOELASTIC PROPERTIES

3.1 ALIGNMENT BY LARGE-AMPLITUDE OSCILLATORY SHEAR

Prolonged application of large-amplitude oscillatory shear results in dramatic changes in the linear viscoelasticity and the orientation of the silicate layers in the shear direction. Application of large-amplitude oscillatory shear leads to a decline in the moduli during shear which gradually decrease with time before reaching a plateau. For the case of the PCLC and PSPI18, qualitatively similar behavior of the modulus during shear was observed, with the shear stress signal almost always remaining sinusoidal. The linear viscoelastic moduli after shear alignment for a PCLC hybrid [9] and a PSPI18 hybrid [11] are shown in Figures 9 and 10 respectively. The temperature-dependent frequency shift factors required to obtain the master curves are, within experimental error, identical to those obtained for the unaligned samples. Small-angle neutron scattering data obtained before and after shear alignment for a $5\,wt\%$ PCL nanocomposite, shown in Figure 11, clearly illustrate the alignment of the nanocomposites with the layers parallel to the shear direction. Examination of the linear viscoelastic results shown in Figures 9 and 10 indicate that both G' and G'' for the aligned nanocomposites are considerably lower than those for the initially unaligned samples. Moreover, the frequency dependence

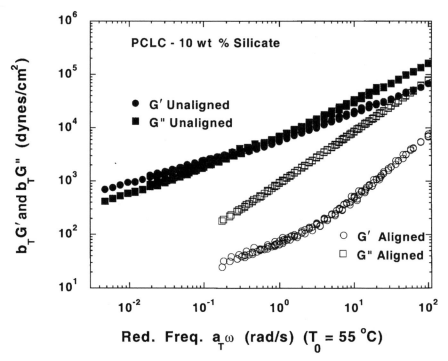

Figure 9 Effect of large-amplitude oscillatory shear on the linear viscoelastic moduli master curves (from data at 70 and 95°C) for a PCL + 2C18M hybrid (10 wt% silicate). Open symbols correspond to the shear-aligned sample while the filled symbols correspond to the unoriented sample, with the squares corresponding to G'' and the circles corresponding to G'. The moduli decrease after shear alignment and have a dramatic effect on the low-frequency response, with a more liquid-like behavior observable

of both G' and G'' for the aligned samples is much stronger than that of the unaligned samples and starts to resemble that of the free homopolymers. These shear alignment results are similar to those observed for ordered block copolymers and small-molecule smectic liquid crystals[33,43–45].

This increased low-frequency dependence of the viscoelastic moduli suggests a breakdown of the percolated silicate network in the shear-aligned sample. In fact, for the case where the tactoids are aligned parallel to the shear direction, percolation would be expected when the effective two-dimensional disk-like objects form a network. For the case of overlapping disks and on the basis of continuum model calculations, it is expected that percolation would occur at a volume fraction of disks of ~ 0.67 [48]. For the case of the intercalated PSPI-based nanocomposites, where tactoids are calculated to be ~ 30 layers in thickness, this calculation would suggest that a critical percolation threshold occurs at ~ 44 wt% of the layered silicate, far

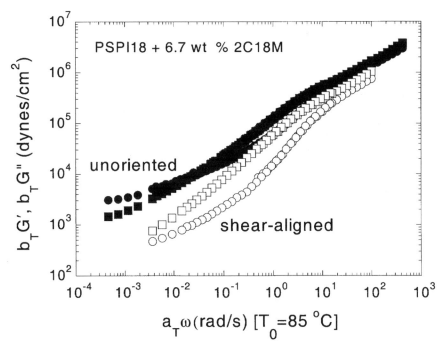

Figure 10 Effect of large-amplitude oscillatory shear on the linear viscoelastic moduli master curves (from data at 85 and 92 °C) for a PSPI18 + 2C18M hybrid (6.7 wt% silicate). Open symbols correspond to the shear-aligned sample while the filled symbols correspond to the unoriented sample, with the squares corresponding to G'' and circles corresponding to G'. The moduli decrease after shear alignment and have a dramatic effect on the low-frequency response, with a more liquid-like behavior observable

above any concentration examined in this study. Thus, in close analogy to their small-molecule counterparts, it appears that the layered silicate-based polymer nanocomposites possess a property-defining mesoscopic structure that can be significantly altered by the application of large shear flow fields.

3.2 NON-LINEAR DYNAMIC MODULI

The non-linear complex viscosity, η^*, probed as a function of strain amplitude at a fixed frequency for the PSPI-based nanocomposites [49] is reported in Figure 12. With increasing silicate content, the transition from linear to non-linear behavior occurs at lower strain amplitudes. It should be noted that the strain amplitude sweep was carried out as consecutive measurements, not each carried out from the quiescent random state. The sharp decrease in the complex viscosity with increased strain

a   b

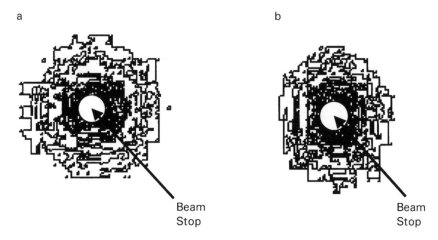

Beam   Beam
Stop   Stop

Figure 11 Small-angle neutron scattering on a 5wt% montmorillonite poly (ε-caprolactone) end-tethered nanocomposite before (prealignment) and after (post-alignment) large-amplitude oscillatory shear. The aligned samples show that the silicate layers preferentially lie parallel to the shear direction

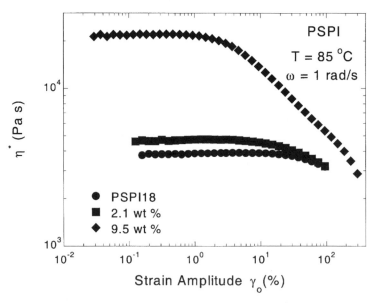

Figure 12 Complex viscosity, η^*, as a function of strain amplitude, γ_0, for PSPI18 nanocomposites of different concentrations at $T = 85\,°C$ and $\omega = 1\,rad/s$

amplitude for the nanocomposites is consistent with the quiescent state mesoscopic structure described earlier and the fragility of this random structure to the application of shear. With increased strain amplitude, the silicate layers would be expected to be better aligned in the shear direction, resulting in a decreased complex viscosity.

On the other hand, similar strain amplitude sweeps performed on an aligned PCL-based nanocomposite [2] provide distinctly different results. Complex viscosity, η^*, and phase angle, $\tan\delta$, as a function of strain amplitude, γ_0, for a representative 10 wt% PCLC at $\omega = 1$ and 3 rad/s ($T = 55\,°C$) during both increasing and decreasing strain amplitude cycles are shown in Figure 13. At all strain amplitudes the force signal was confirmed to be sinusoidal, thereby allowing for interpretation in terms of standard viscoelastic functions. The η^* at low strain amplitudes is independent of γ_0 and is dominated by the viscous response. However, progression to higher γ_0 leads to an increase in η^*, with the elastic component becoming more prominent as seen by the decrease in $\tan\delta$. At the highest strain amplitudes, η^* appears to saturate, with a value much higher than that observed at low strain amplitudes. A slight hysteresis in the transition from high to low viscosity is also observed when the strain amplitude is decreased from a large value to a small value (as compared with the case where the strain amplitude is increased from low to high values). Furthermore, the magnitude of the viscosity increase is considerably greater at low frequencies, which is a result of to the low value of the initial viscosity at low frequencies.

Based on the behavior for four PCL-based nanocomposites (i.e. 3, 5, 7 and 10 wt% silicate), the upturn in the η^* (and the downturn in $\tan\delta$) occurs at a critical strain amplitude, γ_0^c, over a wide range of temperatures and frequencies (for a given silicate loading). Three important features are observed in the rheological response:

(a) The process is reversible (Figure 13a).
(b) There is a critical strain amplitude for the transition that decreases with increasing silicate content (Figure 14).
(c) The elastic component to the rheological response becomes more important with increasing strain amplitude (Figure 13b).

Typically [23], homopolymers and intercalated nanocomposites (Figure 12) exhibit decreasing viscosity with increasing shear rate. Moreover, confined polymer solutions such as those present in a surface force apparatus (SFA) shear thin at a critical velocity, presumably owing to the slip of the confining mica layers [50–68]. These confined polymer solutions have also shown a dramatic increase in normal force beyond a critical Weissenberg number, attributed to a combination of hydrodynamic instabilities and flow-induced brush thickening, as well as shear-induced diffusion [69, 70]. In fact, for a solution brush created by adding 50 wt% toluene (a good solvent) to a 7 wt% silicate PCL nanocomposite, the shear thinning behavior is recovered as shown in Figure 15. The γ_0 at which the brush starts to shear thin decreases roughly linearly with increasing frequency, ω, consistent with the earlier observations of confined polymer solutions in a SFA [50–68]. This suggests that the

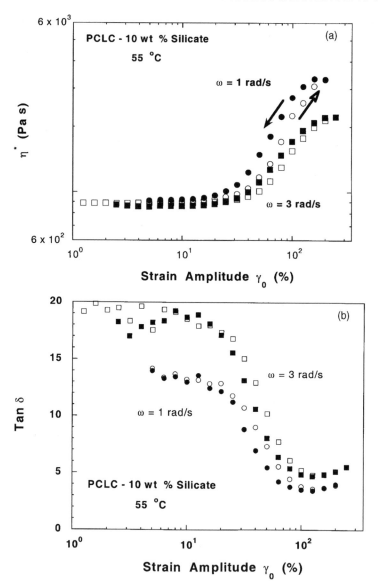

Figure 13 (a) Complex viscosity, η^*, and (b) phase angle, $\tan\delta\,(=G''/G')$, as a function of strain amplitude, γ_0, for a 10 wt % poly(ε-caprolactone) nanocomposite at $T = 55\,°C$ and $\omega = 1$ and 3 rad/s. Open symbols were obtained with increasing strain amplitude and filled symbols with decreasing strain amplitude

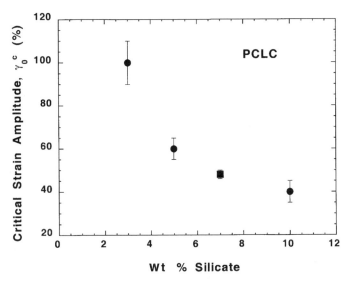

Wt % Silicate

Figure 14 Silicate content dependence of the critical strain amplitude for the onset of strain hardening in the melt brush nanocomposites described in Figures 3 and 4

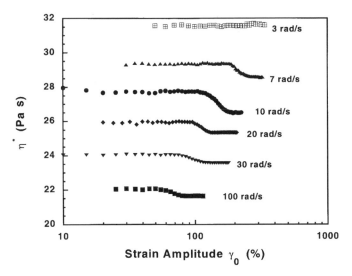

Strain Amplitude γ_0 **(%)**

Figure 15 Complex viscosity, η^*, as a function of strain amplitude for a solution brush, 7 wt % poly(ε-caprolactone) nanocomposite diluted with toluene (50 wt %), over a wide range of frequencies at $T = 60\,^\circ$C. Data shown were obtained with increasing strain amplitude. The solution brush clearly exhibits shear thinning while the melt brush exhibits strain hardening. Note that the abscissa is on a linear scale in Figure 13, while in the melt brush figures it is on a logarithmic scale

product of γ_0 and ω is roughly constant, thereby indicating that there exists a critical velocity for the onset of shear thinning. Additionally, unlike the melt brush system described in Figure 13, little or no hysteresis is observed in the viscoelastic response for this solution brush.

In the aligned end-tethered melt nanocomposites, we observe an increase in viscosity with increasing shear strain amplitude, with the transition occurring at modest strain amplitudes (in the same range where shear thinning occurs in the solution). The brush-like nature of the tethered polymer, where the chains stretch perpendicular to the surface even in the absence of any external forces owing to the high density of grafting at the surface and the resulting lateral crowding, is suggested to be responsible for the observed viscoelastic response. Upon application of shear strain (beyond the γ_0^c identified in Figure 14), the tethered polymers are expected to completely unwind in response to the applied shear [50, 69]. Based on the limitations of a 'melt brush' system, where the tethered polymer has to fill all space, the chains stretch. This stretching occurs at a critical displacement that only depends on the geometrical restrictions of the system, i.e. the distance between tethering surfaces. This is further validated by the observation that the critical displacement decreases with increasing silicate content. An estimation of the interlayer distance, based on the assumption of uniform distribution, suggests an inverse dependence of the interlayer distance with the weight fraction of silicate in the composite, consistent with the results presented in Figure 14. Furthermore, it is expected that, as the spacing between layers is decreased, the equilibrium chain structure is further distorted, and therefore the critical shear displacement required for chain stretching is decreased.

Alternative explanations for the strain hardening observed here include a physical jamming (or network formation) of the layers as a result of the applied shear. This appears unlikely as the layers are oriented parallel to the plates of the rheometer and would be forced to jam in a lateral manner. In addition, the complete reversibility of the strain hardening further suggests that the jamming of layers is not responsible for this effect. Moreover, as shown earlier for the PSPI-based nanocomposites where the linear viscoelastic response of the quiescent structure is dominated by the meso-scopic structure and a percolation of the filler network, no such strain hardening was observed.

3.3 STEADY SHEAR RESPONSE

The steady shear response of layered silicate-based polymer nanocomposites has important consequences for the potential processability of the materials. The melt-state steady shear viscoelastic behavior has been systematically studied for the PSPI-based nanocomposites. The steady state viscosity, η, as a function of shear rate, $\dot{\gamma}$, for these nanocomposites is shown in Figure 16. At low shear rates, the addition of even small quantities of layered silicate results in non-Newtonian behavior and a significant enhancement in the viscosity. Furthermore, for the nanocomposites with

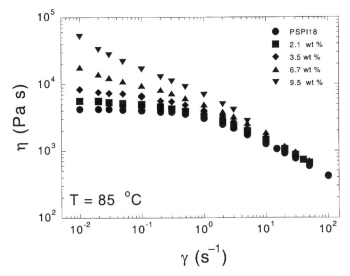

Figure 16   Steady state viscosity, η, as a function of shear rate, $\dot{\gamma}$, for the PSPI-based nanocomposites studied here. The addition of even small quantities of layered silicate at low shear rates results in non-Newtonian behavior and a significant enhancement in the viscosity. For the hybrids with 6.7 and 9.5 wt% silicate, the low-frequency viscosity diverges and is consistent with the pseudo-solid-like behavior described in the section on linear viscoelasticity

6.7 and 9.5 wt% silicate, the low-frequency viscosity diverges and is consistent with the pseudo-solid-like behavior described in the section on linear viscoelasticity. However, at high shear rates, the viscosity and rate of shear thinning for the nanocomposites are comparable with those of the unfilled polymer as a result of the preferential orientation of the layers of the silicate or even anisotropic tactoids parallel to the flow direction.

Based on the steady state shear stress, $\sigma(\dot{\gamma})$, measured as a function of shear rate, $\dot{\gamma}$, a yield stress, σ_0, can be estimated for each of the nanocomposites by fitting to [71]

$$\sigma(\dot{\gamma}) = \sigma_0 + k(\dot{\gamma})^n \qquad (9)$$

where k and n are positive numbers. The estimated yield stresses are shown in Figure 17. It is clear that, for nanocomposites with silicate fractions of 6.7 and 9.5 wt%, there exists a finite yield stress, while for the composites with low silicate loadings the yield stress, within experimental error, is zero. The presence of a yield stress is consistent with the presence of a mesoscopic structure for the 6.7 and 9.5 wt% nanocomposites, where the tactoids of the silicate layers are unable to relax independently and result in the observation of pseudo-solid-like behavior for these hybrids.

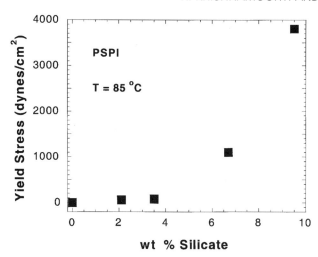

Figure 17   Estimated yield stresses for the PSPl18-based nanocomposites using equation (9). For nanocomposites with silicate fractions of 6.7 and 9.5 wt % there exists a finite yield stress, while for the composites with low silicate loadings the yield stress, within experimental error, is zero

Thus, the steady shear viscoelasticity provides complementary information to the linear and non-linear dynamic oscillatory shear measurements regarding the quiescent structure and the ability of shear to deform and orient these nanocomposites. For low shear rates, the mesostructure remains essentially unaffected by the imposed flow and the viscoelastic response reflects the quiescent structure, particularly for the high silicate content where a yield stress is observed and indicative of a solid-like response. At high shear rates, on the other hand, the mesoscopic structure is oriented by the imposed shear and suggests that the layers, in the absence of tethering and hence mechanical coupling, orient in the flow direction and contribute negligibly to the viscoelastic properties of the nanocomposite.

4 SUMMARY

The linear and non-linear melt-state viscoelastic properties of two series of end-tethered nanocomposites based on poly(ε-caprolactone) and nylon 6 and one series of intercalated nanocomposites based on a disordered diblock copolymer of polystyrene and polyisoprene are reviewed. The linear viscoelastic measurements suggest a significant deviation from liquid-like flow behavior for the nanocomposites, with a pseudo-solid-like character being observed for the hybrids with higher silicate content. This pseudo-solid-like behavior at long times is attributed to the percolation of a three-dimensional filler network structure comprising a random orientation of grains consisting of locally correlated layers. This picture of the quiescent state mesoscopic structure is further validated by the presence of a yield stress in steady shear measurements. Additionally, this mesoscopic structure can be altered by the application of flow and, at high shear rates, results in the layers of silicate orienting in the direction of shear. The non-linear viscoelastic behavior is strongly dependent on the physical connectivity (i.e. interaction) between the polymer and the layered silicate, with end-tethering resulting in strain hardening, while weaker physical interactions lead to a shear thinning character.

5 ACKNOWLEDGEMENTS

The authors thank Profs Giannelis, Mohanty, Graessley and Colby for useful discussions. The help of Jiaxiang Ren in the preparation of some of the figures of this manuscript is also gratefully acknowledged. We would like to acknowledge financial support from the donors of the Petroleum Research Fund administered by the American Chemical Society, the Texas Coordinating Board (ATP) and Exxon Chemical Company (Baytown, TX).

6 REFERENCES AND NOTES

1. Giannelis, E. P., Krishnamoorti, R. and Manias, E. *Adv. Polym. Sci.*, **138**, 107–147 (1999).
2. Krishnamoorti, R. and Giannelis, E. P. accepted by *Langmuir* (2000).
3. Ramsay, J. D. F. and Lindner, P .J. *J. Chem. Soc., Faraday Trans.*, **89**, 4207 (1993).
4. Ramsay, J. D. F., Swanton, S. W. and Bunce, J. J. *J. Chem. Soc., Faraday Trans.*, **86**, 3919 (1990).
5. Magauran, E. D., Kieke, M. D., Reichart, W. W. and Chiavoni, A. *NGLI Spokesman*, **50**, 453 (1987).
6. Dijkstra, M., Hansen, J. P. and Madden, P. A. *Phys. Rev. Lett.*, **75**, 2236–2239 (1995).
7. Pignon, F., Piau, J.-M. and Magnin, A. *Phys. Rev. Lett.*, **76**, 4857–4860 (1996).
8. Pignon, F., Magnin, A. and Piau, J.-M. *Phys. Rev. Lett.*, **79**, 4689–4692 (1997).
9. Krishnamoorti, R. and Giannelis, E.P. *Macromolecules*, **30**, 4097 (1997).
10. Krishnamoorti, R., Vaia, R. A. and Giannelis, E. P. *Chem. Mater.*, **8**, 1728 (1996).
11. Ren, J., Silva, A. S. and Krishnamoorti, R. *Macromolecules* **33**, 3739–3746 (2000).
12. Ferry, J. D. *Viscoelastic Properties of Polymers*, 3rd edition, New York, Wiley, 1980.
13. Messersmith, P. B. and Giannelis, E. P. *Chem. Mater.*, **5**, 1064 (1993).
14. Messersmith, P. B. and Giannelis, E. P. *J. Polym. Sci., Part A, Polym. Chem.*, **33**, 1047 (1995).
15. Malkin, A. Y. *Adv. Polym. Sci.*, **96**, 69 (1990).
16. Usuki, A., Kojima, Y., Kawasumi, M., Okada, A., Gukushima, Y., Kurauchi, T. and Kamigaito, O. *J. Mater. Res.*, **8**, 1179 (1993).
17. Kojima, Y., Usuki, A., Kawasumi, M., Okada, A., Fukushima, Y., Kurauchi, T. and Kamigaito, O. *J. Mater. Res.*, **8**, 1185 (1993).
18. Yano, K., Usuki, A., Karauchi, T. and Kamigaito, O. *J. Polym. Sci., Part A, Polym. Chem.*, **31**, 2493 (1993).
19. Jo, W. H., Chae, S. H. and Lee, M. S. *Polym. Bull.*, **29**, 113 (1992).

20. Han, C. D. and Yang, H.-H. *J. Appl. Polym. Sci.*, **33**, 1199 (1987).
21. Fetters, L. J., Lohse, D. J., Milner, S. T. and Graessley, W.W. *Macromolecules*, **32**, 6847 (1999).
22. Fetters, L. J., Lohse, D. J. and Graessley, W. W. *J. Polym. Sci., Part B, Polym. Phys.*, **37**, 1023 (1999).
23. Graessley, W. W. *Physical Properties of Polymers*, 2nd edition, American Chemical Society, Washington, DC (1993).
24. Gotro, J. T. and Graessley, W. W. *Macromolecules*, **17**, 2767 (1984).
25. Lin, C. C., Jonnalagadda, S. V., Kesani, P. K., Dai, H. J. and Balsara, N. P. *Macromolecules*, **27**, 7769–7780 (1994).
26. Fredrickson, G. H. and Larson, R. G. *J. Chem. Phys.*, **86**, 1553 (1987).
27. Larson, R. G. and Fredrickson, G. H. *Macromolecules*, **20**, 1897 (1987).
28. Ninomiya, K. and Ferry, J. D. *J. Colloid Sci.*, **14**, 36 (1959).
29. Sato, T., Watanabe, H., Osaki, K. and Yao, M.-L. *Macromolecules*, **29**, 3881 (1996).
30. Watanabe, H., Yao, M.-L., Sato, T. and Osaki, K. *Macromolecules*, **30**, 5905 (1997).
31. Rosedale, J. H. and Bates, F. S. *Macromolecules*, **23**, 2329 (1990).
32. Koppi, K. A., Tirrell, M., Bates, F. S., Almdal, K. and Colby, R.H. *J. Phys. II (Paris)*, **2**, 1941 (1993).
33. Larson, R. G., Winey, K. I., Patel, S. S., Watanabe, H. and Bruinsma, R. *Rheol. Acta*, **32**, 245–253 (1993).
34. Kawasaki, K. and Onuki, A. *Phys. Rev. A*, **42**, 3664 (1990).
35. Rubinstein, M. and Obukhov, S. P. *Macromolecules*, **26**, 1740 (1993).
36. Halperin, A., Tirrell, M. and Lodge, T. P. *Adv. Polym. Sci.*, **100**, 31 (1992).
37. Ohta, T., Enomoto, Y., Harden, J. L. and Doi, M. *Macromolecules*, **26**, 4928 (1993).
38. Doi, M., Harden, J. L. and Ohta, T. *Macromolecules*, **26** (1993).
39. Witten, T. A., Leibler, L. and Pincus, P. *Macromolecules*, **23**, 824 (1990).
40. Kojima, Y., Usuki, A., Kawasumi, M., Okada, A., Kurauchi, T., Kamigaito, O. and Kaji, K. *J. Polym. Sci., Part B, Polym. Phys.*, **32**, 625 (1994).
41. Kojima, Y., Usuki, A., Kawasumi, M., Okada, A., Kurauchi, T., Kamigaito, O. and Kaji, K. *J. Polym. Sci., Part B, Polym. Phys.*, **33**, 1039 (1995).
42. Limary, R., Swinnea, S. and Green, P. F. *Macromolecules*, **33** 5227–5234 (2000).
43. Fredrickson, G. H. and Bates, F. S. *Ann. Rev. Mater. Sci.*, **26**, 501 (1996).
44. Colby, R. H. *Curr. Op. Col.*, **1**, 454 (1996).
45. Kossuth, M. B., Morse, D. C. and Bates, F. S. *J. Rheology*, **43**, 167 (1999).
46. The b_T values ranged from 0.97 to 1.03 for the PS-based nanocomposites. This leads additional credence to the substantial b_T values observed for PSPI and PSPI-based nanocomposites and suggests that the concentration fluctuations in those systems are responsible for the large b_T values.
47. Silva, A. S., Tse, M. F., Wang, H.-C. and Krishnamoorti, R. submitted to *Macromolecules* (2000).
48. Isichenko, M. B. *Rev. Mod. Phys.*, **64**, 961 (1992).
49. Ren, J. and Krishnamoorti, R. in preparation (2000).
50. Fytas, G., Anastasiadis, S. H., Seghrouchni, R., Vlassopoulos, D., Li, J.B., Factor, B.J., Theobald, W. and Toprakcioglu, C. *Science*, **274**, 2041 (1996).
51. Klein, J. *Ann. Rev. Mater. Sci.*, **26**, 581 (1996).
52. Rabin, Y. and Alexander, S. *Europhys. Lett.*, **13**, 49 (1990).
53. de Gennes, P. G. *Scaling Concepts in Polymer Physics*, Cornell University Press, Ithaca, NY, 1979.
54. Joanny, J.-F. *Langmuir*, **8**, 989 (1992).
55. Subramanian, G., Williams, D. R. M. and Pincus, P. A. *Macromolecules*, **29**, 4045 (1996).
56. Milner, S. T. *Science*, **251**, 905 (1991).

57. Semenov, A. N. *Langmuir*, **11**, 3560 (1995).
58. Brochard-Wyart, F. *Europhys. Lett.*, **23**, 105 (1993).
59. Brochard-Wyart, F., Hervet, H. and Pincus, P. *Europhys. Lett.*, **26**, 511 (1994).
60. Israelachvili, J. N. and Tabor, D. *Proc. R. Soc. (Lond.) A*, **331**, 19 (1972).
61. Israelachvili, J. N. *Surf. Sci. Rep.*, **14**, 109 (1992).
62. Klein, J. *J. Chem. Soc., Faraday Trans. 1*, **79**, 99 (1983).
63. Overney, R. M. *Trends Polym. Sci.*, **3**, 359 (1995).
64. Reiter, G., Demirel, A. L. and Granick, S. *Science*, **263** (1994).
65. Klein, J., Perahia, D. and Warburg, S. *Nature*, **352**, 143 (1991).
66. Granick, S. *MRS Bull.*, **21**, 33 (1996).
67. Granick, S., Demirel, A. L., Cai, L. L. and Peanasky, J. *Israel J. Chem.*, **35**, 75 (1995).
68. Cai, L. L., Peanasky, J. and Granick, S. *Trends Polym. Sci.*, **4**, 47 (1996).
69. Doyle, P. S., Shaqfeh, E. S. G. and Gast, A. P. *Phys. Rev. Lett.*, **78**, 1182 (1997).
70. Szleifer, I. and Carignano, M.A. *Adv. Chem. Phys.*, **94**, 165 (1996).
71. Larson, R. G. *The Structure and Rheology of Complex Fluids*, Oxford University Press, New York, 1999.

INDEX